Ethics Across the Curriculum—Pedagogical Perspectives

Elaine E. Englehardt · Michael S. Pritchard
Editors

Ethics Across the Curriculum—Pedagogical Perspectives

Springer

Editors
Elaine E. Englehardt
Utah Valley University
Provo, UT, USA

Michael S. Pritchard
Western Michigan University (Emeritus)
Kalamazoo, MI, USA

ISBN 978-3-319-78938-5 ISBN 978-3-319-78939-2 (eBook)
https://doi.org/10.1007/978-3-319-78939-2

Library of Congress Control Number: 2018937694

Printed on acid-free paper

This Springer imprint is published by the registered company Springer International Publishing AG part of Springer Nature
The registered company address is: Gewerbestrasse 11, 6330 Cham, Switzerland

Contents

Editors and Contributors

About the Editors

Elaine E. Englehardt is a Distinguished Professor of Ethics and Professor of Philosophy at Utah Valley University (UVU). She received her Ph.D. from the University of Utah. She has written and directed seven multi-year, national grants in the area of Ethics and Ethics Across the Curriculum. One grant founded Ethics Across the Curriculum at UVU and another funded the Society for Ethics Across the Curriculum and the journal*Teaching Ethics*. She is a SEAC Board Member and with Michael Pritchard was co-author of the journal*Teaching Ethics*. She has authored ten textbooks. She has written numerous peer-reviewed articles and ten chapters. She has served in various administrative positions.

Michael S. Pritchard is Emeritus Professor of Philosophy and former Co-director of the Center for the Study of Ethics in Society at Western Michigan University. He received his Ph.D. in Philosophy from the University of Wisconsin. He received his B.A. from Alma College, Michigan. He teaches courses in Ethical Theory, Practical Ethics, Professional Ethics, 18th Century British Philosophy, and Philosophy for Children. For seven years, he served as co-editor, with Elaine Englehardt of*Teaching Ethics,*the official journal of the Society for Ethics Across the Curriculum. He is a Founding Board Member for SEAC. Among his publications are: On Becoming Responsible (Kansas, 1991); Communication Ethics (Wadsworth, 1994), with James Jaksa; Reasonable Children (Kansas, 1996); Professional Integrity (Kansas, 2007); Ethical Challenges of Academic Administration (Springer, 2010), edited with Elaine Englehardt, Kerry Romesburg, and Brian Scrag; Taking Sides: Business Ethics, 11th edition (McGraw-Hill, 2011), edited with Lisa Newton and Elaine Englehardt; Engineering Ethics (Wadsworth, 5th edition forthcoming in 2013), with C.E. Harris, Elaine Englehardt, and Ray James; and Obstacles to Ethical Decision-Making (Cambridge 2013), with Patricia Werhane, Laura Hartman, Crina Archer, and Elaine Englehardt. He has been awarded several National Science Foundation and National Endowment for the Humanities grants in various areas of ethics including research ethics.

Contributors

Robert Baker earned his Ph.D. in Philosophy from the University of Minnesota and was a Postdoctoral Fellow at Albany Medical College. He serves on the editorial board of*Bioethics*, holds the William D. Williams Professorship in Philosophy at Union College where he directs the college's Ethics Across the Curriculum initiative, and is Founding Director and Professor of Bioethics in the Bioethics Program of Clarkson University and the Icahn School of Medicine at Mount Sinai. A four-time National Endowment for the Humanities awardee and a Hastings Center Fellow, Professor Baker has written, edited, or co-edited numerous books including*The American Medical Ethics Revolution*(Johns Hopkins University Press) and*The Cambridge World History of Medical Ethics* (Cambridge University Press). Oxford University Press published his most recent book,*Before Bioethics: A History of American Medical Ethics*. His current project,*A Theory of Moral Revolutions: A Study of Transformations in the Morality of Abortion, Euthanasia, and Our Treatment of Dying and the Dead*, will be published by MIT University Press.

Michael D. Burroughs is Director of the Kegley Institute of Ethics and Assistant Professor of Philosophy at California State University, Bakersfield. He is also Vice President of the Philosophy Learning and Teaching Organization. He specializes in ethics education, philosophy of education, community engaged philosophy, and social and feminist epistemology. More information on his current projects, research, and teaching can be found on his Web site.

José A. Cruz-Cruz has a B.S. in Computer Engineering, an M.S. in Information Management, and a Ph.D. in Intelligent Business Systems. Presently, he is a Professor of Information Systems in the College of Business Administration at the University of Puerto Rico at Mayagüez. For over eighteen years, he has participated in and led workshops for faculty from all disciplines in Ethics Across the Curriculum (EAC). He was the Principal Investigator of "Collaborative Development of Ethics Across the Curriculum Resources and Sharing of Best Practices", an NSF grant designed to promote the development and sharing of EAC best practices. He was also part of the team that led efforts resulting in ACBSP accreditation for UPRM's College of Business Administration.

Randall Curren is Professor and Chair of Philosophy at the University of Rochester, New York. He was the Ginny and Robert Loughlin Founders' Circle Member at the Institute for Advanced Study in Princeton, New Jersey, for 2012–2013, and was named Professor in the Royal Institute of Philosophy, London, and Chair of Moral and Virtue Education in the Jubilee Centre for Character and Virtues at the University of Birmingham, England, for 2013–2015. He is the author of*Aristotle on the Necessity of Public Education*(Rowman & Littlefield, 2000), first author of*Living Well Now and in the Future: Why Sustainability Matters*(MIT

Press, 2017) and*Patriotic Education in a Global Age*(University of Chicago Press, 2018), and editor of the*Companion to Philosophy of Education*(Blackwell, 2003).

Michael Davis is Senior Fellow at the Center for the Study of Ethics in the Professions and Professor of Philosophy, Illinois Institute of Technology, Chicago, IL 60616. Among his recent publications are: Thinking Like an Engineer (Oxford, 1998); Ethics and the University (Routledge, 1999); Conflict of Interest in the Professions (Oxford, 2001); Profession, Code, and Ethics (Ashgate, 2002); Actual Social Contract and Political Obligation (Mellen, 2002); Engineering Ethics (Ashgate, 2005); Code Writing: How Software Engineering Became a Profession (Center for the Study of Ethics in the Professions, 2009); and Ethics and the Legal Profession, 2nd edition (Prometheus, 2009). Since 1990, he has received four grants from the National Science Foundation to help integrate ethics into technical courses. Email: davism@iit.edu

Aine Donovan is a Professor of Business Ethics at Dartmouth's Tuck School of Business and the Director Emerita of the Dartmouth's Ethics Institute. Her research focus is on all areas of applied ethics, but especially the emerging issues in the intersection of business and medicine. Prior to her appointment at Dartmouth, she was a Professor of Ethics at the United States Naval Academy where she was tasked with creating a new center for military ethics and a required ethics course for all midshipmen. Her current research project at the Tuck School of Business is a collaborative venture that explores notions of the common good and economic development in Mississippi.

Deni Elliott holds the Eleanor Poynter Jamison Chair in Media Ethics and Press Policy at the University of South Florida, St. Petersburg. She is Coordinating PI on the National Ethics Project, which is funded in part by the Andrew W. Mellon Foundation and the Spencer Foundation.

Karin D. Ellison 's research, teaching, and program development activities span the areas of research ethics, ethics education development, and the history of American science and technology. As Director of the Life Science Ethics Program with the School of Life Sciences, she coordinates opportunities for students, faculty, and staff to learn about and reflect on the ethical dimensions of advances in the life sciences and life science research. Working with colleagues at the National Academy of Engineering and ASU on an NSF-sponsored project, she is leading enhancement of life and environmental science ethics education materials for the Online Ethics Center. She also fosters graduate education at ASU through her positions as interim program chair of the Masters in Applied Ethics and the Professions and as Associate Director of the Center for Biology and Society.

Challie Facemire has a Master of Science Degree in Bioethics, Policy and Law from Arizona State University. She works at the University of Phoenix in the School of Advanced Studies in Tempe, AZ.

William J. Frey is a Professor in the College of Business Administration at the University of Puerto Rico at Mayagüez. His teaching and research interests include business and engineering ethics, moral psychology, computer ethics, and research ethics. He is currently Co-principal Investigator with CRWS, "Cultivating Responsible Wellbeing in STEM" (NSF 1449489), an NSF grant devoted to integrating social, ethical, and global issues into UPRM's Science, Technology, Engineering, and Mathematics curriculum.

Joseph R. Herkert is Associate Professor Emeritus of Science, Technology and Society and a Visiting Scholar in the Genetic Engineering and Society Center, North Carolina State University. He was formerly Lincoln Associate Professor of Ethics and Technology in the Consortium for Science, Policy & Outcomes, Arizona State University (retired May 2015). He has been teaching engineering ethics and science, technology and society courses for thirty years. He is editor of*Social, Ethical and Policy Implications of Engineering: Selected Readings*(Wiley/IEEE Press, 2000) and co-editor of*The Growing Gap between Emerging Technologies and Legal-Ethical Oversight: The Pacing Problem*(Springer, 2011), and has published numerous articles on engineering ethics and societal implications of technology in engineering, law, social science, and applied ethics journals. He previously served as editor of*IEEE Technology and Society Magazine*and an associate editor of*Engineering Studies*. He has been a leader in many professional organizations including the Society on Social Implications of Technology (SSIT) of the Institute of Electrical and Electronics Engineers (IEEE), the National Institute for Engineering Ethics, and the Engineering Ethics and Liberal Education/Engineering and Society (LEES) Divisions of the American Society for Engineering Education. In 2005, He received the Sterling Olmsted Award, the highest honor bestowed by LEES, for "making significant contributions in the teaching and administering of liberal education in engineering education."

Karlynd June holds an M.A. in Journalism and Media Studies and B.A. in Mass Communications from the University of South Florida, St. Petersburg and is Project Manager and Research Associate for the National Ethics Project, which is funded in part by the Andrew W. Mellon Foundation and the Spencer Foundation.

Robert F. Ladenson is Professor of Philosophy Emeritus at the Illinois Institute of Technology. He is the creator of ethics bowl and from 1997 to 2009 was the principal organizer and developer of the Association for Practical and Professional Ethics Intercollegiate Ethics Bowl (APPE-IEB).

Christopher Meyers is Professor Emeritus of Philosophy and Director Emeritus of the Kegley Institute of Ethics at CSU, Bakersfield. He is also Adjunct Ethics Faculty at Kern Medical Center, Hospital Ethicist for Adventist Healthcare Bakersfield, and Associate Editor for Journal of Media Ethics. He received his Ph. D. in Philosophy and Academic Certificate in Bioethics from the University of Tennessee and is the author of three books and some four dozen scholarly articles. His hobbies include wandering the Eastern Sierra mountains, woodworking, and standing on the bank of a trout stream, frustratingly watching trout laugh at him as his fly drifts over their head.

Glen Miller is an Instructional Assistant Professor at Texas A&M University. He received his Ph.D. in Philosophy from the University of North Texas. His current research triangulates the history of philosophy, especially ethics and political philosophy, and two emerging areas of ethical concern, the environment and technology. He regularly teaches a large course on engineering ethics and investigates issues in applied, practical, and professional ethics, including bioethics and cyberethics. He is the author of several encyclopedia articles, journal articles, and chapters.

Deborah S. Mower is the Mr. and Mrs. Alfred Hume Bryant Chair of Ethics and an Associate Professor of Philosophy at the University of Mississippi. She specializes in moral psychology, moral education, and applied ethics. She received her Ph.D. in philosophy from the University of Wisconsin, Madison. She is an Active Member of the Association for Practical and Professional Ethics and the current President of the Society for Ethics Across the Curriculum, which promotes scholarship on ethics and moral philosophy as it intersects with moral psychology, citizenship education, and general education as well as the teaching of ethics in all academic disciplines. In 2012, she co-edited*Civility in Politics and Education* (Routledge) with Wade L. Robison. In 2015, she co-edited*Developing Moral Sensitivity*(Routledge) with Phyllis Vandenberg and Wade L. Robison. In 2016, she co-directed (with Phyllis Vandenberg) a National Endowment for the Humanities Summer Institute on "Moral Psychology and Education: Putting the Humanities to Work" (http://www.moralpsychology.co). Twenty-five scholars and seventeen internally recognized faculty studied the moral psychology behind effective moral education, examining how specific areas of the humanities (art, music, the literature, and film) enhance the development of moral reasoning, skills, and action.

Lisa H. Newton is a Professor Emeritus at Fairfield University. She joined the faculty in 1969 and taught courses in ethics, applied ethics, environmental studies, healthcare ethics, and several other areas. She was the Director of the Applied Ethics Center at Fairfield University. She was also the Director of the Program in Environmental Studies at Fairfield University. She received her Ph.D. in Philosophy from Columbia University in New York City. She served for many terms on the Executive Committee of the Association for Professional and Practical Ethics. She is the author of twenty books and hundreds of articles. She received her Ph.D. from Columbia University.

David T. Ozar is Professor Emeritus in the Department of Philosophy, Loyola University Chicago, having taught there from 1972 to 2015 and also taught at various times in Loyola's schools of medicine, nursing, business, law, education, social work, and dentistry. He was Director of Loyola's Center for Ethics, 1993–2006. He earned his Ph.D. in Philosophy at Yale University in 1974. In addition, he has served for more than 30 years as Associate Member of the Professional Staff, Member of the Institutional Ethics Committee, and Consulting Ethicist for NorthShore University HealthSystem in Evanston, Illinois. He has served on committees and consulted for numerous professional organizations in health care and many other professions. He was the Founding President of the American Society for Dental Ethics in 1987 and a Founding Member of the Society for Ethics Across the Curriculum. He is a frequent lecturer to university, professional, and community groups on healthcare ethics, professional ethics, ethics education, and contemporary social issues. In addition to two books, he has published more than 160 articles on these topics in books and professional journals.

Alan A. Preti is the Director of the Institute for Ethical Leadership and Social Responsibility and Associate Professor of Philosophy at Rosemont College in Philadelphia, Pennsylvania. His research interests include theoretical and applied ethics, moral psychology, and moral development. He also has interests in Asian philosophical traditions. He has published the*International Journal of Ethics*and*Philosophy East and West*, and has contributed chapters to*Handbook of Virtue Ethics in Business and Management*(Springer) and*Developing Moral Sensitivity*(Routledge). As an Executive Committee Member of the Society for Ethics Across the Curriculum, he has also served on the Board of Directors of the Greater Philadelphia Philosophy Consortium since 2008.

Wade L. Robison is the Ezra A. Hale Professor of Applied Ethics at the Rochester Institute of Technology. He has published extensively in philosophy of law, David Hume, and practical and professional ethics. His book*Decisions in Doubt: The Environment and Public Policy*(University Press of New England, 1994) won the Nelson A. Rockefeller Prize in Social Science and Public Policy. His latest book is*Ethics Within Engineering: An Introduction.*

Brian Schrag was, for twenty years, the Founding Executive Director of the Association for Practical and Professional Ethics. For eleven years, he served as project director of a collaborative effort of the association for a project, "Graduate Research Ethics Education" (GREE) funded by the National Science Foundation. He has published articles on research ethics, research ethics education, academic ethics, museum ethics, business ethics, civility and community, and allocation of medical resources.

Alan Tomhave is Chair and Associate Professor of Philosophy at Youngstown State University.He is the co-editor of the journal Teaching Ethics.He received his PhD from the University of Missouri in social and political philosophy and Ethics.

Phyllis (Peggy) Vandenberg is a Professor of Philosophy at Grand Valley State University and received her Ph.D. in Philosophy from the University of Utah. She serves on the Executive Board for the Society for Ethics across the Curriculum and is a Regular Presenter at the Association for Practical and Professional Ethics. She co-edited and contributed to the Routledge, 2015, volume,*Developing Moral Sensitivity*(with Deborah S. Mower and Wade L. Robison). She co-directed (with Deborah S. Mower) the 14th International Conference of the Society for Ethics across the Curriculum (SEAC) in 2012 and (with Erik Wingrove-Haugland) the 19th International SEAC Conference, 2017, both at Grand Valley State University. In 2015, she co-directed with Deborah Mower a four-week National Endowment for the Humanities Summer institute,*Moral Psychology: Putting the Humanities to Work*. She is a contributor and reviewer for the Internet Encyclopedia of Philosophy and also has reviewed for Teaching Ethics and the NEH. Her presentations and publications are primarily on the ethical theories of the 18th century Scots: Francis Hutcheson, David Hume, Adam Smith, and Thomas Reid. Her latest pieces are "Relationships and the Spectator Perspectives in Hutcheson, Hume, and Smith" in*Cultura: International Journal of Philosophy of Culture and Axiology*, "Reid and Hume in Agreement on Moral Foundations" in*Journal of Scottish Thought* (University of Aberdeen Press 2011), and "A Humean Look at Feminist's Ethics"*Special European Legacy Edition on Hume (*Routledge, June 2013).

Mark Vopat is Professor of Philosophy and Director of the Dr. James Dale Ethics Center at Youngstown State University. He and his colleague Alan Tomhave are co-editors of the journal Teaching Ethics. His research interests include social and political philosophy, ethics, and applied ethics with particular emphasis on children's rights. He has written and presented papers in a variety of areas of applied ethics including: business ethics, professional ethics, the ethics of boycotting, moral sensitivity and the intersection of libertarianism and Christianity. In 2015, he published a book Children's Right and Moral Parenting with Lexington Press.

Sandy Woodson is a Teaching Professor in the Division of Humanities, Arts, and Social Sciences at the Colorado School of Mines, where she has taught Ethics and Environmental Philosophy since 1999. She is the Director of the Ethics Across Campus Program, the Undergraduate Adviser for the HASS division, and she coaches the Mines Intercollegiate Ethics Bowl team.

Daniel E. Wueste is Professor of Philosophy at Clemson University. He was Founding Director of the Rutland Institute for Ethics; he stepped down as director after fifteen years, in the summer, 2016, and returned to teaching and research. His research and writing focus on issues in three areas: practical and professional ethics, legal philosophy, and social and political philosophy. He has a special interest in what ethicists can learn from legal philosophers and vice versa. His work has appeared in various journals including*Cornell Law Review, Harvard Journal of Law and Public Policy, the Canadian Journal of Law and Jurisprudence, Teaching Ethics, Dimensions of Critical Care Nursing, Critical Reviews in Biomedical Engineering, The Annals of Thoracic Surgery,*and*Research in Ethical Issues in Organizations.*In addition to teaching in the Department of Philosophy and Religion, he teaches in two of Clemson's Ph.D. Programs: Healthcare Genetics and Policy Studies. He is Member/Researcher at the Institute of Human Values in Health Care, Medical University of South Carolina, and a Member of the Executive Board and Treasurer of the Association for Practical and Professional Ethics. He has been a Member of the Society for Ethics Across the Curriculum since its inception and is a Member of its Executive Committee and proudly served two terms as its president.

Qin Zhu is a Research Assistant Professor in the Ethics Across Campus Program and the Division of Humanities, Arts, and Social Sciences at the Colorado School of Mines. He holds a Ph.D. in Engineering Education from Purdue University, USA, and a Ph.D. in Philosophy of Science and Technology from Dalian University of Technology, China. His research focuses on ethical decision-making in engineering, ethics education assessment, and ethics of technology.

Introduction

Elaine E. Englehardt and Michael S. Pritchard

Abstract Traditionally confined largely to programs in philosophy and religion, the teaching of ethics has in recent decades spread virtually across the curriculum of higher education. The contributors to this book discuss the rationale for supporting efforts to teach ethics across the academic curriculum, the variety of challenges such efforts face, and the sorts of benefits faculty and students who participate in ethics across the curriculum endeavors can expect.

Keywords Teaching ethics · Practical ethics · Academic integrity
Institutional programs · Research ethics · Assessing ethics education

Teaching Ethics: Where and Why?

Traditionally confined largely to programs in philosophy and religion, the teaching of ethics has in recent decades spread virtually across the curriculum of higher education. The contributors to this book discuss the rationale for supporting efforts to teach ethics across the academic curriculum, the variety of challenges such efforts face, and the sorts of benefits faculty and students who participate in ethics across the curriculum endeavors can expect.

According to the Society for Ethics Across the Curriculum (SEAC) (2000), EAC refers to "the teaching of ethics in all academic disciplines." EAC is grounded on

The original version of this chapter has been revised: The co-author name and her email address has been corrected to Karlynd June and karlyndjune14@gmail.com. The correction to this chapter can be found at https://doi.org/10.1007/978-3-319-78939-2_26

E. E. Englehardt (✉)
Utah Valley University, Provo, UT USA
e-mail: elainee@uvu.edu

M. S. Pritchard
Western Michigan University (Emeritus), Kalamazoo, MI USA
e-mail: michael.pritchard@wmich.edu

E. E. Englehardt and M. S. Pritchard (eds.), *Ethics Across the Curriculum—Pedagogical Perspectives*, https://doi.org/10.1007/978-3-319-78939-2_1

the notion that the teaching of ethics should not be restricted to one or two courses in the philosophy or religion departments but rather be covered throughout the curriculum. We agree and believe it is important to note that ethics across the curriculum can be understood from at least two fundamental perspectives. First, it can be viewed from within individual academic disciplines. For example, accredited engineering programs are expected to ensure that their students are introduced to the ethical dimensions of engineering. This can involve consideration of ethical issues within particular areas of engineering (e.g., civil, mechanical, electrical, chemical) as distinctive segments of certain courses (e.g., those that focus on design problems), or as a full semester course in ethics in engineering. Similar approaches can be taken in nursing, medicine, law, social work, psychology, accountancy, management, and so on. That is, some emphasis on ethics can be found in a variety of academic disciplines.

However, second, this approach does not ensure that ethical issues will be seen as possibly cutting across different parts of the curriculum, or even all of the curriculum. Many ethical issues require careful attention from the perspectives of several disciplines at once, and in ways that require their joining hands. For example, engineers design traffic roundabouts. However, well designed roundabouts must consider not only the safe, efficient travel of vehicles, but also the safety and welfare of pedestrians who cross the roads near them. A common oversight in the initial design of roundabouts was the failure to take into account the needs of pedestrians who have limited or no vision. Traffic signals that require vehicles to stop before proceeding provide critical sound cues for those with serious visual limitations. Roundabouts do not. So, there is a need for collaboration between traffic engineers and those with expertise in attending to the rights and needs of those with serious visual disabilities.

Recognizing that adequately addressing many ethical issues may require the inclusion of perspectives from a variety of disciplines makes apparent the need for effective communication and reflection across disciplines, not simply within them. This, in turn, suggests that faculty and their students can benefit from special programs that are designed to include participants from a variety of disciplines. Such programs will be a central feature of this book. Some issues, such as those pertaining to academic integrity or environmental sustainability can surface in virtually any discipline. Although some differences may arise in how such issues can best be discussed across different parts of the curriculum, these discussions might be joined in ways that help students, faculty, administrators, and the wider public in higher education better appreciate their shared ethical ground.

EAC provides opportunities for using a variety of classroom methods. Many teachers favor using case studies written by professors, professionals and students. Additionally, many EAC courses are team taught by a professor and a professional, or two professors from different disciplines.

Students

It is clear that in all walks of life, professional or otherwise, college students will encounter challenging ethical problems. As they leave our institutions of higher learning, how well prepared will they be to recognize, analyze, and constructively address these problems? More to the point, what responsibilities, if any, do the institutions from which they emerge have to help prepare students to anticipate and successfully address these problems?

Although, as we have said, traditionally courses in ethics have been readily available in philosophy and religion departments, most colleges and universities do not require students to take such courses. Furthermore, until quite recently, few of these courses have been designed to help students anticipate the sorts of ethical problems that, say, those entering specific vocational areas, such as business, engineering, psychology, social work, or various technical fields, will encounter. Helping students understand and clearly address ethical problems in these more specialized fields requires instructors who can talk about, not only ethics in general (as philosophy and religion courses historically have done), but also ethical problems as they arise concretely in those more particular contexts.

EAC Models

One (but not the only) promising way of helping students become more ethically prepared for their future work is to develop comprehensive "ethics across the curriculum" (EAC) programs. It is not enough to have some sort of exposure to ethics occur in different parts of the curriculum, leaving it to students to figure out for themselves how these separate and independent exposures might be connected. Equally important, the faculty need to be acquainted with one another's interests in ethics to be able to make connections that will help broaden and deepen the focus of their students.

Beginning in 1986 with a series of national grants to then Utah Technical College, Elaine E. Englehardt initiated a series of EAC faculty workshops at her own institution, as well as at a variety other colleges and universities. Some 30 years later, Utah Technical College has become the largest university in the state of Utah, Utah Valley University (UVU). The EAC faculty programs remain a staple feature of UVU.

The UVU workshops appeal to faculty from a variety of disciplines. Faculty attend workshops and seminars to strengthen their understanding of ethics. Each summer a widely recognized ethics scholar is invited to UVU to lead a week-long workshop for interested faculty. Follow-up meetings encourage faculty to develop case studies and to change their syllabi to reflect some ethics content. After participating in these programs, faculty use cases to engage their students with ethical issues. This enables faculty and students to understand how ethics can permeate the disciplines.

As contributors to this volume illustrate, there are many possible models for EAC programs. Although they may be readily replicable, their implementation is not always easy. Of course, a first requirement is to ensure that there is a sufficient number of faculty who are both able and willing to devote their energies to make a given program work well. Also, some resistance may be offered by some in the academic institution or in the surrounding community who either value only one side of an ethics problem or who fear that an EAC program itself will attempt to impose a narrow ethical perspective on students. However, insofar as EAC programs strive to be open to multiple perspectives, the potential gains for faculty and students are substantial. EAC programs can aid faculty in designing syllabi that will improve their students' understanding of ethics by helping them to:

- expand and deepen their moral sensitivities and assist them in examining the nature of their ethical assumptions.
- strive for clarity and consistency in their value frameworks.
- critically examine relevant facts and develop decision-making strategies for resolving ethical issues.
- increase their awareness and understanding of current ethical problems, whether in their areas of specialization or more generally in society.

Part I begins with a brief history of how and why the study of ethics has both broadened and deepened its place in higher education. It addresses fundamental questions about the relationship between theory and practice in the study of ethics, and it wrestles with questions about how philosophers, in particular, can best participate in this expanded role of ethics in education. The following are descriptions of the entries in Part I.

Deni Elliott and Karlynd June's essay traces the development of ethics education in U.S. higher education from 1980–2015. It discusses how ethics education aligns with the moral purposes of higher education, the growth in scholarly literature, and how ethics education has become an integral part of many college and university campuses.

Michael Davis asks what part moral theory should have in ethics across the curriculum? After distinguishing five kinds of ethics across the curriculum, he argues that the part moral theory should have depends on the kind involved.

David Ozar argues that assessing the effectiveness of an ethics across the curriculum program depends on having clear answers to two questions: (1) Who is it that the EAC program is intended to serve?; and (2) What good is the program intended to achieve for them? His essay offers detailed answers to these two questions, with special attention to identifying learning objectives for ethics education programs. It then explains how to use this information to assess an EAC program's effectiveness and describes the value of several other assessment questions relevant to EAC programs.

Deborah Mower's essay presses us to consider what our ethical obligations are in educating students and what psychological capacities we should seek to develop for moral action. Moral sensitivity is a complex discriminative capacity that enables us to interpret situations and determine a morally appropriate course of action. Her essay addresses various concrete strategies to develop moral sensitivity both within a stand-alone professional ethics course as well as within a comprehensive EAC program.

Glen Miller considers that the idea of identity development can orient the diverse elements that constitute professional ethics programs. Identity is crafted by answering qualitative questions that define one's self, which is shaped by what one has done, what one aims to do, and what one holds as significant. Understood in this way, professional ethics encourages students to articulate and develop a coherent moral identity that combines personal and professional intentions, actions, and goals and is informed by standards, practices, and exemplars of the profession.

Alan Tomhave and Mark Vopat, the current editors of *Teaching Ethics*, give a basic overview of the goals, purpose, and contribution of this periodical, the official journal of the Society for Ethics Across the Curriculum (SEAC). They discuss six areas of clear benefit offered by *Teaching Ethics*.

Part II discusses challenges posed by efforts to teach ethics across the curriculum. Elaine Englehardt and Michael Pritchard discuss what teaching practical ethics well requires of faculty, particularly those whose discipline is other than philosophy. However, they also discuss challenges that philosophers face in attempting to keep ethics practical rather than allow it to become absorbed by theoretical considerations whose practical implications are at best unclear.

Christopher Meyers argues that those engaged in the work of practical and professional ethics should adopt what Nick Fotion calls a "weak" approach to ethical theory and reasoning, one that is situated within a practice-driven attitude toward ethics engagement. This approach insists that ethical reasoning should strive to find real solutions to real world problems while also embracing the consequence that such solutions are frequently only tentative. The resulting model embraces key insights from the great ethical traditions and works to merge them into a practical method of ethical reasoning.

Alan Preti explores ways in which John Dewey's moral philosophy anticipated current trends in applied and practical ethics. He introduces Dewey's conception of moral inquiry and then moves to a discussion of contemporary moral decision-making models and accounts of moral imagination. He concludes that practical ethics instructors and their students would be well-served by a Deweyan approach to moral inquiry.

Lisa Newton observes that most students no longer come to college for the reasons that used to draw them, and she urges that the teaching of ethics should reflect this change. More than ever, students are focused on life in the economic world that they will enter immediately upon graduating. For the professionally oriented, there is a rich literature on professional ethics; for the others, there are exciting experiments in experiential learning and internships that can be tailored to incorporate ethically problematic situations like those that students will likely encounter.

Wade Robison maintains that questions about what it is to be a professional of a particular kind can be contested. For example, it is a continuing disagreement in law whether lawyers should be hired guns, doing whatever needs to be done, even pushing against the law, to help their clients, or whether they are guardians of the law, with a higher obligation to justice. He argues that this kind of internecine dispute is a neglected aspect of professional ethics, but worth exploring because different conceptions of a profession have competing ethical implications.

Phyllis Vandenburg writes about challenges academic philosophers face when engaging with those outside their academic discipline. Of special concern is the involvement of philosophers in civic and public affairs where there is a need to discuss issues in practical ethics. This includes making policy in government, corporations, professions, or in any societal settings where there is need for practical determinations of an ethical nature.

Part III examines ethical topics that are especially well suited for discussion across the curriculum, such as research ethics, pre-college ethics education, sustainability issues, what has come to be known as the ethics bowl, and academic integrity.

Brian Schrag addresses the nature of research ethics, and he discusses opportunities in the university provide effective research ethics education. He provides an account of the wide range of ethical issues in research, and of the practical moral recognition and reasoning required to address those issues and their implications for research ethics. He draws on his experiences as leader of an 11 year NSF funded project, "Graduate Research Ethics Education" (GREE), as well as the reported experiences of graduate students and post-doctoral alumni in the program as they later encountered challenges and opportunities to do research ethics education in their own careers.

Michael Burroughs presents a range of approaches to early childhood ethics education, arguing for the benefits of methodological pluralism. He stresses the importance of identifying the many continuities and opportunities for collaboration across the theoretical and practical divisions set up in the field of ethics education. Finally, he calls for collaborative teacher-researcher partnerships in order to develop effective ethics education programming.

Michael Pritchard argues that integrating ethics into science classes requires a kind of critical thinking about values in science that can play a significant role in fostering the reasonableness of students. He offers several reasons for concluding this is an appropriate objective of science education in the schools, even at the middle and elementary levels.

Randall Curren notes that students, faculty, and administrators have all played roles in an expansion of sustainability-related teaching cross the collegiate curriculum in recent years. This has involved faculty in many different fields, and those without a background in ethics struggle with how to address the obvious yet ill-defined normative aspects of sustainability. This essay provides such faculty with some ways of thinking about ethics across the curriculum and outlines an approach to sustainability ethics that may be useful in a variety of curricular contexts.

Robert Ladenson seeks to make apparent that the basic educational objectives of the ethics bowl coincide closely with those of ethics across the curriculum. He describes the format, procedures, and rules of the Association for Practical and Professional Ethics Intercollegiate Ethics Bowl (APPE-IEB) and sets forth an interpretive analysis of the basic educational objectives implicit in the APPE-IEB as it has developed over the past twenty years. The essay concludes that ethics bowl qualifies for inclusion in a "toolkit" of educational approaches for implementing ethics across the curriculum.

Daniel Wueste's essay argues that there is a connection between academic integrity and teaching ethics across the curriculum. The twofold purpose of the essay is to (a) explain the connection and (b) to make the case for the approach it involves. The essay draws on the work of Jonathan Haidt, Oliver Wendell Homes Jr., and, especially John Dewey.

Part IV describes some of the institutional programs that have been able successfully to launch ethics across the curriculum programs. These EAC programs function differently at each institution, according to the needs of the faculty, students, community and administration.

Elaine Englehardt discusses the pioneering efforts of Utah Valley University (UVU) in designing and implementing an Ethics Across the Curriculum program (EAC) with funding help from National Endowment for the Humanities (NEH) and continued support from the Fund for the Improvement of Post-Secondary Education (FIPSE). Her essay explains UVU's advancement from the modest beginnings of a core ethics course to the establishment of a vibrant ethics across the curriculum program housed in UVU's Center for the Study of Ethics.

Karin Ellison, Challie Facemire, and Joseph Herkert explain how the School of Life Sciences at Arizona State University began a systematic expansion of an ethics across the curriculum program as part of its Life Science Ethics Program in the fall of 2015. They present the initial elements of that program, review relevant literature on ethics across the curriculum programs and responsible conduct of research education programs. They conclude with lessons from that literature for ethics education program development.

Robert Baker notes that Wilhelm Dilthey (1833–1911) and other German philosophers classify a certain type of literature as *Bildungsroman*, a concept loosely translated as a "coming of age narrative." Baker gives an account of a *Bildungsroman* of an ethics-across-the-curriculum initiative, offering an account of its coming of age at Union College, a small liberal arts college. As is characteristic of *Bildungsroman,* it tells tales of daunting challenges and draws lessons for the edification of anyone seeking to start an ethics-across-the-curriculum initiative.

Sandy Woodson and Quin Zhu present the historical background and institutional context within which the Ethics Across Campus (EAC) Program at the Colorado School of Mines was created and has evolved. Focusing on the EAC program's major campus and curricular initiatives designed to enhance the education of Mines students, a variety of activities is described. There is also a discussion of some specific challenges faced by the EAC program in its commitment to further cultivate an ethical campus climate for STEM education and research.

Aine Donovan's essay presents an overview of one of the earliest ethics across the curriculum programs, designed specifically to enhance liberal arts teaching and research. She observes that the Dartmouth academic community has long had an emphasis on moral meaning-making, but the threads of that interest were varied and not connected to the mission of the college. The EAC program sought to integrate the mission and the educational objectives into the framework of ethics education as a necessary component of a robust liberal arts education. EAC at Dartmouth, now in its 15th year, has become a much sought-after opportunity for faculty and staff.

William Frey and Jose Cruz's essay describes over three decades of EAC experiences at the University of Puerto Rico at Mayagüez. It outlines strategies for sharing EAC best practices and provides methods for refashioning materials from faculty workshops into EAC teaching modules. The goal is to provide readers with a roadmap for developing a successful EAC program.

Reference

Society for Ethics Across the Curriculum. (2000). Society for ethics across the curriculum. http://www.rit.edu/~w-ethics/seac/.

Part I
The Changing Landscape in Teaching Ethics

The Evolution of Ethics Education 1980–2015

Deni Elliott and Karlynd June

Abstract Ethics education became an integral part of most U.S. institutions of higher education between 1980 and 2015. Growth can be seen in institutional messaging, number of courses in ethics offered throughout the graduate and undergraduate curricula, national recognition of degrees and certificates granted in ethics by the federal National Center for Educational Statistics, creation of campus-wide ethics centers and co-curricular initiatives, and an explosion of peer-reviewed journals in the intersection of disciplinary areas and ethics. Yet, much research is yet to be done. Connections between ethics education and students' civic and moral development remain unclear. The impact of ethics education remains unknown. There is no consensus on what counts as effective ethics education. Student voices are largely absent from discussions on the topic. And conversations relating to curricular and co-curricular ethics education continue to be divorced from analysis of the ethical implications of institutional choices.

Keywords Ethics · Morality · Values · Ethics education · Moral education
Higher education · Moral development · Institutional ethics · Education ethics
Student ethics

Long time observers of U.S. higher education have witnessed a series of shifting trends in mission and purpose. In some periods, the priority is to graduate students with civic responsibility. Then, for a while, it's vocational readiness. Sometimes

---

The original version of this chapter has been revised: The co-author name and her email address has been corrected to Karlynd June and karlyndjune14@gmail.com. The correction to this chapter can be found at https://doi.org/10.1007/978-3-319-78939-2_26

D. Elliott (✉) · K. June
University of South Florida, St. Petersburg, FL USA
e-mail: elliott@mail.usf.edu

K. June
e-mail: karlyndjune14@gmail.com

E. E. Englehardt and M. S. Pritchard (eds.), *Ethics Across the Curriculum—Pedagogical Perspectives*, https://doi.org/10.1007/978-3-319-78939-2_2

stimulating students' intellectual and moral development for their own intrinsic good is in the background. Other times this goal is front and center. Priorities shift one to another and back over time. The stated purpose of higher education reflects political and social expectations of the era as well as the character of the institution and the branding by leadership at a particular moment in time.

Occasionally, an idea takes hold that creates fundamental change in how higher education is understood, how its purposes are achieved, or in how its achievements are measured. An analysis of 35 years of artifacts provides evidence that ethics education is an idea of this type. Ethics education, both as a basis for, and style of, critical inquiry, seems to be here to stay. Between 1980 and 2015, ethics education became embedded in the mission, vision and values statements of institutions of higher education, in written policies, in academic programs leading to certificates and degrees, in the creation of ethics courses across the curriculum, and through co-curricular activities with implicit or explicit focus on ethics. It also found its place in scholarly literatures. Practitioners, policymakers and critics turned attention to ethical issues internal to the university as well. For example, in the late 20th century, faculty research misconduct and student cheating were noted as areas of ethical concern on campus that demanded the institution's response. Faculty conflict of interest policies and disclosure of external support became common in this period. Power inequity was noted as a fatal ethical flaw in faculty-student romances. Full and part-time faculty salaries, unionization of graduate students, along with institutional purchasing and investment choices were recognized as having a previously ignored ethical component.

In this chapter, we examine the most sustained "ethics boom" (Davis1999) in the history of U.S. higher education. This boom was first formally noted and analyzed by a research team convened by The Hastings Center in the late 1970s, resulting in seminal essays and monographs about ethics education at US colleges and universities.[1] The Hastings Center's materials, published in 1980, comprised the first landscape study of the teaching of ethics in US institutions of higher education.[2] Our examination of artifacts launches from this foundation and ends in 2015, as that

[1]Along with multiple presentations at academic conferences, in 1980, the team published a book of collected essays,**Ethics Teaching in Higher Education** (Daniel Callahan and Sissela Bok, editors), and nine monographs on the teaching of ethics,**The Teaching of Ethics in Higher Education** (by The Hastings Center),**Legal Ethics and Legal Education** (by Michael J. Kelly), **Teaching Ethics in Journalism Education** (by Clifford G. Christians and Catherine L. Covert), **Teaching Bioethics: Strategies, Problems, and Resources** (by K. Danner Clouser),**Ethics in the Education of Business Managers** (by Charles W. Powers and David Vogel),**The Teaching of Ethics and the Social Sciences** (by Donald P. Warwick),**Ethics and Engineering Curricula** (by Robert J. Baum),**Ethical Dilemmas and the Education of Policymakers** (by Joel L. Fleishman and Bruce L. Payne) and**Ethics in the Undergraduate Curriculum** (by Bernard Rosen and Arthur L. Caplan).

[2]The Hastings Center study included a systematic survey of literature on the teaching of ethics in American higher education, review of 2000 college catalogs, consultations with more than 1000 teachers of ethics, a summer workshop for 150 participants, using a grounded theory approach to identify common practices and patterns along with problems and issues in ethics education.

is the last year that material could fully be captured at the time of this writing. Specifically, we look at how ethics education supported the stated purposes and implicit values of higher education, examine trends in academic writing relating to ethics education, and conclude with a survey of some of the different ways that ethics education has played out on U.S. campuses, including a comparison of how instructional goals, student outcomes, and pedagogy and assessment have been discussed. While significant attention has been focused on many aspects of ethics education, we also identify areas that are in need of more systematic attention.

The Moral Purpose of Higher Education

Throughout history, higher education has been expected to play a role in developing students' moral and civic capacities, regardless of students' particular fields of study. One can reach back to Plato's utopian dialogue,*The Republic*, or Aristotle's experimental school,*The Lyceum*, to appreciate the long-held view of the importance of education in preparing future leaders who have the character necessary to govern. More recently, 19th century British philosopher, John Stuart Mill identified education as the social construction that made it possible for autonomous individuals to understand that one's own happiness was dependent on the health and happiness of the community within which they lived (Mill1863/1991, p. 166). This realization was an important step in moral growth and development for all citizens, not just that of future leaders. In 1945, then Harvard President James Bryant Conant echoed Mill in prescribing an education that produced both good individuals and good citizens by "balancing free enquiry and critical individualism with the necessity for individuals to 'subordinate their individual good to the common good'" (Keohane2006, p. 99).

Teaching ethics in earlier periods of higher education was meant to promote and reinforce community standards. In 19th century America, for example, often it was the school's president who taught required capstone courses or delivered lectures to reinforce the virtues deemed important for moral leadership in the ministry, government and law.[3] The President as instructor highlighted the significance of the lesson. Professional associations in engineering, journalism and law trace their first codes of ethics to the 1920s, which were given to practitioners as sanctioned statements of values and expectations and provided to students as standards of the professions that they had chosen to follow.

In contrast, in contemporary teaching, ethics "is treated as a subject in which controversy is normal, argument is appropriate, and answers are to be worked out in a shared search for the best reasons." A profession's code of ethics "is not just handed down. It is treated as a historical artifact to be examined, appraised, defended, or condemned" (Davis1999, p. 15). While scholars (Colby et al.2003) and

[3]See Eliot (1869) and Stearns (1908).

stakeholders (Association of American Colleges & Universities) agree that colleges and universities have an educational and civic obligation to unapologetically teach for personal and social responsibility, the effectiveness of implicit or explicit attempts to meet the obligation are largely unknown (Dey and Associates2010).

The values construct of institutions of higher education is not limited to the ethical perspective that graduates might or might not have as they exit from their college years. Institutions of higher education themselves are dependent on shared values to promote student learning and to sustain the environment that supports the simultaneously collaborative and competitive work of seeking new knowledge. Shared values can be extrapolated from expectations for classroom conduct, research procedures, and conventions of residence hall co-habitation. Values weave through the curriculum, co-curricular activities and campus life. These values include honesty, integrity, self-discipline, "mutual respect, open-mindedness, the willingness to listen to and take seriously the ideas of others, procedural fairness, and public discussion of contested issues" (Colby et al.2003, p. 13; Wolff1994, p. 106). Roberts (1999) adds that values central to education include "those associated with the promotion of questioning, dialogue and reflective human activity" (p. 19). Ebels-Duggan (2015) includes "intellectual charity," to the list, noting that,

> The intellectually charitable person approaches new ideas and texts with the presumption that there is something true and worthwhile to be found there. He or she thus refrains from immediate criticism, striving first to understand the positions and to reconstruct them in a way that brings out what seems most plausible. Humility is a corresponding attitude governing one's relationship to one's own view. (p. 82)

While the campus, as a whole, is rich with opportunities for ethics learning (Colby et al.2003, p. 277), ethics education has been assumed to produce more consistent results when it is offered in designated courses than when relying on students' informal socialization to facilitate their civic and moral development, or on the tendency for controversies to surface here and there during the course of classroom discussions. Classes in ethics theoretically provide opportunities for students to critically examine values and ethical issues, to reason about the issues and to examine the justifications for holding particular values (Ozar1977; Davis1990; Whitbeck1995; Matchett2008). Mayhew and King have suggested that a key element for ethics education is that specific courses in the field "encourage perspective-taking or that they provide structured opportunities to practice moral decision making" (Mayhew and King2008, p. 36).

Ethics education typically includes the teaching of substantive content as well as the development of ethical reasoning skills. Content may include philosophical theories that provide the foundation for systematic moral analysis. It is likely to include examinations of the major historical and contemporary controversies within a particular discipline or field of study. Ethical reasoning "requires students to be able to assess their own ethical values and the social context of problems, recognize ethical issues in a variety of settings, think about how different ethical perspectives might be applied to ethical dilemmas and consider the ramifications of alternative actions" (Association of American Colleges and Universities2010). Ethical reasoning is likely to include the teaching of argument construction, logical analysis and fallacies. Colby et al. (2003) see the importance of ethics coursework as:

> working to move students beyond moral relativism, supporting deep understanding of and personal connections with ethical concepts, teaching the skills of moral discourse, promoting the values and themes that are central to the institution's goals for moral and civic education, and supporting transfer of learning to contexts beyond the classroom. (p. 142)

In summary, policymakers, scholars and stakeholders think that it is important that institutions of higher education meet their moral purpose of producing graduates who perceive themselves as having personal, civic and social responsibilities. Courses in ethics, no matter where in the curriculum that they appear, are considered important in helping to achieve this goal (Dey et al.2009). To date, however, there is little evidence that ethics education, however it is delivered in higher education, is responsive to the myriad hopes, assumptions and expectations associated with it.

Ethics Education Within Academic Literature

Thinking and writing about ethics exploded 1980–2015 in scholarly literature and lay discussions alike. Print and broadcast news magazines gave voice to discussions of equity and fairness, political ideology, self-determination, and imperialism. Where past generations might have trusted the government to make choices on their behalf, the contemporary generation demanded to know why. Public philosophers and their books became staples on television talk shows as well as in classes taught and taken by non-philosophers. Well-respected philosophers Sissela Bok, Michael Boylan, Philippa Foot, Martha Nussbaum, Lisa Newton, Michael J. Sandel, and Peter Singer, among others, produced non-fiction works on topics in ethics that sold well in the trade press in addition to their writings that appeared in philosophy journals.*Ethics in America*, a show produced for PBS by Columbia University in 1989, brought prominent lawyers, articulate philosophers, and important contemporary leaders from government, business and media, together to discuss controversial topics of the day. The Socratic questioning and roundtable discussion modeled ethical reasoning and civil dialogue for a national audience.

As journal publication is the coin of the realm for achievement in higher education, we limited our examination of scholarship on ethics education to these peer-reviewed publications. While ethics was becoming a concept discussed at the dinner table, in scholarly literature, it simultaneously morphed into a field with sub-disciplines. Peer-reviewed journals specific to disciplinary or topical areas of practical ethics, such as medical ethics, business ethics, engineering ethics, and environmental ethics grew from fewer than 20 journals prior to 1980 to 145 by 2015. In addition, a journal devoted specifically to pedagogy for ethics education, *Teaching Ethics Journal*, was founded in 2001.

A selection of journals in ethics and journals in higher education were examined here to capture trends and major developments in the field.[4] A birds-eye view revealed an increase in the number of journals that focus on practical ethics alongside an upward trend in scholarship on ethics education published in*The Journal of Higher Education*. On the other hand, traditional journals in moral philosophy, such as*Ethics*, maintained their distance from writings in practical ethics education, as did three out of four prestigious journals in higher education.

The first of the two analyses we conducted tracked ethics education trends in flagship journals of higher education. These journals are important in establishing which aspects of ethics education have been of interest to researchers and readers of scholarship who study higher education. The second analysis examined the most prestigious journals in the two most prolific areas within practical ethics: medicine/ bioethics,*The Journal of Medical Ethics (JME),* and business & economics,*The Journal of Business Ethics (JBE)*. In addition, we examined*The Journal of Moral Education (JMED)*, the premiere interdisciplinary journal in moral education and development, as well as two prominent journals in moral and political philosophy—*Ethics* and*Philosophy & Public Affairs*.

We divided the literature into timed sequences and categories: 1980–1989, 1990–1999, 2000–2009, and 2010–2015 to provide a closer analysis of trends from 1980 forward.[5] A range of themes emerged from our content analysis of articles, which were then categorized accordingly as follows[6]:

1. **University Culture** (expressions of institutions' moral purpose; modeling and reinforcing of core values perceived as necessary for higher education)[7];
2. **Research Ethics** (animal and human subjects protections and research misconduct primarily regarding faculty researchers)[8];
3. **Ethical Responsibilities of Faculty or Administration** (impact of direct faculty and administrator behavior on students)[9];
4. **Student Academic Integrity** (cheating and other forms of academic misconduct primarily regarding student behavior)[10];

[4]Keywords used to search descriptors for journals and articles include "higher education" combined with "ethics" or "ethical" or "moral" in the journal or article title, subject word, or description. Initial search results were then manually culled to include only academic articles (not book reviews, for example) that addressed topics related to ethics education in higher education.

[5]See Sloan (1980), for a summary of literature in ethics education prior to 1980.

[6]Artifacts that could have reasonably been coded in more than one category were placed in a primary category based on title, abstract or other determination early in the article of major focus. Two researchers independently categorized journals and articles in our study, with disagreements discussed and consensus achieved.

[7]See, for example, Besvinick (1983), Thornton and Jaeger (2008) and Wilshire (1987).

[8]See Steneck (1994).

[9]See Scriven (1982).

[10]See Thompson (2006).

5. **Ethics Education Goals & Outcomes** (articulations of expectations for ethics education and assessment, primarily curricular)[11];
6. **Ethics Pedagogy & Teacher Preparation** (examinations of teaching practices and determinations of adequate background for teaching in the field)[12];
7. **Civic Education** (development of student knowledge, skills and motivation for civic engagement, including experiential learning)[13];
8. **Student Moral Development** (development of moral sophistication at the individual student level through interventions both in and outside of the classroom)[14];
9. **Co-curricular Ethics Learning** (institution-supported ethics education that occurs external to formal curricula)[15];
10. **Other** (e.g. articles that reported on surveys of students or other stakeholders, comparisons between corporate or professional practice and the academy, reviews of trends in literature or practice).[16]

Journals in Higher Education

We identified four flagship journals in higher education based on impact factors, citation ranking, and acceptance rates:*Harvard Educational Review (HER)*,*The Journal of Higher Education (JHE)*,*Review of Research in Education (RRE)*, and*Teachers College Record (TCR)*. Of these four, only*JHE* published a significant number of articles (38) in our area of interest.*TCR* published five,*RRE* and*HER* published three each. Of those 49 relevant articles, 21 of them were published in the 19-year-period of 1980–1999. Twenty-eight were published in the 15-year-period of 2000 through 2015, indicating that ethics in higher education is of continuing and growing interest in our period of study for researchers who study it from a higher education perspective.

The first articles in this period both appeared in*JHE* in 1982, “Should there be an academic code of ethics?” (Callahan) and “Professorial Ethics,” (Scriven). Callahan’s (1982) article surveyed the list of ethical issues that confront decision-makers in higher education, determining that, even though there is “certainly good reason to confront, and to grapple with, the long list of ethical problems facing the university,” (p. 341), writing a code of ethics is not the answer. He endorsed that campuses have an ongoing project in which the whole campus community examined the school’s ethical issues. Callahan’s focus is particularly

[11]See Camenisch (1986).

[12]See Tsei (2002).

[13]See Rhoads (1997).

[14]See Meyhew (2012).

[15]See Magolda and Abowitz (1997).

[16]See Lee and Taylor (2013).

interesting in that The Hasting Center project that he co-directed resulted in ten publications, which all focused specifically on teaching ethics within a classroom context. His*JHE* publication is the only publication in this period that focused on university culture and how the institution as a whole could improve its ethical decision-making. Scriven (1982) discussed the need for ethical conduct of faculty in regard to students (p. 313) but also argued that practical ethics should be taught throughout the curriculum, just as writing and critical thinking are taught through many courses (p. 310–11). Three articles were published in higher education journals in the 1980's that specifically addressed the teaching of ethics.[17] Each of them referenced the work of The Hastings Center study team, reinforcing our view that The Hastings Center publications served as a foundation for the field of ethics education.

The categories of articles that appeared in the flagship journals over the 35 years reflect ethical concerns within the broad higher education environment rather than matters that might be of specific interest to instructors teaching ethics in the classroom: University Culture (11), Student Academic Integrity (9), Ethical Responsibilities of Faculty and Administration (8), Civic Education (5) and Research Ethics (5). Student Academic Integrity was the primary category of article in the 2000s (7).

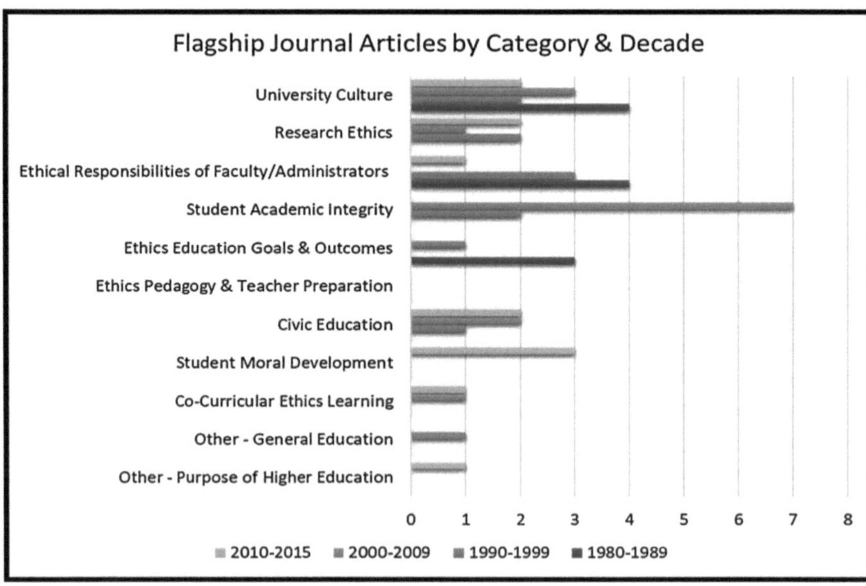

[17]See Camenisch (1986), Stark et al. (1986) and Rivage-Seul (1987).

The Growth of Literature in Practical Ethics Journals

Journals focused on scholarship in disciplinary areas within practical ethics proliferated in our study period as well. Of the 145 disciplinary journals identified in this study (InCites Journal Citation Report2017; UlrichsWeb2017),[18] only 17 began publication prior to 1980, with only two of them beginning publication prior to 1960. The premiere journal in moral philosophy,*Ethics*, began publication in 1888.

Ethics Journals by Discipline	
Bio, Medical	33
Business & Economics	24
Moral Philosophy	20
Political Science	9
Higher Education & Teaching	7
Information Sciences & Technology	7
Social Sciences; Sociology	7
Religions & Theology	6
Environmental	6
Law & Criminal Justice	5
Psychology	5
Communications & Media	4
Military	3
Engineering	2
Public Health & Safety	2
Sports & Games	2
Public Administration	1
Animal Ethics	1
Social Services & Welfare	1
TOTAL	**145**

For further analysis, we examined publications in the*Journal of Medical Ethics (JME)* and the*Journal of Business Ethics (JBE),* the top journals in the two disciplinary areas with the most journals. In addition, we looked at high-ranking journals

[18]While we believe that we captured most of the journals that publish articles in practical ethics, moral education or moral development, no one database seems to have captured all peer-reviewed journals that belong in our study. InCites Journal Citation Reports (JCR) was chosen as a recognized source analyzing citation references within 11,000 + indexed journals including "nearly" 250 disciplines. Ulrichs Web is recognized among librarians as the premier periodical indexing system with more than 300,000 periodicals.

in moral philosophy,*Ethics*, and*Philosophy & Public Affairs* to ascertain what interest ethics education might have had for those reviewers and readers. Finally, we examined*The Journal of Moral Education (JMED),* as moral and civic development is sometimes addressed concurrently with ethics education and the researchers most likely to examine the outcomes of ethics education are moral psychologists. We identified 168 articles from these five journals for analysis and categorized them in the same manner as used in our review of the flagship journal articles.

JBE, which began publication in 1982, published 88 articles related to ethics in higher education in our study period. The earliest published article in our time period relevant to the study was "Ethics in Education: A comparative study," (Lane and Schaupp1989), which we categorized as student moral development. According to the authors, findings included greater competitiveness among the business students when compared to students in other colleges; business students were far more likely to see a need to "step on people" and to "clear their path" to attain their goal (p. 947).

In the 1990s,*JBE* published 12 relevant articles: Ethics Education Goals & Outcomes (2), Civic Education (2), Ethics Pedagogy & Teacher Preparation (1), Research Ethics (1) and Student Moral Development (1). Five articles were categorized as Other and included articles that compared corporations with academic settings, student perceptions with those of workers in business, and student surveys. In the five-year period between 2010 and 2015, the most recent period of review, *JBE* published 52 articles related to ethics in higher education. Ten of those focused on pedagogy and nine of them focused on institutional culture. More than any other journals examined,*JBE* published articles relating to the ethics of the institution in about equal balance with those relating to formal instruction of ethics in business and accounting education.

The*Journal of Medical Ethics (JME),* which began publication in 1975, published 52 articles that include higher education as a keyword from 1980–2015, with "Teaching medical students on the ethical dimensions of human rights" (London and McCarthy1998), the earliest published in our timeframe. Because of the role of biomedical research in higher education, a substantial number of articles (8) related to research ethics, with the highest number of articles (27) focused on formal ethics education for undergraduate or medical students in either the Ethics Education Goals & Outcomes category (16) or Ethics Pedagogy & Teacher Preparation (11).

The longest-running and most esteemed journal in moral philosophy,*Ethics,* published only two articles in the time frame that were directly relevant to our study. One appeared in 1993: "Liberalism and campus hate speech: A philosophical examination," (Altman1993), the other in 2007: "Fair opportunity in education for citizenship," (Anderson2007). Similarly,*Philosophy & Public Affairs* saw only two relevant publications in our study period, "Diversity," (Shaw1999), and "Yes Means Yes: Consent as Communication," (Dougherty2015). We categorized all four of these articles as addressing University Culture, as they addressed ethical issues relating to higher education broadly speaking. But, the primary focus for each of the four articles was the philosophical concept rather than the campus environment: liberalism as political ideology, what counts as fair opportunity, the meaning of diversity and the nature of consent. It is not surprising, then, that ethics—as it plays out in practical ethics courses or in practical matters within the higher education environment—was not of primary interest to reviewers or readers.

The*Journal of Moral Education (JMED),* which began publication in 1971, was included in our review as it has been the premiere interdisciplinary journal in moral development and education for more than 40 years. It is appropriate to include a review of a journal in moral education as "moral education" and "ethics education" are sometimes used synonymously. In addition, moral psychologists speak directly to the practice of ethics education. First they have engaged in discussing whether moral growth and development are appropriate goals for ethics education. Next, they have offered objective measures for assessment of whatever moral development might have taken place within an ethics education attempt. They have also pointed out the many ways that moral development theories can be used in analyzing moral growth in higher education. For example, Schmidt et al. examined a method for promoting cognitive moral reasoning (Schmidt et al.2009; Lee et al.2015) examined other-oriented motivations for moral behavior as compared with motivations that were self-motivated.

This journal published 24 relevant articles in the period under consideration relating to ethics education in higher education. While it is certainly not surprising that almost one-third of the articles were categorized as Student Moral Development (7),*JMED* also published 10 articles categorized as Ethics Education Goals & Outcomes or Ethics Pedagogy & Teacher Preparation, because of the focus on objective assessment of ethics education attempts.

As with the flagship journals, the disciplinary journals together showed a steady increase of relevant publications between 1980 and 2015. The trend between 2010 and 2015 showed a substantial increase in*JBE, JME,* and*JMED.*

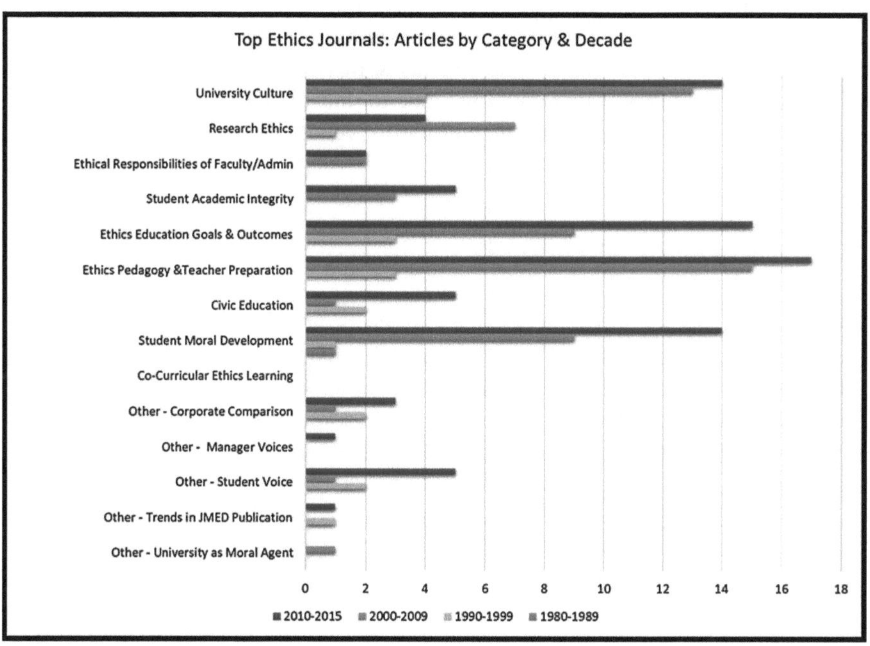

Although *JBE, JME,* and *JMED* all showed an upward trend of publications in ethics education, the role of ethics within higher education cannot be said to be the primary focus for any of the journals examined. For example, the percentage of published articles relating to ethics education within higher education rose from 6.3% in the period between 1981 and 1990 to 10.9% in the period between 2001 and 2011 (Lee and Taylor 2013) in the *JMED.* In the same comparative period, these authors found no significant change relating to K-12 or professional education (p. 414). But, even an increase to 11% of journal articles devoted to the teaching and learning of ethics on college campuses reflects a relatively low level of researcher and reviewer interest.

The study of ethics education specifically spawned journals dedicated to the teaching of ethics: *Teaching Business Ethics* began publication in 1982, but then merged with the *Journal of Business Ethics* in 2004. *Teaching Ethics* (2001) is a peer-reviewed journal that publishes biannually and welcomes articles relating to the teaching of ethics in any learning environment. This is the one continuing journal devoted to ethics pedagogy. One new publication in ethics pedagogy, *International Journal of Ethics Education* began publication outside of our time-frame (2016), but claims to "present a platform for exchange of theoretical and practical experiences with teaching ethics in various educational settings" (Springer Publication 2017). Articles on ethics education also appear in *Teaching Philosophy*, which has been published since 1975.

Our examination showed that journals with a focus on higher education were mostly concerned with ethics education as it affects the university as a whole: university culture and ethical issues that cross disciplinary lines, such as faculty research misconduct and student cheating. Within the disciplinary journals, we saw greater activity in publication on issues relating to teaching content and skills. Only the *JBE* published a balance of articles between those focused on classroom activity with those relating to the university as a whole. The siloes that we see in the thematic foci of journals reflects the fracturing of ethical concerns that persists on college campuses today. More could be done to understand and address how ethics in the classroom connects with ethics in the university environment.[19]

Ethics Education on Campus

The increase in ethics education scholarship has been mirrored by growth of ethics education practice on campuses. In the years of this study, many higher education policy makers, leaders and curriculum planners have been convinced that intentional ethics education has a place in higher education. In some cases, the move toward ethics across the curriculum happened in concert with university-wide or programmatic accreditation. In other cases, ethics stepped outside of the traditional departments of

[19] See especially Keenan (2015).

religion or philosophy through creative team-teaching by professors working cross discipline to provide student their shared expertise. On yet other campuses, administrators or curricular specialists found opportunities to seed ethics education throughout their institution's programs of study and provided faculty development to assist instructors in gaining needed expertise. In this section, we examine a number of the different ways that ethics education has become part of the face of campus.

Evidence for Interest in Intentional Ethics Education

Some institutions have chosen ethics education as the basis for their campus-wide Quality Enhancement Projects (QEP). At the time of this writing, QEPs were required by the Southern Association of Colleges and Schools (SACS), which is one of the six regional accrediting councils in the U.S. for institutions of higher education. That SACS has accepted campus-wide curricular and co-curricular ethics projects as appropriate QEPs is a significant indication of policy-maker acceptance of the importance of intentional ethics education.[20] As an example, beginning in 2010, James Madison University developed a saturation technique for ethics across the curriculum.*The Madison Collaborative: Ethical Reasoning in Action* invited faculty to integrate eight key questions into their courses, regardless of field. The eight questions addressed fairness, outcomes, responsibilities, character, liberty, empathy, authority and rights (James Madison University2017).

In addition, national accreditation councils for specific disciplines, such as the Association to Advance Collegiate Schools of Business (AACSB), Accreditation Board for Engineering and Technology (ABET) and Accrediting Council on Education in Journalism and Mass Communication (ACEJMC) include ethics education as a qualifier for program accreditation.

While a few institutions of higher education, including Harvard, have required undergraduate study in ethics from the time that general education requirements were put in place, other schools began requiring ethics in or near our period of review.[21] Many more schools have added ethics requirements to particular majors. A steadily increasing number of institutions are offering ethics degrees at the undergraduate, master's or doctoral levels.

Beginning in 2000, institutions of higher education have been able to offer certificate and degree programs in ethics recognized by the federal National Center for

---

[20]Schools that developed a campus-wide QEP based on ethics include Barry University, Campbellsville University, Carson-Newman College, Eastern Kentucky University, Georgia Military College, Hardin-Simmons University, James Madison University, Marymount University, Oakwood University, St. Philip's College, Texas Tech University, The Citadel, Virginia Military Institute, Webber International University, William Peace University.

[21]See for example, The University of Montana, which introduced the general education requirement, Ethics and Human Values in 1975 and continues through the time of this writing. Thirty courses are listed as providing general education credits in this area including the intriguingly-named literature course, "Placebos: The Power of Words".

Education Statistics. The Center hosts the Integrated Post Secondary Education Data System (IPEDS), which in turn provides data categories that all post-secondary educational institutions must use in reporting details regarding their institutions, academic programs and students. IPEDS created a standardized system for reporting academic programs that lead to degrees or certifications, the Classification of Instructional Programs (CIP) in 1985. CIP codes relevant to this review are*Ethics* (38.0103, introduced in 2000),*Applied and Professional Ethics* (38.0104, introduced in 2010), and*Bioethics/Medical Ethics* (51.3201, introduced in 2010).

Ethics CIP 38.0103: "A program that focuses on the systematic study of the theory of moral good and its application to various theoretical and practical problems. Includes instruction in ethical theory, history of ethics, belief and value systems, ethical constructs, and applications to specific topics, issues and problems."

Applied and Professional Ethics CIP 38.0104: "A program that focuses on the systematic study of ethical issues in the workplace and public life, and the application of ethical decision-making to the practical problems of society and the professions. Includes instruction in ethical theory; history of ethics; contemporary social dilemmas; methods in applied ethics; and applications including medical ethics, legal ethics, business ethics, environmental ethics, and criminal justice ethics."

Bioethics/Medical Ethics 51.3201: "A program that focuses on the application of ethics, religion, jurisprudence, and the social sciences to the analysis of health care issues, clinical decision-making, and research procedures. Includes instruction in philosophical ethics, moral value, medical sociology, theology, spirituality and health, policy analysis, decision theory, and applications to problems such as death and dying, therapeutic relationships, organ transplantation, human and animal subjects, reproduction and fertility, health care justice, cultural sensitivity, needs assessment, professionalism, conflict of interest, chaplaincy, and clinical or emergency procedures." (National Center for Educational Statistics2017).

The chart below represents the number of institutions offering degrees in each of the CIP codes. The chart begins with 2010–11 as that was the first year that all three CIPS were available:

IHE with Programs

	2010-11	2011-12	2012-13	2013-14	2014-15
Ethics 38.0103	24	20	27	26	30
Applied & Prof. Ethics 38.0104	5	8	11	13	14
Bio/Med Ethics 51.3201	18	22	25	30	26

The overall number of institutions offering degrees in one or more of these CIP codes has grown steadily. The growth pattern is consistent across public institutions, private institutions and religious institutions. For example, across the three CIP codes in 2014–15, 25 public institutions had certificate or degree programs in one or more of the three CIPs, 20 private institutions had programs, and 24 religious institutions were offering certificates or degrees in one or more of the ethics CIP codes. In 2014–15, New England College of Business and Finance became the first for-profit institution of higher education offering a degree in ethics. Schools offering degrees include large public research institutions, religious schools, and private liberal arts colleges.[22]

Goals and Pedagogy in Ethics Education

The 1980 Hastings Center report was a response to concerns about intentional ethics education that surfaced in the 1960s and 1970s. The study team asked themselves what the new focus on ethics education meant: "What are the appropriate purposes of courses in ethics? What kinds of student should such courses try to reach, and at what point in the curriculum? Who should teach such courses, and what training ought they to have?" (p. xiv). The Hastings Center's research team lamented that students' opportunities to formally examine ethical questions in general life or in the professions were "often scant and episodic" (The Hastings Center1980, p. 79). The research team identified appropriate goals in the teaching of ethics as: "stimulating the moral imagination, developing skills in the recognition and analysis of moral issues, eliciting a sense of moral obligation and personal responsibility, and learning both to tolerate—and resist[23]—moral disagreement and ambiguity" (The Hastings Center, p. 80).

Those goals have been echoed (Camenisch1986) and restated in student outcome "behavioral" terms (Elliott2007).

Much has happened to meet these goals and to answer those concerns in the 35-year study period. If we confine ourselves to considering how teachers of ethics talk about their craft, some recommendations from The Hasting Center's team have become standard practice in the field. Based on an aggregation of results from studies performed in 2008 and 2015–16 by Cooper, that included interviews with 80 senior ethics teachers at selective English-speaking institutions in the US,

---

[22]See for example Arizona State University (multiple campuses), Brown University, Carnegie Mellon University, Case Western Reserve University, Epic Bible College, Kansas City University of Medicine and Biosciences, Kennesaw State University, Loma Linda University, New England College of Business and Finance, Northwestern University, Oral Roberts University, Smith College, Utah Valley University, University of Maryland (multiple campuses), Western Michigan University, and Yeshiva University.

[23]In some of the 1980 Hastings Center publications, the term used here is "reduce" rather than "resist".

Canada, UK, Europe, Asia, Australia and New Zealand (Cooper2017), Cooper (2017) noted some consensus: all agreed that "Ethics as a discipline of moral reasoning should be taught consistently in colleges and universities" (p. 67).

Concern that instructors might indoctrinate students into their own ideological views, expressed from the 1960s and articulated in The Hastings Center's report, seems to have dissipated. Cooper reported "little difference by those teaching in the U.S., British, Asian, Canadian, and Australasian institutions. Hence, despite the uniqueness of the tutorial teaching system used at Oxford and Cambridge, and of the Confucian, Taoist, and other traditions in the Pacific, this study did not reveal appreciable national and cultural differences in attitudes toward the teaching of ethics" (Cooper2017, p. 66). Pluralism was further promoted by an expansion beyond what Cooper called an "all but rigid reliance upon the classical canon of revered deceased philosophers" (p. 70) to include contemporary voices of women and a diversity of cultures along with secondary texts to accompany primary classical readings.

We can see consistency over the years in what counts as adequate ethics education. Matchett (2008) utilized work from Ozar and Rest in specifying knowledge and skills to be achieved in ethics education to include: values, principles and ideals, conflicts among those, and facts relevant to ethical decisions in specific areas in the knowledge arena. Skills include: multiple perspective taking, formulating logical arguments, employing conceptual tools such as ethical theories, and accurately applying justifiable standards that are reasonable to expect within the professional or social role examined (Matchett2008, pp. 32–33). Cooper's (2017) sample reported teaching rigorous moral reasoning, critically informed decision-making, taking deeper perspective on important issues and adopting a more philosophical or transcendent approach to life and ethical dilemmas (p. 68).

Scholars who caution against expecting too much change to occur within a single semester still adopt goals reflective of The Hastings Center's goals: one scholar suggests that a realistically attainable and significantly valuable goal for ethics courses is to get "students to go beyond demonizing and to downgrade their own intuitions" (Murphy2014, p. 426). Even this basic blow to student subjectivism fits The Hasting Center's appropriately ambiguous goal of helping students learn both "to tolerate—and to resist—moral disagreement and ambiguity" (The Hastings Center1980, p. 80).

In contrast to ethics teaching within departments of philosophy or religion, in which courses in ethics might lead students through the examination of particular philosophers, texts, traditions or theories, ethics courses taught across the curriculum have been focused on controversies. The issues that first led clinicians to consult with philosophers, such as brain death, distribution of limited resources including cadaver organs for transplant, and obligations to treat severely disabled neonates were now offered to undergraduate, graduate and medical students for analysis. Undergraduate and graduate students as well in business ethics and social responsibility courses learned processes for analyzing the ethical implications of outsourced labor and obligations to employees. They learned that ethical considerations might lead one to a outcome different from when the economic bottom line was used as the only criterion for success. Engineering students learned behind the scene details of disasters including Chernobyl and the Challenger explosion to help them appreciate the tension between "thinking like an engineer" as compared with "thinking as a manager." The latter frame of reference implies that in important ethical values that govern the engineering profession were set aside in these cases for the priorities expressed by other powerful stakeholders. According to Cooper's findings, "the teaching of ethics should be a catalyst to both intellectual growth and to deeper understanding of moral choice" (2017, p. 70).

Pedagogy thought to best accomplish those goals is based on active student learning, such as discussion/debate (DuBois and Burkemper2002; Dean and Beggs2006) with the goal of steering students away from a polarized analysis of good/bad, right/wrong answers for ethical controversies in favor of an appreciation of the complexity of assumptions and justifications evident in any moral choice. Students were taught to consider the adequacy of means along with the justifiability of ends. Teachers of ethics were encouraged to focus on a range of moral permissibility, so that students were introduced to thinking of choices as morally prohibited, morally required, morally permitted or morally ideal (Gert2005) instead of simply right or wrong. One pedagogical method offered is for educators to present ethical issues within what Whitbeck called a "design problem model." This model represents potential outcomes as expressing more than one potential good, rather than providing two alternatives or closed-choice multiple alternatives (Whitbeck1995, p. 302).

Goals for the teaching of ethics have been further dissected to affirm the importance of students practicing ethical thinking skills as compared with the less-engaged learning of how to apply elements of argumentation and recognition of the relevant points of disagreement in controversies. For example, Kenneth Goodpaster distinguished these two pedagogical approaches as the difference between*praxis*, which he calls the "salient element" of ethics teaching and*poiesis* (Goodpaster1982).

Instructor Preparation for Ethics Education

A common pedagogical concern of potential ethics educators and assessment specialists is the interdisciplinary preparation required to teach ethics well. Hasting Center scholars agreed that some deep understanding (the equivalent of a Master's degree) should be required both in moral philosophy and in the area of analysis (Bok, p. 30; Callahan,1980, p. 77). Practical ethics instructors should be able to teach critical thinking skills, important components of major ethical traditions and theories, and how to build good arguments and how to evaluate the construction of arguments offered. Instructors who are practitioners within the field of analysis or who have credible expertise in the field provide important modeling. "Each time a teacher in a professional school raises a question of professional ethics, she is an example of a member of her profession concerned about its ethics" (Davis1990,

p. 36). Along with the need for interdisciplinary knowledge, teachers of ethics need good facilitation skills, the courage to allow students to discuss controversial matters, and the ability to steer student conversation to more sophisticated and theoretical levels rather than allowing it to devolve into polarized positions on a particular case. According to Matchett (2008), "many non-ethics faculty have fairly limited ideas about how they might lead a productive discussion about ethical issues related to their course subject matter," as well as being unsure about how to assess student ethical thinking (p. 26).

The Hastings Center study team set a high bar for qualifications for the teaching of ethics: "As an ideal, those teaching applied and professional ethics—where knowledge of one or more fields is necessary—ought to have the equivalent of one year of training in the field in which they were not initially trained" (Callahan and Bok1980, p. 301) or team-teaching which is cost prohibitive in many institutions. Finding sustainable methods for "training up" instructors so that they feel adequately prepared to address ethical issues within the context of their fields of expertise has remained a significant challenge.

The proliferation of graduate certificates and degrees in ethics may provide an answer in the long run, as students complete the ethics qualification in addition to disciplinary training. In the meantime, professional associations and campuses have stepped into assist fledgling ethics instructors. For example, the Association for Practical and Professional Ethics (APPE) has hosted a half-day Graduate and Early Career Scholars' Seminar in Teaching Ethics as part of its annual meeting beginning in 2010. The Center for the Study of Ethics at Utah Valley University (UVU) provides an annual five-day seminar for UVU faculty who want to teach stand-alone courses in ethics or incorporate the teaching of ethics into their regular curriculum offerings (Utah Valley University, n.d.). Such attention to faculty needs is a "best practice" for ethics across the curriculum (Matchett2008, p. 36).

Moral Psychology, Assessment, and Ethics Education

The Hastings Center team said, "Courses in ethics ought not explicitly to seek behavioral change in students. They should seek to assist students in the development of those insights, skills, and perspectives that set the stage for a life of personal moral responsibility, manifesting careful and serious moral reflection" (1980, pp. 80–81). Discussions regarding the distinction between facilitating true moral growth and teaching content and skills have become more nuanced over time. Some scholars have sought to distinguish instructional outcomes that can be measured from "pedagogical hope," (Elliott2007, p. 40) for how a student might turn out to be. Cooper's interviewees, the majority of whom taught in philosophy departments, were opposed (46%) or unsure (17%) about whether ethics as "moral improvement" or "character development" should be taught in higher education ethics courses" (Cooper2017, p. 67). Yet many ethics scholars who consider disciplinary areas outside of philosophy as their home departments discuss ethics education and moral and civic development in the same breath.

The connection between moral psychology and ethics education as addressed by scholars seems intuitive, but research has confirmed the close connection of cognition with moral judgment (Kohlberg1981,1984; Rest1986). Various instruments have been developed to measure a research subject's moral competence or stage of moral development.

One test, the Personal Ethical Threshold (PET) was developed for assessing moral behavior in the face of situational pressure (Coleman et al.2015, p. 26). Another, the Moral Competence Test (MCT) assumed that personal preferences impact moral judgment and identified consistencies in applying those preferences in solving moral problems (Biggs and Colesante2015, p. 499). But, by far, the most discussed assessment technique used to evaluate student moral growth and development in this period of study was the Defining Issues Test (DIT), developed by James Rest based on Lawrence Kohlberg's stage-based theory of moral development. The test has been used in attempts to identify progress in students' moral reasoning in higher education as an outcome of interventions ranging from a semester-long ethics course to the full four undergraduate years.

Challenges to the use of moral development testing for ethics education include the inadequacy of the moral development theory that serves as the test's foundation as well as concerns about the test's ability to capture the complex processes at work, much less the multiple pedagogic goals in ethics teaching and learning (Thoma et al.2016). Other theories of moral development that are compatible with Kohlberg's theory and flesh out the notion of moral sophistication, specifically those of Carol Gilligan and William Perry, were not considered in the development of the DIT (Elliott2007). Scholars argued that the use of Kohlberg's theory as the sole basis for the instrument created a fundamental bias based on Gilligan's charge "that Kohlberg's view of morality and moral development is a decidedly masculine construction, culminating in abstract principles and rights, and that it ignores a more contextual and relational understanding of morality focusing on responsibility to persons" (Camenisch1986, p. 506).

Kenneth Goodpaster (1982) identified other concerns with the DIT: the problem of time lag and the inappropriateness of instructors treating students as research subjects. There may be a significant time lag between an ethics course and when students recognize the importance of what they learned or until they apply the tools well after the class has ended. They may not encounter a profession-based problem until they have completed years more of pre-professional training. Goodpaster also pointed out that DIT measures only one expression of moral sophistication while philosophers encourage a pluralistic approach to application and use of ethical theory. Finally, Goodpaster argued that the DIT misses an important distinction: the ethics instructor works with students to achieve change rather than treating them as human subjects in which the results of an intervention can be measured. Goodpaster (1982) argued that teachers should not be dispassionate in the same way that experimenters should. Teachers should be fair in how they treat a group of students, but should be deeply invested in the success of each.

Evaluation of student progress in practical ethics courses has generally followed best practices for humanities disciplines, which rely on conventional methods such as classroom discussions, tests and writing assignments. Based on the assumption of Callahan's goals for the teaching of ethics, particular student abilities emerge as providing justifiable basis for evaluation: quality of arguments for moral views; mastery of theories and principles of ethics; identification of moral issues; and ability to argue both sides of a position (Caplan1980, pp. 148–149). This approach centers on assessment of students' ethical reasoning and critical thinking skills. Sometimes assessment tools also include a component that addresses values clarification.[24] It is fair to say that methods for comprehensive assessment of student learning in ethics, those that measure ethical reasoning, have yet to be developed.

Co-curricular Growth in Ethics Education

Alongside the increase in ethics degrees and courses, some schools have created co-curricular support for the institutional ethics mission, including ethics centers and student ethics competitions. While scholars noted that integration of ethics education within campuses or between curricular and co-curricular attempts is rare (Colby et al.2003) the elements are present on many campuses for this to occur. Creative administrators often have the elements to build ethics across the curriculum programs from existing curriculum and co-curricular programs and to support faculty development and networking and the campus-wide unification of ethics education. Ethics centers were created on many college and university campuses, at least in part, to meet some of these goals.

A 2017 study of seventy-five ethics centers found that most ethics centers were campus wide and commonly provided activities that crossed disciplinary lines including lectures, fellowship opportunities, and student competitions (Safatly et al.2017, p. 156). As with other ethics initiatives, most centers began operations in the late 20th century, with the number of centers picking up in the 2000s[25] (Safatly et al.2017). Data is not available to provide a full accounting of the number of college and university-based ethics centers, institutes and initiatives. While there are certainly additional ethics centers not affiliated with APPE, that association claims more than 150 institutional members, which are almost all ethics centers within institutions of higher education. Most APPE-affiliated centers have campus-wide, community-wide or even nation-wide focus, while others are disciplinarily grounded (APPE).

Of the ethics education efforts offered through co-curricular activities and community service,[26] the Intercollegiate Ethics Bowl (IEB) has most successfully

[24]See, for example, theAACU LEAP VALUE rubric for Ethical Reasoning.

[25]The study found at 15% of the centers examined were established in the 1970s, 14% in the 1980s, 32% in the 1990s and 39% between 2001 and 2010.

[26]See, for example, King and Mayhew (2002) and Coleman et al. (2015).

engaged the highest number of undergraduate institutions in a shared activity. The IEB was established in 1993 at theIllinois Institute of Technology by Professor Robert Ladenson and the Center for the Study of Ethics in the Professions (CSEP). The competition became national in 1997. Since then, the IEB has been hosted annually by APPE in conjunction with the annual meeting (APPE2017; Ladenson2001) and has added schools hosting undergraduate teams every year. By 2001, thirty-two teams representing colleges and universities across the United States participated. And by 2015, more than 250 campuses fielded teams that participated annually in the IEB, involving more than 1000 student team members, graduate student coaches and faculty sponsors in 10 regional bowls, with 36 of the best teams invited to compete in the national competition (APPE2017).

Ladenson (2001) has credited the IEB with providing three contributions in the development of students' capabilities in ethical judgment and reasoning: (1) developing a framework of analysis for addressing ethical issues in an intellectually well-organized manner; (2) providing opportunities to acquire valuable background information on ethical issues of special importance to them in light of their respective interests, concerns, and career aspirations; (3) fostering the capacity for ethical understanding over a broad range of important subjects.

The IEB received the 2006 American Philosophical Association/Philosophy Documentation Center's 2006 prize for Excellence and Innovation in Philosophy Programs (APPE2017). The competition has been endorsed by a multitude of college instructors who claim to have seen significant growth in students' ethical reasoning after participating in an ethics bowl competition (Ladenson2001; Borrego2004; Connolly2009; Meyer2012; Merrick et al.2016). Students have also consistently reported that their ethical reasoning has become more systematic from participation and that they have learned to more carefully consider alternative perspectives (Meyer2012).

The IEB has spawned the two-year college bowl competition, a national high-school competition as well as served as the model for disciplinary-based bowls along and many in-class explorations. At the time of this writing, more than 500 case presentations were available for use, free of charge, through the ethics bowl archives (Ladenson2001).

Conclusion: Missing Elements in the Ethics Education Discourse

Our examination of the current ethics boom illustrates that, even though ethics education has made substantial and enduring changes in the practice of higher education, there are still urgent improvements to be made.

While students are reportedly in agreement with other stakeholders as to the importance of ethics as a component of their undergraduate education (Coleman et al.2015), what that means to them and how and when students believe that ethics education works best is not often explored. Students learn how to reason about prescribed matters of ethical concern, but their voices in the development of curricular and co-curricular attempts to teach ethics are largely absent. This means that we have limited understanding about whether ethics education is addressing the ethically demanding situations that students actually face or anticipate facing, and whether current offerings do so in a way that resonates with them.

In recent years, higher education scholars and policymakers have begun to ask students to share their perceptions of their college experience. Since 2000, institutions throughout the U.S. in partnership with the National Survey on Student Engagement (NSSE) have queried over five million students about their college experience within and beyond the classroom (Kuh2003).[27] Yet such studies of student perceptions of their learning experiences and environments have rarely been connected to ethics coursework (Dey and Associates2009). While instructors solicit feedback on ethics courses via course evaluations, they do not typically ask whether ethics offerings have enabled students to navigate the ethically demanding situations that they actually face or anticipate facing. Aligning ethics programming with students' needs today is all the more important given the ethical quandaries students are likely to face in their personal, civic, and professional lives. Higher education must prepare students to succeed even as the world as we know it changes, with predictions of a less organizationally-bound and more entrepreneurial workforce; the ubiquity of social media and the ethical quandaries it raises about interpersonal communications and civic engagement; and changing social norms (e.g., sexual norms; shifting conceptions of privacy). Is ethics education meeting students' needs in light of the ethical dilemmas these shifts present? To date, this question remains open.

And whereas institutions have focused more on some ethical obligations—such as protection for members of the campus community from sexual assault, discrimination and harassment, protections for human and animal subjects in school-sponsored research, investigations for suspected research misconduct, and enforcement of students' academic honesty—broader expectations of institutions' moral agency are largely absent from the conversation.[28] Institutions model moral choices for their students through policies regarding resource use and sustainability, vendor choice, investments, and the ratio of full time faculty to adjuncts to name but a few. Rarely are students invited to participate in this institutional ethical decision making.

The then president of Harvard, Derek Bok, said, "If a university expects to overcome the sense of moral cynicism among its students, it must not merely offer courses; it will have to demonstrate its own commitment to principled behavior by making a serious effort to deal with the ethical aspects of its investment policies, its

---

[27]The annual national survey of student engagements (NSSE) examines student self-reports on "items that represent outcomes that characterize interpersonally effective, ethically grounded, socially responsible, and civic minded individuals" (Kuh2003).

[28]See Keenan (2015).

employment practices, and the other moral dilemmas that inevitably confront every educational institution" (Bok1976, p. 29). Yet, the moral aspects of institutional choices have remained a relatively unstudied subject for individual institutions and in the literature, especially in regards to choices impacting employees and external entities.

James Steve Counelis (1993) reported the result of his search in the late 20th century for "empirical studies on the moral behavior of those who comprise the American university community" (p. 75). He claimed that his review of bibliographies from professional ethics literature, encyclopedic works on higher education, institutional research reports on academic management, and "a computer search on moral behaviors of university boards of trustees, their administrators and faculty" yielded no results (Counelis1993). While it is clear that Counelis missed existing literature on the subject of institutional morality in higher education and that more literature has been published since his review,[29] empirical studies on the role that ethics plays in academic institutional decision-making seem non-existent.

In this chapter, we have laid out evidence from scholarly literature as well as from policy and practice in U.S. higher education that documents the firm rooting of ethics education in U.S. colleges and universities. Still lacking, however, are clear indications of what ethics education is currently accomplishing, and how—beyond proliferation of journals, journal articles, courses and degrees—it can best serve all stakeholders and society, thus enhancing the mission of higher education. We argue that attention to students' own articulation of needs in the 21st century, a better understanding of instructional goals supported by comprehensive assessment measures, and cohesive institutional commitments to professional ethics would all constitute productive next steps.

Acknowledgements By Deni Elliott and Karlynd June with thanks for research assistance and feedback from National Ethics Project co-investigators Jess Miner and Anne Newman. This chapter was produced in part from research funded by the Spencer Foundation and Andrew W. Mellon Foundation.

References

Altman, A. (1993). Liberalism and campus hate speech: A philosophical examination.*Ethics*, 302.
Anderson, E. (2007). Fair opportunity in education for citizenship.*Ethics*, 595–622.
Association for Practical and Professional Ethics. (2017, July 12). Retrieved from Association for Practical and Professional Ethics,http://appe-ethics.org/about/.
Association of American Colleges & Universities. (2010, June).*Ethical reasoning VALUE Rubric.* Retrieved July 21, 2017, from Association of American Colleges & Universities:https://www.aacu.org/ethical-reasoning-value-rubric.
Association of American Colleges & Universities. (n.d.).*Core commitments: Guiding philosophy.* Retrieved November 22, 2016, fromhttps://www.aacu.org/core_commitments.

[29]See, for example, Englehardt et al. (2010) and Cahn (1992).

Besvinick, S. L. (1983). Integrity and the future of the university.*The Journal of Higher Education, 54*(5), 566–573. 10.2307/1981629 Retrieved fromhttp://www.jstor.org/stable/1981629.
Biggs, D. A., & Colesante, R. J. (2015). The moral competence test: An examination of validity for samples in the United States.*Journal of Moral Education,44*(4), 498–516.
Bok, D. (1976). Can ethics be taught?*Change,8*(9), 26–30.
Borrego, A. M. (2004). Ethics bowls exercise students' moral muscles.*The Chronicle of Higher Education*.
Cahn, S. (1992).*Morality, responsibility, and the university*. Philadelphia: Temple University Press.
Callahan, D. (1980). Goals in the teaching of ethics. In D. Callahan & S. Bok (Eds.),*Ethics teaching in higher education* (pp. 61–80). New York: Plenum Press.
Callahan, D. (1982). Should there be an academic code of ethics?*The Journal of Higher Educaiton*, 335–344.
Callahan, D., & Bok, S. (1980).*Ethics teaching in higher education*. New York: Plenum Press.
Camenisch, P. F. (1986). Goals of applied ethics courses.*The Journal of Higher Education,57*(5), 493–509.
Caplan, A. L. (1980). Evaluation and the teaching of ethics. In D. Callahan & S. Bok (Eds.),*Ethics teaching in higher education* (pp. 133–150). New York: Plenum Press.
Colby, A., Ehrlich, T., Beaumont, E., & Stephens, J. (2003).*Educating citizens, preparing America's undergraduates for lives of moral and civic responsibility*. San Francisco: Jossey-Bass.
Coleman, P. D., Atkinson, J. K., & Waduge, S. (2015). Should college education play a significant role in teaching students about ethics?*Journal of Leadership, Accountability and Ethics,12*(4), 24–30.
Connolly, P. (2009). History of the Ethics Bowl. In P. Connolly (Ed),*Ethics in Action* (pp. 505–509). United Kingdom: Blackwell Publishing.
Cooper, T. (2017). Learning from ethicists, Part 2: How ethics is taught at leading institutions in the pacific region.*Teaching Ethics,17*(1), 23–91.
Counelis, J. S. (1993). Toward empirical studies on university ethics: A new role for institutional research.*The Journal of Higher Education,64*(1), 74–92.
Dartmouth College. (n.d.). Retrieved July 30, 2017, from Institute for the Study of Applied and Professional Ethics:http://www.dartmouth.edu/~ethics/about/history.html.
Davis, M. (1990). Who can teach workplace ethics.*Teaching Philosophy,13*(1), 21–38.
Davis, M. (1999).*Ethics and the University*. New York: Routledge.
Dean, K. L., & Beggs, J. M. (2006). University professors and teaching ethics: Conceptualizations and expectations.*Journal of Management Ethics,30*(1), 15–44.
Dey, E., & Associates. (2009, September 30).*Civic Responsibility: What is the Campus Climate for Learning?* Retrieved from Association of American Colleges & Universities:http://www.aacu.org/publications-research/publications/civic-responsibility-what-campus-climatelearning.
Dey, Eric L., & Associates. (2010).*Developing a moral compass: What is the campus climate for ethics and academic integrity?* Retrieved July 28, 2017, from Association of American Colleges & Universities:https://www.aacu.org/sites/default/files/files/core_commitments/moralcompassreport.pdf.
Dey, E. L., Antonaros, M., Ott, M. C., Barnhardt, C. L., & Holsapple, M. A. (2009).*Developing a moral compass: What is the campus climate for ethics and academic integrity*. Association of American Colleges and Universities.
Dougherty, T. (2015). Yes means yes: Consent as communication.*Philosophy & Public Affairs,43*(3), 224–253.
DuBois, J. M., & Burkemper, J. (2002). Ethics education in U.S. medican schools: A study of syllabi.*77*(5), 432–437.

Ebels-Duggan, K. (2015). Autonomy as intellectual virtue. In H. Brighouse & M. McPherson (Eds.),*The aims of higher education, problems of morality and justice* (pp. 74–90). Chicago: The University of Chicago Press.

Eliot, C. (1869).*Inaugural address as the President of Harvard college. Educational reform: Essays and addresses*. New York: The Century Company.

Elliott, D. (2007).*Ethics in the first person: A guide to teaching and learning practical ethics*. Lanham, MD: Rowman & Littlefield.

Englehardt, E., Pritchard, M., Romesburg, K., & Schrag, B. (2010).*The ethical challenges of academic administration*. Springer.

Gert, B. (2005).*Morality, it's nature and justification*. New York: Oxford University Press.

Goodpaster, K. E. (1982, February).*Is teaching ethics 'making' or 'doing'?* pp. 37–39.

Harvard University. (n.d.). Retrieved July 30, 2017, from Edmond J. Safra Center for Ethics: https://ethics.harvard.edu/.

InCites Journal Citation Report. (2017, July 12). Retrieved from Thomson Reuters:https://jcr-incites-thomsonreuters-com.ezproxy.lib.usf.edu/JCRJournalHomeAction.action?SID = A2-Kp3O2cfPLILCHcwAz1zlBX1ZBesgaJM8-18x2dMkPmblR3wFca4tn0JGiq6Ax3Dx3DLU63EqzfAm6YtXtbZIKYfwx3Dx3D-iyiHxxh55B2RtQWBj2LEuawx3Dx3D-1iOubBm4x2FSwJjjKtx2F71AaQx3Dx3D&Src.

James Madison University. (2017).*The madison collaborative: Ethical reasoning in action*. Retrieved 2017, fromhttp://www.jmu.edu/mc/index.shtml.

Keenan, J. F. (2015).*University ethics: How colleges can build and benefit from a culture of ethics*. New York: Rowman & Littlefield.

Keohane, N. O. (2006).*Higher ground: Ethics and leadership in the Modern University*. Durham, NC: Duke University Press.

King, P., & Mayhew, M. J. (2002). Moral judgment development in higher education: Insights from the defining issues test.*Journal of Moral Education,31*(3), 247–270.

Kohlberg, L. (1981).*The philosophy of moral development*. San Francisco: Harper & Row.

Kohlberg, L. (1984).*The psychology of moral development*. San Francisco: Harper & Row.

Kuh, G. D. (2003).*Conceptual framework and overview of psychometric properties*. NSSE: Indiana University, NSSE.

Ladenson, R. F. (2001). The educational significance of the ethics bowl.*Teaching Ethics, 1*(1), 63–78.

Lane, M., & Schaupp, D. (1989). Ethics in education: A comparative study.*Journal of Business Ethics*, 943–949.

Lee, C.-M., & Taylor, M. (2013). Moral education trends over 40 years: A content analysis of the Journal of Moral Education (1971–2011).*Journal of Moral Education,42*(4), 399–429.

Lee, C.-T., Padilla-Walker, L. M., & Nelson, L. J. (2015). A person-centered approach to moral motivations during emerging adulthood: Are all forms of other-orientation adaptive?*Journal of Moral Education,44*(1), 51–63.

London, L., & McCarthy, G. (1998). Teaching medical students on the ethical dimensions of human rights.*Journal of Medical Ethics*, 257.

Magolda, P., & Abowitz, K. (1997). Communities and tribes in residential living.*Teachers College Record,99*(2), 266.

Matchett, N. (2008). Ethics across the curriculum.*New Directions for Higher Education*, 25–38.

Mayhew, M. J., & King, P. (2008). How curricular content and pedagogical strategies affect moral reasoning development in college students.*Journal of Moral Education,37*(1), 17–40.

Merrick, A., Green, R. C., & Hester, D. M. (2016). Introducing the medical ethics bowl.*Bioethics Education*, 141–149.

Meyer, T. (2012). The intercollegiate ethics bowl: An active learning experience.*Marketing Education Review*, 215–224.

Meyhew, M. J. (2012). A multilevel examination of the influence of institutional type on the moral reasoning development of first year students.*The Journal of Higher Education,83*(3), 367–388.
Mill, J. S. (1863/1991). Utilitarianism. In J. Grey (Ed.),*John Stuart Mill, On liberty and other essays*. New York City: Oxford University Press.
Murphy, P. (2014). Teaching applied ethics to the righteous mind.*Journal of Moral Education,43* (4), 413–428.
National Center for Educational Statistics. (2017).*Classification of instructional programs (CIP)*. Retrieved August 23, 2017, from Integrated Post-Secondary Education Data System,https://nces.ed.gov/ipeds/cipcode/browse.aspx?y=55.
Ozar, D. T. (1977). Teaching philosophy and teaching values.*Teaching Philosophy,2*(3–4), 237–245.
Rest, J. (1986).*Moral development: Advances in research and theory*. New York: Praeger.
Rhoads, R. (1997). In the service of citizenship: A study of student involvement in community service.*The Journal of Higher Education,68*(6), 277–297.
Rivage-Seul, M. K. (1987). Peace education: Imagination and pedagogy of the oppressed.*Harvard Educational Review,57*(2), 153–169.
Roberts, P. (1999). A dilemma for critical educators?*Journal of Moral Education,28*(1), 19–30.
Safatly, L., Itani, H., El-Jajj, A., & Salem, D. (2017). Ethics centers' activities and role in promoting ethics in universities.*Ethics and Education,12*(2), 153–169.
Schmidt, C. D., McAdams, C. R., & Foster, V. (2009). Promoting the moral reasoning of undergraduate business students through a deliberate psychological education-based classroom intervention.*Journal of Moral Education,38*(3), 315–334.
Scriven, M. (1982). Professorial ethics.*The Journal of Higher Education*, 307–317.
Shaw, G. (1999). Diversity.*Philosophy & Public Affairs,28*(2), 85–104.
Sloan, D. (1980). The teaching of ethics in the American undergraduate curriculum 1876-1976. In D. Callahan & S. Bok (Eds.),*Ethics teaching in higher education* (pp. 1–57). New York: Plenum Press.
Society for Ethics across the Curriculm. (n.d.). Retrieved July 29, 2017, from Society for Ethics across the Curriculum,http://www.rit.edu/~w-ethics/seac/.
Springer Publication. (2017).*International Journal of Ethics Education*. Retrieved from Springer International Publishing,http://www.springer.com/philosophy/ethics+and+moral+philosophy/journal/40889.
Stark, J. S., Lowther, M., Hagerty, B. M., & Orezyk, C. (1986). A conceptual framework for the study of preservice professional programs in colleges and universities.*The Journal of Higher Education,57*(3), 00001.
Stearns, W. N. (1908). Moral and religious training in the universities and colleges in the United States.*Religious Education,2*(6), 201–210.
Steneck, N. H. (1994). Research universities and scientific misconduct.*The Journal of Higher Education,65*(3), 310–330.
The Hastings Center. (1980).*The teaching of ethics in higher education*. Hastings-on Hudson, NY: Institute of Society, Ethics and the Life Sciences.
Thoma, S. J., Bebeau, M. J., & Narvaez, D. (2016). How not to evaluate a psychological measure: Rebuttal to criticism of the defining issues test of moral judgment development by Curzer and colleagues.*Theory and Research in Education,4*(2), 241–249.
Thompson, C. (2006). Unintended lessons: Plagiarism and the University.*Teachers College Record,108*(12), 2439–2449.
Thornton, C. H., & Jaeger, A. J. (2008). The role of culture in institutional and individual approaches to civic responsibility at research institutions.*Journal of Higher Education,79*(2), 160–182.
Tsei, L. (2002). Fostering critical thinking through effective pedagogy.*The Journal of Higher Education, 2*, 740.
UlrichsWeb. (2017, July 12).*UlrichsWeb*. Retrieved fromhttp://ulrichsweb.serialssolutions.com.ezproxy.lib.usf.edu

Utah Valley University. (n.d.).*Ethics across the curriculum*. Retrieved July 30, 2017, from Center for the Study of Ethics:https://www.uvu.edu/ethics/emphasis/eac/index.html.
Whitbeck, C. (1995). Teaching ethics to scientists and engineers: Moral agents and moral problems.*Science and Engineering Ethics Journal,1*(3), 299–308.
Wilshire, B. (1987). Can the university defend the values upon which it stands?*Journal of Higher Education,58*(3), 249–260.
Wolff, R. P. (1994). The myth of the neutral university. In R. L. Simon (Ed.),*Neutrality and the academic ethic* (pp. 103–109). Lanham, MD: Rowman & Littlefield.

Moral Theory in Ethics Across the Curriculum?

Michael Davis

Abstract The problem of the relation of curriculum to practice cannot be dismissed as "*merely* philosophical". The problem arises each time a teacher chooses to teach one thing rather than another. It certainly arises when a teacher wishing to teach "ethics" decides to teach moral theory—or decides not to. Which brings me to the subject of this Chapter: *What part should moral theory have in ethics across the curriculum?* After distinguishing five kinds of ethics across the curriculum, I argue that the part moral theory should have depends on the kind of ethics across the curriculum involved. This chapter is a contribution to a debate that has raged, or at least simmered, for at least a decade.

Keywords Morality · Professional ethics · Business ethics · Social ethics
Academic ethics · Research ethics

There is an old joke, "Those who can, do; those who can't, teach." Like most jokes that survive long enough to become old, this one contains a truth. Trouble with practice can help a teacher appreciate what his students must overcome to learn what he teaches. Those who learn easily make poor teachers.

Because this joke is often made to insult teachers, it is worth noting that its surface claim has been refuted many times. The first recorded refutation seems have been about 600 BCE. Thales, one of the "seven sages of ancient Greece", was asked why, if he was so smart, he wasn't rich. He responded by predicting a good olive harvest, reserving all the local olive presses at pre-harvest prices, and then renting them out for much more when demand peaked during the harvest, thus proving both that those who can, need not, and that those who teach, can also practice what they teach. (Aristotle 1964, p. 31.)

Of course, behind that old joke is a problem, the relation between what teachers teach (facts, skills, and theories) and what practitioners need (wisdom). The problem is primarily philosophical, though. In practice, teachers seem to teach a reasonable

M. Davis (✉)
Illinois Institute of Technology, Chicago, USA
e-mail: davism@iit.edu

E. E. Englehardt and M. S. Pritchard (eds.), *Ethics Across the Curriculum—Pedagogical Perspectives*, https://doi.org/10.1007/978-3-319-78939-2_3

amount of what practitioners need. The practitioners generally do better after being taught. Were it otherwise, few societies would spend much on schools. Yet, the problem of the relation of curriculum to practice cannot be dismissed as "*merely* philosophical". The problem arises each time a teacher chooses to teach one thing rather than another. It certainly arises when a teacher wishing to teach "ethics" decides to teach moral theory—or decides not to. Which brings me to the subject of this Chapter: *What part should moral theory have in ethics across the curriculum?* My thesis is: It depends. This chapter is a contribution to a debate that has raged, or at least simmered, for at least a decade. (See, especially, Harris 2009a; Davis 2009a, 2010, 2014; Harris 2009b, 2010; Gert 2010; Englehardt and Pritchard 2013, 2015).

Introduction

Between 1990 and 2009, the Center for the Study of Ethics in the Professions at Illinois Institute of Technology (IIT) offered summer workshops (funded by the National Science Foundation) to help faculty integrate "ethics" into courses across the curriculum.[1] During the first three years, the Center trained IIT faculty. Having figured out how to do that, the Center began offering the workshop to faculty outside IIT, including faculty from as far away as France, Japan, New Zealand, the Philippines, and Sudan. Of the workshop's seven days, only half of one was devoted to moral theory. The theme of that half day was that moral theory was too complex to teach, much less use in class, outside a philosophy course. That theme was part of a general strategy to make teaching ethics in technical courses seem as much as possible like teaching the technical material in such courses. The Center feared that non-philosophers, especially engineers and scientists, would not try to integrate ethics into their courses if they thought doing so required teaching moral theory.

The Center called what it was doing "ethics across the curriculum" because that was what two young professors of mechanical engineering called it in 1989. They had come to the Center for help with including some engineering ethics in their engineering courses. They thought that the Center would have something it could pull off the shelf to insert into, for example, Thermodynamics. The Center did not have anything like that but it set about inventing something like what the two engineers were asking for.

No one now recalls where the two got the term. I had not heard it before they used it, but "writing across the curriculum" was already current. I initially supposed "ethics across the curriculum" to rely on the obvious analogy between "writing" and "ethics". I did not then appreciate how "ethics" might differ from "writing", especially how much less practical than writing ethics might be (depending on what

[1]1991–1995 (DIR 9014220); 1997–1999 (SBR-9601905); 2000–2003 (SES-9985813), and 2006–2009 (EEC-0629416). For an assessment of this undertaking at one university, see Davis et al. (2016).

"ethics" was thought to mean). Indeed, almost a decade passed before I realized that there were several kinds of ethics across the curriculum competing for the name. A few more years passed before I realized how important it is to be clear which kind is under discussion.

I shall now distinguish five kinds of ethics across the curriculum before narrowing my focus to the one kind that many teachers seem to assume should include the study of moral theory to improve conduct. No doubt there are other kinds, but these five seem to be the most common today, what we may call: (1) *morality*-across-the-curriculum; (2) *moral-theory*-across-the-curriculum; (3) *social-ethics*-across-the-curriculum; (4) ethics-*from*-across-the curriculum; and (5) *professional-ethics*-across-the-curriculum. Let us consider the five in that order.[2]

First Four Kinds

By "morality" I mean those standards of conduct that all of us (at our most reasonable) want everyone else to follow even if that means having to follow them ourselves. (For a defense of something like this definition of morality, Gert 2005, esp. pp. 9–14.) Among those standards are the (prima facie) rules "Don't lie", "Don't cheat", and "Keep your promises", principles such as "Help the needy" and "Return good for good", and ideals such as truthfulness or equal justice under law. One kind of ethics across the curriculum understands "ethics" as a mere synonym for "morality" (in this sense). Morality-across-the-curriculum may exist in one of at least three forms (though, more often, in a mixture of them).

One form emphasizes adherence to a written code, such as an "honors code" or rules of a religion (understood as a statement of universal standards). A second form emphasizes the cultivation of certain virtues (morally good dispositions to act in certain ways) rather than the following of any particular standards (though it is expected that "virtuous conduct" will satisfy moral standards). These first two forms of morality-across-the-curriculum seek to guide conduct; the third, what we may call "moral literacy" (or "value clarification"), does not (at least not directly) seek to guide conduct. It is a largely intellectual undertaking. Morality is treated as

[2]For an earlier discussion of this distinction, see Davis (2004). Lawrence Hinman has published an "EAC Grid" also distinguishing five kinds of "academic" ethics across the curriculum (the left side of the grid). Though useful for some purposes, especially planning, its five-kind distinction is different from mine. Its five kinds are: (1) centralized required ethics courses; (2) specialized ethics courses within philosophy departments; (3) team-taught ethics courses with philosophers and non-philosophers; (4) ethics-component in non-ethics courses with philosopher as guest lecturer; and (5) course component taught by non-philosophers. These five kinds are (as I shall partially show below) possible subdivisions in all of my kinds except the fourth. The "non-academic" side of his grid (five kinds of service learning in the middle and three kinds of community service on the left) are not strictly ethics across the *curriculum* at all. They belong to a larger category, what we might call "ethics across the *campus*". For more, see ethics.sandiego.edu/Presentations/EAC/Tokyo/Implementing_EAC.pdf (accessed November 9, 2016).

a subject worthy of scholarly investigation across the curriculum. There is, nonetheless, the expectation, or at least the hope, that improving a student's knowledge of morality will improve her conduct. To know morality is to love it; to love it, to seek to do what it asks.

A university (or other institution of higher education) can teach (or, at least, try to teach) morality in one of these three ways without having a program deserving the name "ethics across the *curriculum*", for example, through honors courts, public lectures, regular chapels, or other extra-curricular activities. What makes a program of moral uplift deserve the name "ethics across the curriculum" is that the code, the virtues, or the moral literacy are integrated into classrooms all across the curriculum —not merely for study but in an attempt to improve student conduct both inside the classrooms and out. The best examples of morality-across-the-curriculum are probably to be found in nineteenth-century American colleges of liberal arts (well before "ethics across the curriculum" was a name for it), but we can find something quite close in some "Christian" colleges today. For example, Bob Jones University understands the whole institution, including its classrooms, as a single "place where Christ would be the center of all thought and conduct."[3] But today, morality-across-the-curriculum, even in the form of moral literacy, seems to be much more at home in primary and secondary schools (though not under the name "ethics across the curriculum"). I have discussed its strengths and weaknesses elsewhere (See Davis 2003a).

By "moral theory" I mean (roughly) the attempt to understand morality as a reasonable practice, especially the attempt to understand morality in a way that claims one theory to be better than any other. *A* moral theory is a systematic presentation of definitions, distinctions, and arguments designed to show that morality, understood in a certain way, is reasonable, that is, something reason permits, recommends, or requires (depending on the theory). Though moral theory (unlike morality) is generally considered the monopoly of the philosophy department (with an occasional "ethicist" in a department of religion, political science, psychology, or the like privileged to participate), the purpose of

[3]"Throughout his travels, Dr. Bob Jones Sr. saw students whose faith was shaken during college, and he recognized the need for a thoroughly Christian school to train America's youth. His vision was to establish a training center for Christians from around the world that would be distinguished by its academic excellence, refined standards of behavior, and opportunities to appreciate the performing and visual arts. At the same time, Dr. Jones's intent was to make a place where Christ would be the center of all thought and conduct." From Bob Jones University website, www.bju.edu/aboutbju/history/index (accessed July 10, 2004). See also the statement on academic education: www.bju.edu/academics/ed_purpose (accessed July 10, 2004). This statement makes it clear that BJU does not consider its ethics to be "sectarian" (whatever the rest of us may think). The only "true virtue", it is necessarily universal. Unfortunately, both websites seem to have disappeared. Though BJU still promises to "infuse every course with a biblical worldview and strive to offer the best academic experience of any Christian university", its website no longer claims that that "biblical worldview" to be the one true standard. http://www.bju.edu/ (accessed November 16, 2016).

moral-theory-across-the-curriculum is to make "philosophical ethics" (that is, moral theory) available more widely.[4] One way to do that is simply to require all students to take a course in moral theory (a course typically called "Ethics"). While such a course is undoubtedly a contribution to moral literacy, it has two disadvantages.

First, being required, it is likely to burden the philosophy department both with large classes and with students who resent the requirement. Most philosophy departments resist teaching large classes, arguing that teaching philosophy properly means assigning one or more papers a term and grading them with some care. Many philosophy departments also resist teaching *required* courses.

Second, insofar as the requirement is not part of a larger ethics program, an ethics course is unlikely to connect with the rest of the curriculum. Ethics-as-a-field-of-philosophy is spread across the curriculum only insofar as students carry what they learn into other courses. To expect students to carry much moral theory into their other courses (more or less) without assistance is to expect a lot, probably more than most students are capable of. After all, even faculty outside philosophy seem to have trouble integrating ethics into their courses. While this way of doing ethics across the curriculum leaves the philosophy department's monopoly over ethics intact, it does that at the cost of leaving the connection between moral theory and moral practice largely unexplored. (For a typical example of this sort of ethics across the curriculum, see Sia 2001).

Some universities have therefore tried, in addition or instead, to insert moral theory into courses across the curriculum. The simplest way to do that is to have a moral theorist serve as a guest lecturer for a day or two. A guest lecturer will, of course, have trouble making many connections between moral theory and the course she happens to be visiting on a particular day. ("Today is November 12; so, I must be in Computational Biology. Should I mention utilitarianism?") Deep knowledge of a course's content—indeed, even mere familiarity—is hard for any one person to maintain across much of the curriculum, and in all but the largest departments, only one or two philosophers will feel competent to serve as a guest lecturer on moral theory in even a few courses outside their department. That is one problem with guest-lecturing across the curriculum, spreading the guest too thin. Another problem is that not much of use can be said about moral theory in an hour or two of class. Moral theory is also spread too thin. (After all, most of us who teach

---

[4] I put the term "ethicist" in quotation marks for three reasons. One is that it generally now seems to be used in a context like this, that is, where there is a claim to expertise in "ethics" for someone not a philosopher. It is a polite way for philosophers to let a non-philosopher into the club while signaling that she does not quite belong. The other reason for the quotation marks is that this modern usage departs somewhat from the term's original use, itself (according to the *Oxford English Dictionary*, 1989) barely a hundred years old: one who "supports ethics or morality rather than religion" (as, perhaps, in the Ethical Culture Society). For some people, the word may still call up anti-religious associations. Third, for others, the term has the opposite associations, suggesting a claim to knowledge or virtue rather than philosophy's claim merely to pursue one or the other. An ethicist must be an exemplar of ethics (good conduct) much as a cleric must be an exemplar of piety. So, while the headline "Philosopher caught in adultery" invites a shrug or no reaction at all, the headline "Ethicist caught in adultery" is always good for a laugh.

a course in moral theory feel forty-five hours, a semester-long course, to be too little to cover what we should.)

Having a moral theorist serve as co-instructor in courses across the curriculum provides at least a partial solution to these problems, especially if the moral theorist co-teaches for several semesters, that is, long enough to learn the subject to which moral theory is to be applied, understand what the students will need in the way of moral theory, and develop a sense of how much moral theory the students can absorb (and when in the semester they can best absorb it). There is no algorithm for making moral theory useful in a particular non-philosophy course, whether accounting, engineering, history, linguistics, or the like, no substitute for trial and error.

While co-teaching a course can solve the two problems just identified (lack of deep knowledge and lack of integration), it can do that only by generating another. Generally, co-teaching a course doubles its cost (even if one of the teachers does it "for free"), an effect administrators are likely to worry about. Co-teaching moral theory *across the curriculum* is also beyond the resources of most philosophy departments as they are now—and also beyond what a prudent administration is likely to have in its plans. So, a number of universities (for example, Marquette) have undertaken to export moral theory to faculty across the university, hoping to decentralize the teaching of (basic) moral theory. The philosophers offer a summer workshop in moral theory. Having graduated from the workshop, non-philosophers can, it is supposed, use moral theory in their teaching and research (with, perhaps, assistance now and then from the philosophers) (Ashmore and Starr 1991).

My impression is that, generally, these workshops have been more successful in getting faculty across the curriculum to research morality-related questions than in getting them to include moral theory in their own courses. Moral theory demands too much time to win a place in their already crowded courses. The workshops have, however, had the happy side-effect of enlarging substantially the number of people in the university on good terms with philosophy. The workshops seem to reduce substantially the academic isolation that philosophy typically suffers in American universities.

The third kind of ethics across the curriculum is concerned with *social* ethics rather than morality in general or moral theory in particular (though both morality and moral theory may have a place in the analysis of the problems with which social ethics deals). By "social ethics" I mean those questions and would-be answers concerned with social arrangements for which morality does not, or at least does not yet, provide a decisive answer but to which it contributes, or might contribute, considerations relevant to resolution. Whether "we" should steal is not a question of social ethics (in this sense). Stealing is (prima facie) morally wrong. But what to do about stealing, whether "we" should respond to stealing with better locks, medication, a stipend, punishment, reconciliation, vocational training, or the like is (in part at least) a question of social ethics.

Social-ethics-across-the-curriculum may take the form of a single course in which important social issues—for example, abortion, gender assignment, human rights, immigration, privacy, sustainable development, and war—are discussed.

Such a course in "applied" or "practical" ethics may be called "Social Ethics", "Applied Ethics", "Moral and Social Values", "Moral Problems", "Social Issues", or the like. In many philosophy departments, the voluntary version of this course has become a popular alternative to the traditional Introduction to Philosophy. Though philosophers often refer to the course as "baby Ethics", it is seldom a watered-down version of the course in moral theory. Indeed, courses in social ethics often have little or no moral theory. They are instead an attempt to bring philosophy, all of philosophy, to bear on current "hot topics". What the course emphasizes is philosophical method, not moral theory. That is why it is a good introduction to philosophy. (It is this sort of ethics across the curriculum that is the subject of Keller 2002.)

Having a single course in social ethics required across the curriculum has the same administrative problems as requiring all students to take the philosophy department's course in moral theory. For that reason, and perhaps others, universities have found other ways to have social-ethics-across-the-curriculum. The most common of these seems to be courses typically titled "X and Society" (where the "X" stands for the name of the appropriate department, program, or other administrative unit). So, for example, a biology department might have a course in Biology and Society; the computer science department, Computers and Society; the law school, Law and Society; and so on (more or less) across the curriculum. If taught in the philosophy department, such courses will typically have "Ethics" in the title (rather than "Society"), for example, Biology and Ethics, Computer Ethics, or Legal Ethics.

Such courses taught outside the philosophy department will generally differ from Social Ethics in at least three ways. First, they will have a smaller range of social issues. So, for example, while Social Ethics might include any issue appropriate to Computers and Society (or Computer Ethics), Computers and Society could not take up many issues, such as capital punishment or gun control, appropriate to Social Ethics. Second, an X-and-Society course taught outside of the philosophy department will generally emphasize information rather than method. It will not be a good introduction to philosophy. Indeed, even the corresponding course taught in the philosophy department (for example, Computer Ethics) will not be a good introduction to philosophy. Third, the X-and-Society course will be even less likely than Social Ethics to include any moral theory. The corresponding course taught in philosophy will, however, be likely to include "moral theory" of a sort (the "big ideas" discussed below).

Because administrators resist the multiplication of courses generally, and are especially adverse to courses in different departments that seem to belong to some larger category (as X-and-Society courses seem to), universities have developed two "less wasteful" versions of social-ethics-across-the-curriculum. One version chooses a single social issue, appropriate to the institution, to serve as a theme across the curriculum. The Colorado School of Mines offers a good example. The theme, "stewardship of the earth", appears in its Mission Statement, is central to a required first-year course, and is thereafter raised now and then in courses across the curriculum (from Mechanics of Petroleum Production to Senior Design). [See http://www.mines.edu/AboutMines (accessed November 9, 2016).]

Another way to spread social ethics across the curriculum is to have each course in the curriculum contain some moral reflection on what might be done concerning social issues related to the course. So, for example, a genetics course would be an ideal place to discuss social issues in genetic testing: Should such tests be required? Who should be allowed to see the results? And so on. There is no (substantive) theme across the curriculum, only a reoccurring thoughtfulness about social issues. What this way of spreading social ethics across the curriculum lacks in unity, it gains in flexibility. That flexibility seems more appropriate to a liberal arts college or large university, places of deliberate diversity. Unity of theme seems more appropriate in a professional school (such as the Colorado School of Mines) or a small religious university (such as Bob Jones), places of (relative) homogeneity. Social-issues-across-the-curriculum can be quite independent of philosophy; its reoccurring thoughtfulness need include no moral theory.

Nonetheless, getting students to engage in moral reflection on social issues, whether on a campus-wide theme or just on the content of a particular course, is not something faculty in many disciplines outside philosophy do easily. I once attended a session at the annual meeting of the Association for Practical and Professional Ethics in which a philosopher from St. Olaf College described an impressive program for (what he called) "ethics across the curriculum". The "ethics" was primarily (but not exclusively) social ethics. His presentation included a brief video of a discussion in a science class. The instructor posed a question of the appropriate sort but then sat quietly while the students discussed it. He gave no guidance, not even asking further questions. Though the discussion was lively and the students obviously enjoyed it, it seemed to go nowhere. I could not tell from the video what the students were supposed to have gained from the discussion. I was not surprised to learn (in response to a question I put) that the course would include *no graded* assignment or exam question related to the ethical issue discussed that day, a sign to students both that what happened on that day had little to do with the course and that the professor knew it. The philosophy department had prepared the instructor to identify issues, that is, had given him a workshop *in* social ethics but had not taught him *how* to teach social ethics. In my experience, few philosophers have any idea how hard it is for others to do what philosophers do almost without thought. Any workshop for social-ethics-across-the curriculum should include some instruction in how to lead discussion, how to prepare exams and homework assignment in social ethics, and how to grade them.[5]

Ethics-*from*-across-the-curriculum, my fourth kind of ethics across the curriculum, brings faculty (or students) together from across campus to discuss some

[5]For some idea of what can be done to help faculty outside philosophy lead such a discussion, see Davis (1993); or, in a slightly revised version, Davis (1999), pp. 111–142.

controversial issue in which morality is an important consideration. Typically, a philosopher will lead the discussion. The content of this kind of ethics across the curriculum overlaps social-ethics-across-the-curriculum (though it may include some issues of "personal ethics"—for example, "Should I commit suicide?"—that seem beyond social ethics, however the term is stretched).

Ethics-from-across-the-curriculum is a method, not an academic subject. The point of bringing faculty (or students) together is to provoke discipline-specific insights that may shed light on the issue from different perspectives, without necessarily leading to a solution. The insights of a particular discipline may come in the form of an analysis or summary of empirical research (typical of the sciences), in performance of a creative work (typical of theater, music, art, or creative writing), in interpretation of a crucial text (typical of literary critics, religious scholars, and lawyers), and so on. What student are supposed to take from such events, apart from insights into this or that particular issue, is a sense of how different, enlightening, and useful the insights of different disciplines can be—especially when compared. It is an invitation to seek insight into ethical issues from across the curriculum. Moral theory is typically absent. (I owe the concept of ethics from across the curriculum to Ken Alpern.)

Generally, the consideration of a controversial issue will be at a public event that students attend voluntarily. In that case, it is (strictly speaking) not ethics across the curriculum. It is extra-curricular, that is, "ethics across *campus*". But similar events can be staged in one or more classes. Even then, ethics-from-across-the-curriculum is ethics *across the curriculum* only in an extended sense, for at least two reasons. First, at any given time, only a small part of the curriculum will be represented, say, four departments out of twenty-three. Several years may pass before all the departments rotate through, if the issue chosen for discussion allows for such a rotation. While the aspiration is ethics-across-the-curriculum, the reality generally falls well short. Second, few students are likely to attend all, or even many, of these discussions, since the (cross-curricular) events will (given the limits of faculty time) occur only in a few classes. Students are, then, unlikely to benefit fully from the cross-curricular nature of the events.

Nonetheless, ethics-*from*-across-the-curriculum may be a worthwhile undertaking, especially if combined with one or more of the other forms of ethics across the curriculum distinguished here. One long-term effect of faculty participating in such discussions (whether on-stage or in the audience), with a philosopher as guide, may be to teach the non-philosophers among them something of the art of leading such discussions, tempting them to try such a discussion in their own classes. Another long-term effect may be to help (moral) philosophers to identify faculty on campus whom they can usefully invite into their own classes—not only scientific experts but also actors, poets, critics, and the like.

Ethics-Across-the-*Professional*-Curriculum

The fifth kind of ethics across the curriculum distinguished earlier, ethics-across-the-professional-curriculum, is distinct from the other four in two ways. First, it is about *professional* ethics (and, as I shall soon explain, *institutional* ethics), not about morality in general, moral theory, or social ethics. Second, its focus is the professional (or other career-specific) curriculum, not the entire curriculum. Professional ethics crosses the entire professional curriculum only insofar as each department, program, or school has its own professional (or institutional) ethics component, whether distributed throughout its technical courses or taught as a stand-alone course.

To make clear how much ethics-across-the-professional-curriculum differs from the four other forms of ethics across the curriculum distinguished here, I must explain what professional ethics is—and, especially, how it differs from morality, moral theory, and social ethics.

By "profession", I mean a number of individuals in the same occupation voluntarily organized to earn a living by openly serving a certain moral ideal in a morally permissible way (a discipline) beyond what law, market, morality, and public opinion would otherwise require. (For an extended defense of this definition, see Davis 2009b.) Professions organize all, or part, of a single occupation in a certain way. A profession's ethics are the special standards defining the (morally permissible) way the would-be profession is to pursue its moral ideal. These standards are (more or less) arbitrary in the way promises are—and morally binding much as promises are (see, for example, Davis 1987). Ordinary morality sets limits on professional ethics without determining the content. One cannot deduce professional ethics from morality or moral theory. Even those who know morality or moral theory must learn a profession's ethics one standard at a time.

Ordinarily, I use "professional ethics" in this way. Here, however, I want to use the term in a somewhat wider sense—one including what I have elsewhere called "institutional ethics" (see Davis 2003b). What distinguishes institutions from professions proper is that institutions are a form of organization in which more than one occupation works. So, for example, engineering ethics concerns the special standards of one profession, engineering, but business ethics concerns an institution, business, in which accountants, engineers, lawyers, nurses, and other professionals work—along with many who are not members of any profession (clerks, janitors, laborers, mechanics, truck drivers, and so on). Business ethics is one kind of institutional ethics. Research ethics is another; academic ethics, another; and so on. Medical ethics is institutional ethics if it is understood as, say, concerned with the special standards that hospitals and other medical facilities have adopted, or should adopt, but professional ethics—strictly speaking—if understood as concerned with the special standards that physicians have adopted or should adopt.

And, of course, medical ethics is *social* ethics if understood as (primarily) concerned with how "we" (government and the public) should govern medicine-related organizations, medications, or the provision of medical care.

Though the distinction between professional and institutional ethics is generally important, it is not important here. What distinguishes them both from morality, moral theory, and social ethics is their parochialism. Both professional ethics and institutional ethics apply only to certain people, not to everyone or even to society at large. Business ethics applies to people in business and no one else; engineering ethics, to engineers and no one else; and so on. Professional ethics (the field of study) includes not only special problems but also the special (morally permissible) standards that should help to resolve those problems. While a profession's (or institution's) ethics may interest everyone, its standards only govern members of the profession (or institution). For anyone not a member of the profession (or institution) in question (or not thinking as a member of it), professional (and institutional) ethics will seem a sort of social ethics—and the chief question will be whether "we" (society at large) should allow such organizations the freedom most other organizations have to regulate members or restrain that regulation in this or that way.

Professional ethics can be taught in a stand-alone course, as part of "technical" courses (that is, courses in the appropriate discipline), or in both ways. A stand-alone course may be required or optional. It may be "in house" (that is, taught by faculty in the discipline) or "outside" (that is, taught in the philosophy department, religion department, or the like). Whether in-house or outside, a stand-alone course in a profession's ethics may include some "moral theory"—in practice, as much as a week or two out of a fifteen week semester. The technical courses will generally include issues, information, and perhaps a decision-procedure, but not much, if any, "moral theory".

Whether ethics-across-the-*professional*-curriculum should count as ethics-across-the-curriculum proper seems an open question, the similarity of name notwithstanding. Certainly, ethics-across-the-professional-curriculum is a way to get "moral theory" into parts of the curriculum otherwise hard to reach. There are nonetheless at least two reasons to think ethics-across-the-professional-curriculum is not, strictly speaking, ethics across the curriculum. First, professional ethics is not a single subject, though the ethics of one profession may resemble the ethics of another more or less. Architecture ethics differs substantially from business ethics; research ethics, from engineering ethics; and so on. Except for academic ethics, professional ethics divides the curriculum rather than crossing it. Few courses in professional ethics include much comparison of the special standards of one profession with the special standards of another (though such comparison might help students appreciate how the ethics of their own profession resembles others in some ways while differing in others). Second, the subject of professional ethics is not primarily moral theory or its application but certain special standards and *their* application. Moral theory is directly relevant to professional ethics, if it is relevant at all, only to evaluation of those special standards. Evaluation of a profession's standards is likely to be a small part of any stand-alone course in a profession's ethics and not to come up at all when a little professional ethics is inserted into a technical course. What professional ethics courses share across the curriculum are problems of interpreting standards, bringing the courses closer (in that respect at least) to philosophy of law than to moral theory.

Conclusion

My purpose here is not to argue that one kind of ethics across the curriculum is better than another. I approve of all five, though I think one kind may be better than another for a certain purpose, in a certain context, or limited in certain ways. Often much is to be gained by combining two or more. My purpose here is, rather, to identify a conceptual problem for those kinds of ethics across the curriculum that rely on moral theory to teach practical ethics (that is, to teach individuals or groups how they should conduct themselves). I have distinguished the five kinds of ethics across the curriculum to make possible distinguishing those that have this conceptual problem from those that do not. From what I have said so far, it should be obvious that the stand-alone course in professional (or institutional) ethics is the locus of the problem. *Morality*-across-the-curriculum, *social*-ethics-across-the-curriculum, and ethics-*from*-across-the-curriculum seldom, if ever, have much to do with moral theory. *Moral-theory*-across-the-curriculum, in contrast, does not seek to resolve practical problems but to make faculty and students outside philosophy conversant with moral theory, leaving it to them to make the theory practical. Moral-theory-across-the-curriculum leaves our conceptual problem for another day.

Since moral theory is (more or less) the attempt to *understand* morality, including the special standards of professional ethics, as a reasonable practice, moral theory ("philosophical ethics") is not morality but *about* morality. It is "pure" rather than "applied" or "practical". A moral theory is not a decision-procedure but a way of understanding morality, including arguments for and against it. All moral theories on offer, including the most popular, are imperfect. They are all subject to serious objections. There is no decisive reason why we should guide our conduct by any particular theory rather than by one of its main competitors. Indeed, there is reason to think that we should *not* guide our conduct even by the correct moral theory (supposing we knew which that was). All the plausible theories now on offer are complex enough to be hard to use as guides to conduct, for example, to determine whether a certain code of ethics for engineers should have a stronger provision concerning sustainable develop or whistleblowing than it now has. The correct theory is likely to be at least as complex and therefore as hard to apply. That alone makes problematic the very idea of *applying* moral theory to most *practical* problems, that is, to those problems that come with significant limits of time, knowledge, and resources. Like Andrew Marvell, practitioners are likely to say,

> Had we but world enough and time....
> [Applying moral theory] were no crime....
> But at my back I always hear
> Time's wingèd chariot hurrying near.

Why then has "moral theory" seemed so necessary to ethics-across-the-professional-curriculum, for example, necessary enough to claim at least part of a chapter in most textbooks in professional ethics?

There may, I think, be at least three reasons for that, what we may call: (1) the philosophers' academic monopoly on the word "ethics", (2) philosophy-as-big-ideas, and (3) philosophy-as-wisdom. Upon examination, none of these reasons supports the claim that moral theory is necessary for a course in professional ethics, though each helps to explain why some people might think so.

Professional ethics has many names, including (depending on discipline) "professional responsibility", "professionalism", "professional conduct", "responsible conduct", "social responsibility", and "integrity". In many universities, only the philosophy department is allowed to use the word "ethics" for what it teaches. The justification for its control over that word is that "ethics" implies moral theory and only philosophers have the credentials to teach moral theory. To include "ethics" in the title of a course outside the philosophy department would, in effect, be false advertising. Thus, a course entitled "Architecture Ethics", or even "Ethical Issues in Architecture", is a philosophy course—at least prima facie. In contrast, a similar course entitled "Professional Responsibility of Architects" or "Professionalism in Architecture" is an architecture course—at least prima facie—a course only architects should teach. Since there are many ways to say "ethics" (in the special standards sense) without using the word, philosophy's control of the word is easy for other disciplines to accept. But once philosophers have justified their right to teach a course like Architecture Ethics by pointing to their monopoly over the word "ethics", they would seem to be obliged to include some moral theory in the course. They cannot justify their control over the word "ethics" if they rely on the special-standards sense. That sense would give the appropriate professional (or institutional) department grounds for claiming the right to use "ethics" for its courses too.

Where to put professional ethics (name aside) remains a difficult problem for administrators—as well as for anyone else who wants to teach students the ethics of their profession. Some professions, such as engineering, now tend to cede the teaching of their profession's ethics, at least in the stand-alone courses, to philosophy. Other professions, such as law, seem to have made the subject their own. For now, at least, professional ethics remains an amphibian, living both in philosophy, because we do not yet have a good enough understanding of profession as a reasonable undertaking, and in the specific profession, because understanding the reasonableness of a particular profession requires knowing a good deal about that profession, knowledge members of the profession typically have and philosophers typically do not. The philosophers who do a good job of teaching the ethics of a specific profession have had to learn much about that profession. Much of what they teach is how to practice that profession, something largely outside philosophy.

If we admit that moral theories are too complex to serve as practical guides to conduct, we must ask what students in a course in professional ethics *apply* after they receive a week or two of instruction in “moral theory”. The answer seems to be not the moral theories themselves but the “big ideas” associated with them: universalizability, utility maximization, virtue, care, respect for persons, or the like. A course in professional ethics will typically include several of these “big ideas” (if it includes any). Students will be instructed to use all of them or at least several in making decisions. Thus, the “big ideas” are, in one way at least, fundamentally different from the moral theories they resemble. Each moral theory claims to be the only true one; the others are false (except for any that can be shown to be an alternate formulation of the true one). The “big ideas” make no such claim. The application of “big ideas” is, therefore, not the application of moral theories but the application of heuristics resembling the theories in certain (often superficial) ways. Typically, though, the resemblance is enough to sustain the claim that the philosopher is teaching enough moral theory to use “ethics” in the course’s name.

At least since Socrates, philosophy has been defined as the “*pursuit* of wisdom”. Though Socrates claimed that he was the wisest person in Greece because he alone knew that he knew nothing, the common assumption still seems to be that those who pursue wisdom must eventually catch it by the tail at least, learning much more than that they know nothing.

One subject philosophers, especially moral philosophers, know is moral theory. When they discuss moral theory, they seem wise as well as learned. They therefore seem to have something to offer decision-makers in the professions, the wisdom that an understanding of morality confers. The problem is that there is no reason to believe that *understanding* morality does confer wisdom, much less that the wisdom in question is relevant to the profession in question or that it can be passed from teacher to student. While I don’t think there is any (decisive) reason to believe moral philosophers to be less wise than the average engineer, lawyer, or physician, there is, as far as I know, also no (decisive) reason to believe they are wiser.

What philosophers, like other teachers of professional ethics, can usefully offer students, in addition to much information about their profession, is a decision-procedure (a method well short of an algorithm). That procedure may include some “big ideas” of moral theory. So, for example, today’s most popular text in engineering ethics includes a chapter, “A Practical Ethics Toolkit”, offering a procedure: *Determining the Facts*, *Clarifying Concepts*, *Determining How Concepts Apply*, and so on, ending with four *Tests or Application Procedures*. These “tests” (“Utilitarian Thinking”, “Respect for Persons Approach”, “The Self-Defeating Approach”, and “The Rights Approach”) are big ideas drawn from moral theory. (Harris et al. 2014, pp. 24–50) The book includes no moral theory (strictly speaking) but makes enough of a show of it to support the claim that any course using that textbook is teaching (philosophical) ethics.

Thus, professional-ethics-across-the curriculum does not, it seems, require moral theory or even seem likely to benefit from it—except for external reasons of policy.

Ethics across the curriculum does not seem to have a problem with the gap between moral theory and moral practice because it does not seek to bridge the gap.

Acknowledgements My thanks are owed to Elaine Englehardt and Mike Pritchard for helpful comments on earlier drafts of this Chapter.

References

Aristotle. (1964). In E. Barber (Ed.), *The politics of Aristotle*. New York: Oxford University Press.
Ashmore, R. B., & Starr, W. C. (Eds.). (1991). *Ethics across the curriculum—The Marquette experience*. Milwaukee: Marquette University Press.
Davis, M. (1987). *The moral authority of a professional code* (pp. 302–337). Authority Revisited: NOMOS XXIX. New York: New York University Press.
Davis, M. (1993). Ethics across the curriculum: Teaching professional responsibility in technical courses. *Teaching Philosophy, 16*, 205–235 (September 1993).
Davis, M. (1999). *Ethics and the University*. London: Routledge.
Davis, M. (2003a). What's wrong with character education? *American Journal of Education, 110*, 32–57 (November 2003a).
Davis, M. (2003b). What can we learn by looking for the first code of professional ethics? *Theoretical Medicine and Bioethics, 24*, 433–454.
Davis, M. (2004). Five kinds of ethics across the curriculum: An introduction to four experiments with one kind. *Teaching Ethics 4*, 1–14 (Spring 2004).
Davis, M. (2009a). The usefulness of moral theory in practical ethics: A question of comparative cost. *Teaching Ethics*, 69–78 (Fall 2009a).
Davis, M. (2009b). Is engineering a profession everywhere? *Philosophia, 37*, 211–225 (June 2009b).
Davis, M. (2010). The usefulness of moral theory in teaching practical ethics: A reply to Gert and Harris. *Teaching Ethics, 11*, 51–60 (Fall 2010).
Davis, M. (2014). Professional ethics without moral theory: A practical guide for the perplexed non-philosopher. *J Appl Ethics Philos, 6*, 1–9.
Davis, M., Laas, K., & Hildt, E. (2016). Twenty-five years of ethics across the curriculum: An assessment. *Teaching Ethics 16*, 55–74 (Spring 2016).
Englehardt, E. E., & Pritchard, M. S. (2013). Teaching practical ethics. *Int J Appl Philos, 27*, 161–173.
Gert, B. (2005). *Morality: Its nature and justification* (Revised edition). New York: Oxford University Press.
Gert, B. (2010). The usefulness of a comprehensive systematic moral theory. *Teaching Ethics, 11*, 25–38 (Fall 2010).
Harris, C. E. (2009a). Is moral theory useful in practical ethics? *Teaching Ethics, 10*, 51–68 (Fall 2009a).
Harris, C. E. (2009b). Response to Michael Davis: The cost is minimal and worth it. *Teaching Ethics, 10*, 79–86 (Fall 2009b).
Harris, C. E. (2010). A reply to Bernard Gert. *Teaching Ethics, 11*, 39–50 (Fall 2010).
Harris, C. E. et al. (2014). *Engineering ethics: Concepts and cases* (5th ed.) (Wadsworth, 2014).
Keller, D. R. (2002). The Perils of communitarianism for teaching ethics across the curriculum. *Teaching Ethics, 3*, 49–76 (Fall 2002).

Pritchard, M. S., & Englehardt, E. E. (2015). Ethical theories and teaching engineering ethics (Chapter 8). In Sethy, S. S (Ed.), *Contemporary Ethical Issues in Engineering* (pp. 111–120). IGI Global.

Sia, S. (2001). Teaching ethics in a core curriculum: Some observations. *Teaching Ethics 2*, 69–76 (Fall 2001).

Identifying Learning Objectives and Assessing Ethics Across the Curriculum Programs

David T. Ozar

Abstract Assessing the effectiveness of an Ethics Across the Curriculum (EAC) program depends on having clear answers to two questions about the aim of the program: (1.) Who is it that the EAC program is intended to serve? and (2.) What good is the program intended to achieve for them? While EAC programs come in many shapes and sizes (See the surveys of types of EAC programs in Davis 2018 and University of San Diego 2009.), almost all would answer these questions in the same way. Their goal is to benefit the institution's students, that is, to enhance the students' learning of ethics in some beneficial ways. The first three sections of this article will focus on the best way to identify such Learning Objectives for an EAC program and how to use these Learning Objectives to assess the program's effectiveness in benefitting the students. The final section will discuss some additional assessment questions about EAC programs, especially for programs with more limited resources that cannot assess their program's effectiveness in the best way.

Keywords Ethics · Teaching ethics · Ethics across the curriculum
Assessment · Learning objectives · Moral awareness/sensitivity
Moral reasoning · Moral motivation

D. T. Ozar (✉)
Department of Philosophy, Loyola University Chicago, Chicago, USA
e-mail: Dozar@luc.edu

E. E. Englehardt and M. S. Pritchard (eds.), *Ethics Across the Curriculum—Pedagogical Perspectives*, https://doi.org/10.1007/978-3-319-78939-2_4

Determining Learning Objectives for Ethics Education in General

If humans' moral life was simple, it might make sense to think that one can assess the success of any ethics education program, including Ethics Across the Curriculum programs, by determining whether the learners conduct themselves in ways that are more morally/ethically correct than they did before the educational intervention.[1] But humans' moral life is not simple and impacting humans' conduct is even more complex. However, we do know that acting in morally/ethically correct ways, and especially doing so dependably, requires the development of at least four specific components of personality as well as achieved skill in exercising them. So an ethics education program that can successfully enhance learners' abilities in one or more of these four components and the relevant skills will be improving their moral lives.

The developmental psychologist, James Rest, and his colleagues (Rest 1983; Bebeau and Monson 2012) have described these four components of the moral life and demonstrated that growth in them can take place under the influence of life experiences and other persons' contributions, including ethics education programs. With some interpretive additions of my own, Rest describes these four components as:

1. Awareness or sensitivity to what is morally/ethically at stake in a situation;
2. Reasoning and Other Reflective Skills leading to judgments that determine what ought to be done in the situation[2];
3. Motivation/Conviction/Character; that is, the particular set of values/principles/ideals (among the many that are possible) that ordinarily directs a person's action and that, ideally, is developed by the person into habitually choosing and living in accord with these values/principles/ideals;
4. Implementation; that is, the practical and emotional ability to overcome situational barriers in order to carry out the course of action that a person has judged ought to be done and is motivated to do.[3]

---

[1]Note the use of the expression "moral/ethical" here and in what follows. The terms "ethical" and "moral" have many subtly different meanings in ordinary speech and the meanings of these terms are rarely carefully distinguished from one another. For clarity in this essay, I shall stipulate that the terms "ethical" and "moral" will be treated here as synonyms and used interchangeably to indicate that we are talking about something that ought or ought not to be done. No additional content will be implied when either of these terms is used. More complicated distinctions within this arena of discourse are best made explicitly rather than buried in unexpressed connotations of these or other terms. The expression "ethical/moral" will be used from time to time below to emphasize this point.

[2]The expression "other reflective skills"—later abbreviated to "reflection"—is used here along with the term "reasoning" in order to emphasize the role of narrative judgments along with more deductive or syllogistic components of judgment in careful moral/ethical reflection.

[3]Rest and Bebeau include ego-strength, as a factor in overcoming barriers to action, in the fourth component while these seem to this author to be better combined with the other aspects of identity-formation in the third component, i.e. as part of a person's Motivation/Conviction/Character. In this way the fourth component can be reserved for the person's abilities and skills is

All four of these are essential components of careful moral/ethical thinking and effective moral/ethical decision-making. Contributing to learners' growth in any of them is therefore a worthy aim for an ethics education program, and assessing the effectiveness of such a program will consist above all in assessing learners' improvement in the relevant component(s) of the moral life and the skills relevant to their exercise. How such improvements can be assessed will be discussed below. But three other general points about Learning Objectives for every kind of ethics education need to be made first, all of which are relevant to developing appropriate Learning Objectives for a well-designed EAC program.

First, many EAC programs are focused on the ethical performance of a specific social role, for example, a professional role like medicine or nursing or law or else some other service role or occupation for which there are standards (or at least minimum standards) of appropriate conduct that are widely accepted. In addition, many EAC programs in the general education curriculums of two-and-four-year post-secondary institutions employ cases focused on ethical issues in the conduct of persons in various social roles. In both kinds of EAC settings, there is a fifth component of humans' moral life that might become the educational focus; and it is important to ask if it is a sufficient goal for an ethics education program. This fifth component is *information* about the accepted standards of conduct relevant to a particular social role.

It is obvious that, in order to do careful moral/ethical thinking and effective moral/ethical decision-making about appropriate conduct in a specific social role, a person needs to know the accepted standards of conduct for that role. Acquiring this information is therefore important because growth in each of Rest's four components, if the focus is on one or more social roles, depends on one's having learned this information. But no one's moral life or conduct will be improved simply by the acquisition of this information. Merely transmitting this information successfully to the learners—even though it is something that can be fairly easily assessed—will not make it any more likely that their conduct will be any different unless there is also learning in how to use this information to shape conduct, i.e. unless one or more of Rest's four components are among the programs Learning Objectives as well.[4]

That is, even in settings in which learning the accepted standards of a given social role is an important Learning Objective for the students, students should also be guided—usually by the use of cases—to correctly determine what is morally/ethically at stake in terms of these standards (and possibly other relevant moral/ethical considerations as well) in concrete situations, i.e. enhance their Awareness/Sensitivity, and perhaps—depending on their Baseline for Reasoning/Reflective

---

overcoming specifically *situational* barriers to the person's doing what he or she judges ought to be done and is motivated to do. For a more detailed description of Rest's four components, see Bebeau (2006).

[4]Obviously, therefore, an assessment that only verifies that the learners can correctly state the standards of conduct in a published code of ethics, for example, is not an assessment of ethical learning.

Skills (see below)—to appropriately determine what ought to be done in these situations.[5] In other words, those teaching in such programs should be actively working to enhance students' abilities to exercise one or both of these components rather than resting satisfied with whatever levels of ability the students might happen to bring. Otherwise there would be no justification for claiming that the improvement of the students' moral lives is a Learning Objective of the program.

The second general point is that determining which of Rest's four components—or possibly some combination of them—should be the focus of a specific ethics education program will depend on the time available, the skills of the teachers and other instructional resources, and the structure of the program. This is obvious, and it will be discussed more fully below.

Third, determining what ought to be the Learning Objectives of a given ethics education program, including Ethics Across the Curriculum programs, will also depend on the learners' developmental starting point, or Baseline, in regard to the component(s) whose improvement are the focus of the program.

With regard to Implementation, for example, if a case involving a nurse was being discussed by general education students in a two-and-four-year college (as contrasted with nursing students), even if they were provided with the relevant standards of conduct for nurses, they would be unlikely to know enough about the structure of a hospital's nursing service to know to deal with practical ethical problems of Implementation that nurses face. Providing additional information could make a case that depended on such factors a useful teaching tool. But the point is that one has to ask in advance of an educational effort what the typical student is likely to be bringing to the learning situation. This is the Baseline question and it is essential that it be answered very realistically when determining the intended aims of an educational program.

There is obviously little to be gained if a program's intended Learning Objectives for a particular group of students are so far in advance of the students' Baseline that—given the limits on time, teachers' skills, and other resources—they are beyond what is realistically achievable. With regard to moral/ethical Reasoning/Reflective skills, for example, students who have befitted from a course in moral theory[6] can ordinarily be expected to understand and offer answers of some

[5]Note the words "correctly" and "appropriate" in this sentence. At a minimum, teachers in such programs need to guide students to fill in the gaps in their awareness of what is morally/ethically at stake and call the students' attention to the kinds of skills—e.g. looking at the situation from the perspective of everyone affected by it—that will enhance their Sensitivity/Awareness. Similarly teachers may be able to enhance the skills of Reasoning/ Reflection that the students have brought with them by guiding them to test the reasonableness of their judgments about what ought to be done—e.g. by adjusting various facts about the situation and asking students to determine what ought to be done under those conditions.

[6]See the description of such a course in Davis (2018). Note also the phrase "benefitted from" in this sentence. Some of students who have received credit for a course in moral theory may not have benefitted from it; that is, they may not have achieved the growth in Reasoning/Reflective skills that it was intended to achieve. From the point of view of their Baseline, such students are little different from students who have not taken a course in moral theory at all.

complexity when normative significance of different kinds of moral/ethical considerations about some issue need to be weighed comparatively. Students who have had no formal training in moral theory may not even notice, much less be able to describe or reflect usefully on the differences between different kinds of moral/ethical considerations.[7] Therefore, insofar as the Learning Objectives of a given ethics education program depend on the students' exercise of moral/ethical Reasoning/Reflective skills, it is essential to have a concrete understanding of what skills of this sort they are bringing with them and how much they can reasonably be expected to advance in these skills, given the skills of the teacher and the time and other learning resources that are available.[8]

In the same way with regard to Awareness/Sensitivity, two-and-four-year post-secondary general education students who have had no previous focused ethics education can ordinarily be expected to have a significant, i.e. conduct affecting, awareness of such personally relevant values/principles/ideals[9] as: life; health; pleasure and the absence/limitation of pain; and personal autonomy, as well as of some social-relational values/principles/ideals like cooperation, equality, and non-interference with others' choices. But unless they have studied the standards of conduct of specific social roles, they may be blind to role-specific values/principles/ideals or miss the special kinds of importance that even the more common values/principles/ideals might have in typical role-structured situations. Therefore, insofar as the Learning Objectives of a given ethics education program depend on the students' moral/ethical Awareness/Sensitivity in order to determine what is morally/ethically significant in a given situation, it is essential to have a concrete understanding of what skills of this sort they are bringing with them, i.e. their Baseline, and how much they can reasonably be expected to advance in these skills, given the skills of the teacher and the time the other learning resources that are available.

---

[7]By "different kinds of moral/ethical considerations" I am referring to the different kinds of moral/ethical data processed by and the differing ways of processing employed in: moral/ethical thinking that is rule-based versus moral/ethical thinking that is rights-based, or virtue-based, or case-based, or based on comparing the harms and benefits of various possible actions' outcomes. Courses in moral theory typically focus on and asses the importance of the differences between some or all five of these approaches to moral/ethical thinking.

[8]For more detail on typical Baselines for EAC programs, see Ozar (2001).

[9]The phrase "values/principles/ideals" is used here as a placeholder for the whole range of ways in which a situation can be morally/ethically significant and thus for the whole range of moral/ethical "data" about a given situation that would then, ideally, be effectively processed in moral/ethical Reasoning/Reflection.

Learning Objectives for Ethics Across the Curriculum Programs

With this general understanding of how the Learning Objectives of any ethics education program ought to be chosen, we can now focus specifically on appropriate Learning Objectives for Ethics Across the Curriculum programs.

It is worth mentioning that, of the many programs labelled "Ethics Across the Curriculum," some have probably been established more for institutional reasons rather than primarily to improve the moral lives of students. Some may have been established, for example, to please a donor, to satisfy accreditation requirements, to improve students' scores on the ethics-related questions on board exams, or to respond to public relations' concerns that ethics should appear to play a more prominent role in the institution's educational efforts. Most EAC programs, however, have probably been established in the hope that students' moral/ethical lives will be improved.

For the reasons already stated, for an EAC program's effectiveness to be assessable, its aims need to be identified and articulated much more carefully. They need to be identified and articulated in concrete terms related to what the learners will be able to do or do better that they could not previously do at all or as well. To do this, the EAC program's Learning Objectives need first to specify which of the four components of the moral life just mentioned are to be improved. Then one needs to identify the concrete activities by the learners that will indicate to what specific degree of achievement the learners will grow beyond their identified Baseline for whichever of the four components are the program's focus.[10] In other words, one does not yet have an adequate answer to the Learning Objectives Question—"What is it that, at the end of the learning experience, the students will be able to do (or: do better) that they were not able to do (or: do a well) before?"—until one also has a concrete answer to the Assessment Question: "How will you be able to tell that the students have become able to do this? What is it that the students will be invited to do that will demonstrate their growth?"

Suppose the aim is to enhance learners' moral/ethical Awareness/Sensitivity and the typical learner is a two-or-four-year post-secondary general education student who has had no previous focused ethics education. One plausible Learning Objective for learners with such a Baseline would be that the learners recognize and become articulate about the differences between the values/principles/ideals they are already aware of (e.g. life, health, pleasure, the absence/limitation of pain, personal autonomy, cooperation, equality, non-interference, etc.) and the ways in which these can "conflict," i.e. point a person toward decisions and actions in different directions. Another plausible Learning Objective for this group might be

[10]The rubrics for grading assessments of learners' progress often identify a certain level of performance as the "target learning" for the program, and then indicate levels of performance that are "*advanced,* i.e. beyond the target" and "*making progress*, i.e. growth, but not yet at the target level" and possibly "*no progress*, i.e. no significant change from Baseline," and so on.

that the students grow in awareness of other values/principles/ideals that are commonly referred to in adult discourse—e.g. truth, beauty, justice, friendship—and of their connection to or conflicts with values/principles/ideals already known to them. Note that, if this were the program's Learning Objective, then the "other values/principles/ideals" that are its focus should be specifically identified and the teaching strategies employed should be specifically suited to enhancing these values/principles/ideals. A third plausible Learning Objective might be the students' identification of and practice in using the skills that enhance Awareness/Sensitivity. One such skill, for example, is inquiring about a situation from the perspective of everyone who might be affected (by various actions that might be taken in response to the situation) rather than only the most obvious persons.

Teaching strategies to facilitate such growth might include engaging the students in guided reflection and/or discussion and/or writing of their own reflections on (not too complicated) cases in which known values/principles/ideals conflict or in which new-to-the-student values/principles/ideals are at stake and must be identified and dealt with or in which parties not immediately obvious would nevertheless be significantly affected. Relevant works of literature—or of other arts—might also serve this purpose, provided that the guidance of students' reflection/discussions/writing focuses them on expanding or deepening their moral/ethical Awareness/Sensitivity (rather than, for example, literary or artistic form). The effective use of such teaching strategies will depend, as in all teaching, on the teacher's skill in continually assessing students' relevant growth as such exercises go forward, i.e. not only in a summative assessment at the end of the learning unit.

For students with a Baseline in moral/ethical Awareness/Sensitivity similar to those just mentioned, but who are entering (or aspiring to enter) a particular social role, an EAC program focused on Awareness/Sensitivity could aim at introducing them to the role-specific values/principles/ideals that practitioners of that role need to be specifically aware of. An EAC program for undergraduate journalism students, for example, might use (not too complicated) cases and "war stories" in which professional journalists need to attend to journalism's role-specific values/principles/ideals, such as impartiality, protecting confidentiality and privacy, properly balancing harming to specific individuals with benefit to the public, and preserving the public's trust of the profession.[11]

Students in a graduate level professional program or career path will ordinarily have begun observing practitioners of the relevant social role more carefully than younger learners and will therefore often have a more developed Baseline of Awareness/Sensitivity for what counts as professionalism in the role even if they have not formally studied the role's articulated ethical standards. Thus, while their Baseline of Awareness/Sensitivity in regard to general moral/ethics values/principals/ideals may differ little from that of other persons their age, it will ordinarily be more advanced than the Baseline of younger learners with regard to the values/principles/ideals of the relevant social role. Therefore, an appropriate

[11]See Society for Professional Journalists (2014).

Learning Objective for Awareness/Sensitivity for these more advanced students might be, for example, that they become aware of/sensitive to still more subtle conflicts between the relevant role-specific values/principles/ideals and between these and the more common values/principles/ideals as well.

In a similar way, an EAC program whose learning objective was specifically to enhance the Reasoning/Reflective skills would be far more likely to be effective if the learners' Baseline included the benefits of a moral theory course (assuming the learners had in fact obtained these benefits) than would be the case for students who had not taken such a course. Moreover, for the reasons explained by Michael Davis (Davis 2018; see also Ozar 2001), it is rare that an EAC program for students in the latter group has the time or the educational resources to significantly improve students' Reasoning/Reflective skill. One possible exception might be an EAC program whose target is that the students learn how to identify and differentiate and, ideally, deliberately create examples of moral/ethical thinking that involve the most common kinds of moral/ethical considerations.[12] But to be effective, as will be stressed later, such a program would need to be taught by someone skilled in moral theory. In fact, if properly taught, an EAC program of this sort would likely have the additional benefit of enhancing the learners' Awareness/Sensitivity to the differences between different kinds of moral/ethical considerations, thus achieving growth in two of Rest's components at once.

With regard to enhancing students' Implementation skills, there are generally two kinds of situational barriers to a person's carrying out the course of action that a person has judged ought to be done and is motivated to do. These are emotional barriers—most often fear or despair prompted by characteristics of the person's situation—and practical barriers. Enhancing students' ability to overcome practical barriers—for example, knowing enough about the structure of a hospital's nursing service to effectively do what one ought in some situation—is a matter of providing them with relevant information or perhaps with information about how to obtain the information they need. Both for social roles like nursing and for many aspects of ordinary life, having or obtaining relevant information will often be necessary for a person to carry out the course of action that a person has judged ought to be done and is motivated to do, and thus such information will contribute to the person acting as he or she ought. But as has already been explained, merely acquiring information does not by itself improve anyone's moral life; the person must connect the information to Rest's the other three components for it to be morally/ethically relevant. So teachers in EAC programs focused on enhancing Implementation information should be actively working to enhance students' abilities to exercise the other components of the moral life rather than only providing them with Implementation information and resting satisfied with whatever levels of ability the students might happen to bring regarding the components of the moral life.

The other main situational barriers to carrying out what one ought to do are emotions like fear and despair that are prompted by characteristics of the situation

[12]See for example the five kinds of moral/ethical thinking named in Footnote 6.

the person is in. It is certainly the case that humans can assist one another in overcoming fear, despair, and other action-blocking emotions. But this is rarely doable in groups and it typically requires a personal relationship between the two parties, or at least a high level of understanding and expressed compassion on the part of the helping party. In other words, it is unlikely for an EAC program to be effective in enhancing students' ability to deal with barriers to action of this sort. A possible exception would be an EAC program aimed at enhancing students' emotional Implementation abilities by putting them in touch with narratives (real-life or literary) of persons who overcame such barriers in the hope that students' would respond by building patterns of imitation of these persons. This form of moral/ethical learning/teaching will be discussed in the following treatment of moral/ethical Motivation.

What about Motivation? It is obvious that a person can be aware of everything that is morally/ethically significant about a situation and reason appropriately about what ought therefore to be done about it, but still do something else. That is, their action may be directed by values/principles/ideals that are different from those their Reasoning/Reflection identified as morally/ethically salient. Can ethics education affect which values/principles/ideals have the most influence on how a person acts, i.e. which are dominant when multiple values/principles/ideals are motivationally affecting a person?

One of the principal ways that a person adopts a value/principle/ideal as motivationally relevant,—or reinforces an already adopted value/principle/ideal—is by contact with and then imitation of another human who is motivated by it. Thus ethics educators can put learners in contact with the stories of persons who are admirable exemplars of the values/principles/ideals the learners are intended to adopt as motivational directives for action in their own lives. In addition, the more the learners' Baseline in Motivation/Conviction/Character includes a sense of who they hope to be and, ideally, an awareness that development in this direction involves a life-long process of self-formation and is ordinarily linked to imitation of positive exemplars (and rejection of the ways of negative exemplars), the more likely they are to also grow in this respect through reflective exercises and constructive comparisons of themselves with such exemplars, as well as shared reflection with others similarly situated. In addition, effective self-formation depends on the skills of self-assessment and appropriate self-correction and self-commendation and repetition, which can themselves be learned from observation and imitation of those who exercise them. So there are important ways in which enhancing learners' Motivation/Conviction/Character can be an appropriate learning objective for ethics education.[13]

Specifically with regard to EAC programs, encounters with admirable exemplars could be the content of such activities. But the learners' Baseline needs to indicate a

[13]It is worth noting that imitation itself, though it arguably comes naturally to humans in some measure, itself has skill-aspects since it can be improved through self-reflection and directed attention. This suggests that its skill-aspects may also be a focus of educational efforts under the proper circumstances.

match between the learners' interests and the particular values/principles/ideals being exemplified; otherwise imitation of what they observe will be less likely. Therefore, encounters of this sort as part of an EAC program are most likely to be effective when the students are engaged in learning about a particular social role, whether as undergraduates or as graduate or professional students, and the exemplar is a practitioner of that role. Particularly with more mature graduate or professional students, pairing such encounters with structured reflective exercises and/or shared reflection in small groups is likely to enhance the intended learning. But for general education students whose motivational structures are likely to be very diverse, EAC activities aimed at enhancing the learners' Motivation are less likely to be widely effective.

Assessing Learners' Achievement of EAC Programs' Objectives

The best assessment of the effectiveness of an EAC program is obviously determining whether the learners are achieving the program's Learning Objectives. There are four tasks which are essential to assessing learners' achievement of any ethics program's Learning Objectives. The first is to determine the learners' Baseline for the components of the moral life that are the program's focus. General estimates of learners' Baseline are often part of the planning process; but truly accurate assessment of effectiveness will depend on actual data about the learners' Baseline, i.e. some form of pre-learning instrument. (All such instruments, like ethics education programs themselves, are difficult to design in the abstract. Both general assessment instruments and those designed for a specific learning situation need to be experimented with in practice and adjusted and fine-tuned as experience dictates.)

The second task is identifying and articulating the program's Learning Objectives for its students concretely enough that one can explain how the students' learning will be assessed after the fact and do so in concrete terms, i.e. by describing what the students should be able to do to demonstrate they have achieved the intended learning. For example: Given a properly constructed case, the students will be state which of the role-specific values/principles/ideals of journalism are at stake in the case (Awareness/Sensitivity), or: given a properly constructed case, the students will explain why the journalist who is preparing an article in the case ought to give priority to protecting the privacy of person *A* over the possible benefits to the public of reporting fact *X* about person *A* to them (Reasoning/Reflection).

The third task is the obvious one of assessing the learners' abilities in the relevant component(s) of the moral life at the conclusion of the program and comparing the results with the learners' Baseline abilities before the learning experience began.[14] The question of what to do with the results of this final assessment vis-à-vis the learners will be discussed in the next section.

Fulfilling these three tasks, especially for the first time, is obviously a lot work and it requires a specific set of skills, either already available or acquired through proper training. Of course, it is conceivable that an EAC program that did not do this work and proceeded with whatever resources happened to be available might still have some beneficial impact on the participants. But there is no basis for confidence that such a benefit has occurred without actually assessing participants' achievements on the basis of identified Learning Objectives, and the way to do that is by carrying out these three tasks. This is the reason for saying that this is the best, most dependable method of program assessment for EAC programs. Fortunately, once a program's Learning Objectives have been identified and properly (concretely) articulated and methods for determining learners' Baseline abilities and assessing their progress have been designed, they can typically be used again and again. That is, unless there is need for modification, the Learning Objectives can be used directly, and the original assessment instruments can be used as is or used as models by changing the cases employed, for example, and adjusting the rubrics accordingly.

One reason the three tasks are a lot of work is that many educators in post-secondary institutions have not needed—until recently for some, or not ever for others—to be able to identify and concretely articulate their students' Baseline abilities or the Learning Objectives they aim at in their teaching. So these tasks are additional work for these faculty. It is worth noting, however, that a sizeable body of literature on curriculum design has developed in the last three decades to assist teachers in identifying Learning Objectives (and by implication students' Baselines) and most of these are concrete enough to be readily accessible to post-secondary faculty.[15]

A second and often more significant reason why these tasks are hard work is that many people, including many who teach in post-secondary institutions, have not reflected with any care on the complexity of human conduct. Many people imagine that human conduct is or at least ought to be something uncomplicated, with actions simply arising either from one's habits or one's desires without much processing needed. On such a view, situations in which careful thinking is required about what ought to be done are considered atypical and, since what is simple is conceived to be easier, something of a burden.

---

[14]This assessment is obviously a "summative" assessment, i.e. it aims to determine how much of the intended learning took place from the beginning to the end of the program. It is taken for granted here that the teacher will be constantly assessing the effectiveness of his or her teaching by many different means and adjusting it accordingly.

[15]Because the first focus of this literature is on what the students will be able to do as a result of the learning experience, i.e. on the *outcomes* of the learning—with the assignments and other teaching strategies derived from this starting point rather than the other way around—the approach in this literature is sometimes called "Backward Design" or "Outcome-Centered." See for example Wiggins and McTighe (1998), and Ozar (1994).

Therefore, some understanding of the complexity of human conduct is necessary if Learning Objectives for ethics education are going to be usefully identified, much less effectively striven for. That is, those who design ethics curricula, including EAC programs, need to become familiar with a model of (the complexity of) human conduct, whether Rest's or some other. This is a second reason why carrying out these three tasks is a lot of work.

A still more fundamental challenge is the conviction, not at all uncommon among academics, that our students' moral lives are not susceptible to improvement by educational means. This conviction is amazing because, if a colleague who articulates it were asked if he or she has grown in his or her moral life since leaving high school (or pick any other time-frame), the colleague would almost certainly say that he or she has certainly grown morally. Why then should one believe that educational efforts to assist such growth—if carefully designed and based on sound understandings of how the moral life works—cannot possibly be effective?[16]

Finally, it is worth asking whether there is value in seeking students' own assessments of the value of an EAC program. The answer to this question is yes, provided the students are asked to assess the program—i.e. the overall program, the effectiveness of its teacher(s), and the various teaching strategies used in the program—specifically in terms of their effectiveness in assisting the students in achieving the concrete Learning Objectives of the program. This will of course require the students to understand what the program's Learning Objectives are and what would count as improvement in regard to them. As a side benefit, and as many teachers have found, providing students early on with clear information about the concrete Learning Objectives of a course, unit, or in this case an EAC program, often facilitates students' learning. But it is not unusual for students doing assessments of an educational program to be distracted by reactions to the teacher, the assignments, other students, or the setting and to express positive or negative reactions to these in their assessment. Moreover, since the students may not be as perceptive of their growth as a carefully designed assessment instrument, judging whether an EAC program is achieving its goals should not depend solely on student assessments. All three of the tasks identified above are necessary and should be the primary measure of an EAC program's effectiveness.

Other Assessment Tasks for EAC Programs

The previous section discussed the best kind of assessment of an EAC program, i.e. a retrospective one after the learning experience has been completed. But there is an earlier point in time when assessment is needed; namely beforehand, when an EAC

[16]For solid empirical evidence that such learning can be achieved through educational programs and for examples of very sophisticated assessment tools (including significant evidence of the tools' validity and inter-grader reliability), see Bebeau (2006), Bebeau and Monson (2008, 2012), and the research cited in these articles.

program is in the process of being created (or re-created) and then when a completed proposal for establishing (or modifying) such a program is being evaluated by the relevant decision-makers. There are four prospective assessment questions that the persons in these positions should be asking to determine how likely it is that the proposed EAC program will be effective.

The first question is whether the Learning Objectives that have been identified for the proposed EAC program have been articulated concretely enough that the Assessment Question can be concretely answered. If it is not clear how the students' achievement of the program's Learning Objectives will be assessed—if it is not clear how the students will be invited to demonstrate that they have achieved the Learning Objectives—then a dependable assessment of the program's effectiveness will not be possible when the learning experience has been completed. But this means that no dependable prospective judgment is available about the program's likely effectiveness. Under such circumstances, if the relevant resources are available, decision-makers should tell the creators/proposers of the programs to revise the proposal until the Assessment Question has been adequately answered and, if necessary, the articulation of the Learning Objectives appropriately revised.

The second question that creators/proposers of EAC programs should have clearly answered is what the learners' typical Baseline is regarding the component (s) of the moral life that is the EAC program's focus and what evidence is available to support this judgment. Since the goal of an EAC program to improve the learners' moral life in some way, the claim that achieving the program's Learning Objectives will be an improvement for the learners obviously depends on a judgment about their Baseline. Therefore, for a proposal to be considered complete and for creators/proposers to claim it will be effective and decision-makers to approve it as likely to be effective, a clear statement of the learners' Baseline is also essential.

The third question concerns the skills that the proposed faculty will need to lead the learners from their Baseline to the intended Learning Objectives. At a minimum, the faculty seeking to enhance learners' Awareness/Sensitivity, for example, must themselves be already not only aware of/sensitive to the values/principles/ideals that are the focus of the learning experience, but comfortably articulate about them and their role both in the kinds of cases that will be examined by the students, but also in other relevant situations so they can give additional, accessible examples of the role of these values/principles/ideals in the students' own lives and/or in relevant role-specific situations if that is the program's focus. Similarly, faculty who would be teaching in an EAC program aimed at enhancing students' Reasoning/Reflective skills or Motivation or Implementation would need all of the resources just mentioned and the additional knowledge and skills relevant to those Learning Objectives as well. In addition, teaching ethics, especially in an EAC setting, cannot depend heavily on lecture; the faculty who teach in EAC programs need to be not only comfortable, but already experienced in using discussion techniques and other interactive teaching strategies in the classroom and writing and other reflection-requiring assignments outside of it.

In addition, regardless of the specific Learning Objectives being aimed at, since ongoing assessment of the learners' on-going growth towards the Learning

Objectives is necessary for effective teaching, the faculty must also be familiar with how to assess these forms of moral/ethical growth, not to mention that it is likely they who will supervise and possibly create the summative assessment of the learners' moral/ethical growth at the conclusion of the program.

Faculty who cannot demonstrate that they already have these kinds of knowledge and skills should not be teach in the EAC program without extensive training (with its own Learning Objectives and assessments of effectiveness). If no training is available for faculty and no already knowledgeable and experienced faculty are available to teach in the proposed EAC program, then this is a clear sign that the program is likely to be ineffective in achieving its Learning Objectives even if the answers to the first two prospective questions are completely adequate.

The fourth prospective assessment question concerns the other educational resources available for the program. Among the most important of these are class-size and appropriately configured spaces in order to facilitate rather than inhibit discussion and other interactive teaching strategies during class. Even faculty skilled in conducting class discussions about, for example, relevant ethics cases note important numerical thresholds beyond which more and more students can "hide" from participating and, as the numbers increase, decline to attend even passively to the discussion. One such threshold seems to be in the 12–15 student range, after which "hiding" becomes much easier; another in the 28–30 range, beyond which a significant number of students can fairly easily attend to other things than what is going on in the classroom. These numbers probably vary from institution to institution and teacher to teacher. But the point is that proposals for EAC programs should pay specific attention to class size and the kinds of spaces available. If the classes are too large or the rooms not conducive to effective discussion and other interactive teaching strategies, this will be a reason to wonder if the EAC program, even if well designed in every other respect, will achieve its intended Learning Objectives.

One additional assessment question concerns what is best considered a "side benefit" of a well-designed and effective EAC program. It is a "side benefit" because it would not, by itself, be an appropriate reason to establish an EAC program; namely, the benefits the faculty might gain from teaching in an EAC program. For example, if the faculty have the knowledge and skills mentioned above, on which the program's impact on students depends, they are likely to learn a fair amount themselves from the experience. Their own moral/ethical Awareness/ Sensitivity is likely to be enhanced from attending to the range of comments that the students make about, for example, cases being discussed. Many teachers will also find their skills at moral/ethical Reasoning/Reflection growing because of their need to articulate appropriate questions or offer appropriate guidance to students about Reasoning/Reflection regarding the cases being discussed, etc. It is also possible that ideas about Implementation that arise from the students' discussion of cases will enrich a teacher's understanding, and that expressions of moral/ethical Motivation by the students might inspire the faculty member in new ways.

Finally it is important to add two more kinds of assessment question, especially for EAC programs with such limited resources that they cannot assess their program's effectiveness in the best way that was described in Section 'Assessing Learners' Achievement of EAC Programs' Objectives'. For there is a kind of moral/ethical growth by the students that has not yet been mentioned. One potential benefit to the learners in an EAC program that is well taught is that, first, they might grow in Awareness/Sensitivity to the value of clear and careful moral/ethical thinking by observing its benefits when the teacher employs it and is able to be articulate about it and, second, they may grow in their own Motivation to do clear and careful moral/ethical thinking by observing (and potentially imitating) the teacher's commitment to doing it. In both respects, although students' ability to do exercise the skill-components of the moral life more effectively may not be being improved, their increased affective grasp of their importance—as something with moral/ethical significance so that they ought to take it seriously and/or as a valuable component of the life of a well-developed human being – is something valuable.

Many EAC programs operate on very limited budgets and some provide EAC learning experiences that are so diffuse across an institution's departments/programs/units that the program cannot be reasonably described as having Learning Objectives of the sort described in the previous sections. It is a legitimate question to ask if programs of this sort have no positive impact on the students' moral lives, and this is a question that decision makers who control relevant resources should ask seriously. But if the students are benefiting from such a program in the manner just described, then it is positively impacting their moral lives even if it is not able to be effective in the ways explained in the previous sections.

Obviously, however, the key to providing benefits of this sort to the students is well-trained and committed faculty. Though they would probably find it inappropriate to deliberately present themselves as role models to their students, they may nevertheless be significantly impacting students' Awareness/Sensitivity of the value of careful moral/ethical thinking and of the Motivation to undertake it that is modeled in faculty members' efforts to practice it. This is an assessment question well worth trying to answer for those EAC programs, especially those with very limited resources, who do not assess their effectiveness in the best way that was described in Section 'Assessing Learners' Achievement of EAC Programs' Objectives'.

It is also possible that the institution itself might function as a kind of role model for its students a somewhat similar way. That is, it is possible that an institution's students may grow Awareness/Sensitivity to the value of the moral life from observing that the institution—in creating a well-designed, well-resourced, well-taught EAC program—is itself affirming the value of such growth in the moral life and perhaps expressing a collective commitment to its students' growth in the moral life (Motivation), especially if the commitment to an EAC program is genuinely linked to the institution's mission for its students, i.e. rather than for the sake some other institutional goal. The impact of the institution's choices on the students' moral lives will of course depend a great deal on the students' views of the institution as a moral organization (rather than merely a business entity) and/or of its administrators as genuinely mission-focused (rather than merely managers of resources); and obviously these views will be formed in very complex ways and will only be partially affected, if at all, by the institution's official statements. Assessing an EAC program's impact on the students in this way may therefore be valuable, but it also involves the risk of the students asking, if this is the institution's only contribution of the growth of students' moral lives, why the institution has not tried to contribute in more robust ways.

A final question about assessing an EAC program takes this discussion in a different direction. For it is important to ask, regarding the robust kinds of data that the best forms of assessment would produce, what uses might be made of the results of this data besides their role in judging the effectiveness of the program itself. This has been the only use of these results that has been mentioned so far, but two other possible uses of these results deserve examination. One is to provide these results to the students so they can note their own growth (or lack of it) in the relevant components of the moral life and use this information to guide self-reflection and, ideally, further growth. Educators who have used the results of assessments of moral/ethical growth in these ways—though most often in settings in which more robust programs of ethics education were in place rather than only EAC programs —have found communicating the results of assessments to students, especially in combination with one-on-one sessions with the students or guided small group discussions to assist self-reflection, to be an effective means for additional moral/ethical learning and self-formation.[17] This is therefore something that should be given serious consideration.

A second possible use of assessment results, and one that may seem natural to many academics, is to make them a component of the grading process. Admittedly the rationale for doing so is widely employed in other educational settings; namely, that the degree to which learners accomplish the Learning Objectives of a given educational effort ought to be represented in the official report of their progress, i.e. their course grade. But there are several reasons for doubting that this rationale justifies using assessment results as part of the grade for students in an EAC program. One reason is that most EAC units that are inserted into other courses are too brief to produce major advances in any of the four components of the moral life under consideration here. While there may be assessable advances if the course is well-designed, well-resourced, and well-taught, these advances are unlikely to have major impact on the students' conduct and therefore there is a justice/fairness question that needs to be carefully answered before students' grades—which are major contributors to their life-credentials for the future—are impacted by them. A second reason is that many institutions that have EAC programs are not able to provide the time or funding necessary to fully train faculty, or even to engage the resources necessary to properly design the programs themselves, so that holding the students responsible for what the programs achieve—which is what incorporating results of learning assessments into grading implies—is not justified. This use of

[17]See Bebeau and Monson (2008, 2012).

EAC assessment results is therefore something that should not be undertaken without addressing these very serious ethical questions about the practice.

Conclusion

There are too many different kinds of Ethics Across the Curriculum programs for any concrete recommendations to be offered that would apply to all of them. But any human project worth undertaking is worth assessing to see if it is achieving its goals, not to mention the desire of institutions to know that their resources are being put to effective use. Even if resources to assess an EAC program in the best way are not available, such program should nevertheless be clear about what it is aiming at for its students—i.e. its Learning Objectives—and should find an appropriate way to determine whether the students are benefitting according. That much is true for every EAC program worth having.

References

Bebeau, M. (2006). Evidence-based character development. In N. Kenny & W. Shelton (Eds.), *Lost virtue: Professional character development in medical education* (pp. 48–86). Oxford UK: Elsevier.

Bebeau, M., Monson, V. (2008). Guided by theory, grounded in evidence: A way forward for professional ethics education. In L. Nucci & D. Narvaez (Eds.), *Handbook on moral and character education* (pp. 557–582). Hillsdale NJ: Routledge.

Bebeau, M., & Monson, V. (2012). Professional identity formation and transformation across the life span. In A. Mc Kee & M. Eraut (Eds.), *Learning trajectories, innovation and identity for professional development* (pp. 135–163). Dordrecht, The Netherlands: Springer.

Davis, M. (2018). Moral theory in ethics across the curriculum (in this volume).

Ozar, D. T. (2001). Learning outcomes for ethics across the curriculum programs. *Teaching Ethics, 2*(1), 1–27.

Ozar, L. A. (1994). *Creating a curriculum that works: A guide to outcomes-centered curriculum decision-making*. Washington, D.C.: National Catholic Educational Association.

Rest, J. R. (1983). Morality. In P. H. Mussen, J. Flavell, & E. Markman (Eds.), *Handbook of Child Psychology (Vol. 3)—Cognitive development* (4th ed, pp. 556–629). New York: Wiley.

Society of Professional Journalists. (2014). *SPJ code of ethics*. Indianapolis, IN: Society for Professional Journalists.

University of San Diego. (2009). Implementing Ethics across the Curriculum. ethics.sandiego.edu/Presentations/EAC/Tokyo/Implementing_EAC.pdf. Accessed May 31, 2017.

Wiggins, G., & McTighe, J. (1998). *Understanding by design*. Alexandria VA: Association for Supervision and Curriculum Development.

Increasing the Moral Sensitivity of Professionals

Deborah S. Mower

Abstract The 'educators' charge' is to answer what our priority should be in educating college and university students. Despite shifts in terminology, a long-standing view is that educators should develop the moral sensitivity of their students to prepare them for both their private, civic lives as well as their professional roles and positions in institutions and society. In this chapter, I explain the development of the concept of moral sensitivity, and that it is a complex discriminative or diagnostic expertise that functions as guide for moral action: a sophisticated ability to 'see' or diagnose the right course of action based on extensive learning and practice. Ethics Across the Curriculum (EAC) Programs are uniquely poised to develop moral sensitivity because they provide (1) diverse perspectives and resources from a variety of fields, (2) the ability to focus on developing specific domains of capacities that underlie moral sensitivity, (3) the opportunity to practice and develop moral sensitivity over the span of several years, and (4) the opportunity for directed training of moral sensitivity under the expert and watchful eye of professors and other professionals. Institutions without a robust EAC program may also be able to initiate the moral sensitivity of their students through the use of a specially designed stand-alone Professional Ethics course. The educators' charge is to develop moral sensitivity, which can be accomplished across a range of institutional opportunities.

Keywords Ethical sensitivity · Moral sensitivity · James Rest
Empathy · Sympathy · Compassion · David Hume · Adam Smith
Aristotle · Practical wisdom · Four component model

D. S. Mower (✉)
University of Mississippi, Oxford, USA
e-mail: dsmower76@gmail.com

E. E. Englehardt and M. S. Pritchard (eds.), *Ethics Across the Curriculum—Pedagogical Perspectives*, https://doi.org/10.1007/978-3-319-78939-2_5

Introduction

Students within our courses, degree programs, and universities are with us for a short period of their lives and educational careers. In the short time that we have with them, what is—or what should be—our utmost priority when it comes to teaching ethics? We might call this 'the educators' charge.' Answers to this charge have changed dramatically over time. According to Sloan (1980), in the nineteenth century, the emphasis within undergraduate moral philosophy courses was to provide a synthesizing framework for moral thought and action that combined the insights of a university education into a moral guide for life. He notes that such courses "sought to equip the graduating seniors with the *ethical sensitivity* and insight needed if they were to put their newly acquired knowledge to use in ways that would benefit not only themselves and their own personal advancement, but the larger society as well" (2, emphasis added). Rather than a comprehensive moral guide, The Hastings Center's Summary Recommendations (1980) on the teaching of ethics in higher education calls for the development of a range of capacities: "The general purpose of the teaching of ethics ought to be that of stimulating the moral imagination, developing skills in the *recognition and analysis of moral issues*, eliciting a sense of moral obligation and personal responsibility, and learning both to tolerate and to resist moral disagreement and ambiguity" (300, emphasis added).

Despite the apparent discrepancy between the call for a comprehensive moral guide and the development of specific capacities, what is common is the emphasis on moral sensitivity. At its briefest description, moral sensitivity can be described as the complex diagnostic expertise to interpret a situation morally and to respond in morally appropriate ways. Although much of normative theory myopically emphasizes the role of moral reasoning, moral reasoning depends on one's having seen, noticed, or realized the moral import of a situation. Blum (1991) notes that "one of the most important moral differences between people is between those who miss and those who see various moral features of situations confronting them" (701). Moral sensitivity is the developed capacity to 'see' morally, and it is gaining the attention of educators. Founded in 1915, the Association of American Colleges and Universities (AAC&U) is composed of thirteen hundred institutions as current members. As a body, the AAC&U shapes the educational goals and learning outcomes of its member institutions. One of the general educational outcomes focuses on personal and social responsibility (Swaner 2004, 2005), one of the five fundamental dimensions of which includes moral sensitivity (AAC&U 2014). It is clear that many would answer the educators' charge that our utmost priority when it comes to teaching ethics is to develop moral sensitivity. In this chapter, I first explain the concept and background history of the concept of moral sensitivity, followed by an exploration of how it can be developed within Ethics Across the Curriculum (EAC) Programs. However, because not all colleges and universities have such programs, I detail how educators can still attempt to meet this charge through a specially designed Professional Ethics course.

Moral Sensitivity

Developmental psychologist James Rest (1982) coined the term 'moral sensitivity' to refer to the capacity to morally interpret a situation. Rest conducted an overview of the psychological literature and noticed four phenomena studied as part of moral development and identified as part of mature moral functioning. As it later came to be known, his Four-Component Model (1986) detailed moral sensitivity along with moral judgement, moral motivation, and moral character. Each of the four components contribute to and are integrated for moral action, involving both affective and cognitive processes. Rest and his colleagues expanded on his earlier definition, describing moral sensitivity as:

> the awareness of how our actions affect other people. It involves being aware of different possible lines of action and how each line of action could affect the parties concerned. It involves imaginatively constructing possible scenarios, and knowing cause-consequence chains of events in the real world; it involves empathy and role-taking skills. (1994, 23–24)

In essence, for Rest, moral sensitivity is the perceptive capacity to recognize moral issues within situations.

Despite the fact that the term 'moral sensitivity' originated in psychology, ethicists have long known of the importance of moral sensitivity as a capacity for moral action. Aristotle (1985) describes the importance and role of practical wisdom, or *phronesis*, for understanding and interpreting moral situations and for comprehending the appropriate course of action in a given situation. Such practical wisdom is not the stuff of reasoned, considered judgment or abstract, theoretical knowledge, but a complex and developed expertise gained through extensive life experience in learning a variety of social scenarios and in thinking about and interacting with particular persons. Knowing how to ride a bicycle is akin to practical wisdom in that it is a complex ability that involves managing speed, balance, forward motion, steering, and impulsive movement. One learns how to ride a bicycle only by engaging in riding, and the knowledge gained comprises a skill or ability that is difficult to put into words but transferrable to novel experiences; one builds on knowing how to ride a bicycle in learning to ride a unicycle. Practical wisdom is precisely that: an ability that depends on knowledge gained through experience and practice.

The Scottish sentimentalists Hutcheson (1694–1746), Hume (1711–1776), and Smith (1723–1790) also understood the importance of moral sensitivity for moral action, and detailed aspects of its origin, operation, and development. Hutcheson (1993) described a capacity that he called 'benevolence' as the source of our motivation to engage with and value the welfare of others, and its role in counterbalancing our own self-interest and discerning the nature of our obligation in situations. Hume described a capacity that he called 'sympathy'[1] which enables

[1]Hume referred to the capacity of moral sensitivity as 'sympathy' in the *Treatise of Human Nature*, but in later writings—most specifically in the *Enquiry Concerning the Principles of Morals*—he referred to it as 'benevolence.'.

us to spontaneously identify and comprehend the situation of another which he called ‘fellow feeling’—whether in real life or presented to us in the form of fiction, travel stories, or news reports—and to have a felt, evaluative response of either approbation (approval) or disapprobation (disapproval). These felt responses are moral sentiments: moral evaluations of the action taken (e.g., of a particular individual), factual conditions within the situation (e.g., the type and level of harm), potential outcomes (e.g., given the description of hypothetical situations or future states of affairs), or the issues at play within the situation (e.g., conflicting obligations). Moral sentiments, as moral evaluations, depend upon the moral values one holds as a member of a moral community as well as a complex body of knowledge gained through extensive life experience. Much like the form of knowledge that comprises the ability to ride a bicycle, the moral values and social experiences one has comprise one’s capacity for moral sensitivity.

Hume (1751/1957, 1888/1978) and Smith were particularly interested in the development of moral sensitivity, and much like Aristotle, recognized that the capacity developed over time and through experience by engaging with others and learning more about social situations, factual conditions, potential outcomes, and the complexity of issues. In particular, Hume and Smith were interested in social interactions as a form of moral correction for development. Each made use of the metaphor of a mirror in that we observe the moral sentiments of others in response to our action, and another’s approbation (approval) or disapprobation (disapproval) serves to either reinforce or initiate change in our own responses and actions. An important part of moral development involves learning the norms of a community, and praise and censure are effective means for transmitting messages about such norms as well as attitudes about the moral appropriateness of specific actions given those norms. As one’s capacity of moral sensitivity develops over time, one is capable of increasingly nuanced identifications and interpretations of situations, yielding practical wisdom about the appropriate course of action in a given situation. Smith refers to this developed capacity of moral sensitivity as the ‘impartial spectator’—an internal discriminative capacity or developed expertise to interpret a situation, recognize the moral complexity within it, and to comprehend the appropriate course of action—which functions as a guide that lives “within the breast” (1750/1759, 3.2.32, 185). The capacity of moral sensitivity, as it matures, functions as guide for moral action: a sophisticated ability to ‘see’ or diagnose the right course of action based on extensive learning and practice.

Given the above descriptions, we can understand moral sensitivity as a complex discriminative or diagnostic expertise. As a complex expertise, it develops from the interaction of a set of psychological capacities that cross the emotional/affective, perceptual, cognitive, and normative domains.[2] Within the emotional/affective domain, moral sensitivity harnesses empathy to identify and interpret the mental

[2]For additional analyses of the concept of moral sensitivity see Jordan (2007) and the literature review by Weaver et al. (2008). For extensive discussion of the concept and multi-factorial development of moral sensitivity, see Mower et al. (2015).

states of others[3] and emotional responses such as compassion[4] for identified harms. Within the perceptual domain, moral sensitivity harnesses basic vision as well as attention and schemas as perceptive filters for salient features of situations and events. Within the cognitive domain, moral sensitivity harnesses knowledge and concepts, both theoretical (e.g., knowledge of laws and the concept of 'patient') and normative (e.g., 'bodily integrity' and 'autonomy'). And within the normative domain, moral sensitivity harnesses the norms of particular communities as the grounds for value and motivation (e.g., professional codes and the value of privacy). Although some of the underlying capacities within these domains may be socially inculcated, others, such as basic visual perception, are innate; despite their ultimate origin, all develop through experience and maturation and are integrated into this complex ability.

Moral Sensitivity and EAC Programs

Given that moral sensitivity is a complex expertise for perceiving, interpreting, and knowing how to respond in morally appropriate ways to a situation, it is an essential capacity for all moral action. It affects whether and how we notice others or if they need help, whether and how we engage with and interact with them, the nature and degree of our aid (if necessary), evaluations of our obligations and conflicting aims, and our level of tact and delicacy in dealing with emotionally fraught situations. Moral sensitivity helps us navigate complex social interactions both within the private realm of our personal and civic relationships as well as the professional realm of our interactions with students, colleagues, other professionals, and the public at large.

Because moral sensitivity undergirds all moral actions, whether private or professional, educators should strive to develop moral sensitivity. Viewed simplistically

---

[3]Although many situations require the accurate identification and comprehension of the mental states of others, not all do: some situations involve the analysis of potential risks and harms or mere knowledge of patterns (e.g., the long-term psychological effects of verbal abuse). Empathy is an important underlying capacity for moral sensitivity, but it is neither necessary nor sufficient for its operation. As a consequence, although empathy is an important component to study and cultivate, it should not be equated with moral sensitivity. See Smith et al. (2011) and Steuber (2006) for discussion of empathy as a capacity for accurate knowledge of the mental states of others as well as Coplan and Goldie's (2014) edited volume for extensive discussion about the role of empathy.

[4]Similarly, while compassion is an important underlying capacity for moral sensitivity, it is also neither necessary nor sufficient for the operation of moral sensitivity. Although compassion can be a helpful motivating force for action and support the ability to see the needs of others, accurate diagnoses of morally appropriate actions do not require such an emotive response. Further, a developed capacity of moral sensitivity may need to dampen the effects of compassion to avoid bias and compassion fatigue for moral appropriateness. For a contrasting view on the necessity of compassion within the professions, see Maxwell (2008).

and at the minimal level of educative responsibility, educators transmit knowledge to their students. But viewed carefully, and with an eye to real-world effects and the full range of educative responsibility, educators shape their students into the inheritors of bodies of knowledge who are responsible for maintaining the traditions and standards of academe, the pursuit of truth and knowledge in research, the values and methods of particular fields, and the unique function and service that their field provides to clients and within the broader society. Consequently, educators have a responsibility to develop the knowledge of their students, whether that entails knowledge of legal restrictions, professional standards, or specialized knowledge of their fields (e.g., the level of risk for an off-label use of a medicine). Correlatively, educators also have the responsibilities to develop the skills of their students, whether that includes highly specialized skills such as the ability to run a computer design program or skills of moral sensitivity in interacting with their clients, colleagues, other professionals, and members of the public.

Ethics Across the Curriculum (EAC) programs provide a unique resource for the development of moral sensitivity within young professionals because they provide (1) diverse perspectives and resources from a variety of fields, (2) the ability to focus on developing specific domains of capacities that underlie moral sensitivity, (3) the opportunity to practice and develop moral sensitivity over the span of several years, and (4) the opportunity for directed training of moral sensitivity under the expert and watchful eye of professors and other professionals. One of the most important benefits of a robust EAC program is the diversity of perspectives and resources that they provide to faculty and students.[5] The range of events offered by many EAC programs affords an opportunity to focus on complex moral questions and problems at a more sustained level than what can often be achieved within a specific course or degree program. For example, many courses can be co-taught by experts with both ethics expertise and technological skill from another discipline, classes across disciplines can be designed in sequence to build on related knowledge and skills, and prominent figures addressing ethical questions surrounding the issues can be brought in as visiting scholars, speakers, or consultants. Such collaborative teaching and study within an EAC program allows for extended examination over time by making a particular issue the focus of several classes or events related to that issue.

Perhaps most importantly, an EAC program provides a deeper examination of complex moral problems and questions by making use of diverse resources from a variety of fields. For example, an EAC program at a technological school may focus on not only the ethical issues surrounding the design and product safety for the consumer, but also utilize (1) resources from business management and counseling to address worker employment issues, (2) experts from education and psychology to examine the effects of the new technology on memory, attention, or learning,

[5]See for example Englehardt (2018), Davis (2018), Robison (2018), Donovan (2018), Frey and Cruz (2018), Newton (2018), Baker (2018), Herkert and Ellision (2018), and Woodson and Zhu (2018).

(3) research from biologists on the environmental effects of factory production and how to increase recycling of component parts after the demise of the items, and (4) research from sociologists and economists on the impact of the factories or the international market on local communities. By making use of the resources of a variety of fields and the expertise of professors within those fields, students gain a deeper level of understanding of the complexity of the issue and the range of information that needs to be brought to bear to on it. The simultaneous breadth and depth that can be achieved through an EAC program develops moral sensitivity by providing greater access to the complex and nuanced knowledge needed to appraise moral questions confronted within situations. Although we develop and train students to be professionals within particular fields, gaining greater awareness of the potential resources from other fields, and gaining greater knowledge of the methods and research employed within those other fields provides young professionals with a greater range of options for how to respond to a given situation. In a certain sense, such an educative opportunity opens up new possibilities or new horizons for appropriate moral response by providing students with a broader vision.

The second primary benefit an extensive EAC program provides allows the ability to focus on developing specific domains of capacities that underlie moral sensitivity. In an extensive EAC program in which students participate over the course of their university education, educators can focus on developing the emotional/affective, perceptual, cognitive, and the normative domains. In the emotional/affective domain, moral sensitivity harnesses empathy to identify and interpret the mental state of others and emotional responses such as compassion for identified harms. Within a robust EAC program, courses within English might focus on novels, short stories, or historical fiction that detail the effects on individuals due to technology, research, or particular practices within the field. For example, *The Immortal Life of Henrietta Lacks* (Skloot 2011) details biomedical research using the cells of a woman who had given no consent for her tissue use in research, the effect on her family after a breach of privacy, and the long-term impact of financial loss and poverty for the woman and her family despite a thriving multi-million dollar industry based on the use of her cells for vaccine and cancer research, among others. Similarly, *Love at Goon Park* (Blum 2004) details psychological research on primates, and the benefits of that research for supporting attachment theory in psychology and the demonstrable devastating effects of isolation on any social animal, humans included. Theatre classes might study or perform plays that explore the human impact of certain kinds of professional practice. For example, Michael Frayn's (1998) play *Copenhagen* explores questions surrounding research on the atomic bomb for the Nazi regime and ethical questions about personal relationships to other scientific colleagues, the possible loss of human life, and the reach of scientific responsibility. Theatre classes also might provide greater insight into the lives and experience of those different from us to induce empathy. *The Laramie Project*, a play by Moisés Kaufman (2000), uses extensive interviews to confront anti-LGBT attitudes and to bring to life the real-life characters involved in the murder of University of Wyoming student Matthew Shepard. Finally, courses in the medical humanities could make use of service learning or poverty simulations so that students gain a detailed understanding about many of the persons whom they will serve and the conditions and quality of their lives. For example, poverty simulations provide students with realistic experiences of the challenges faced, the precariousness and reduced range of options, the high level of psychological stress, and the negative impact on health while living in poverty.

In the perceptual domain, moral sensitivity harnesses basic vision, attention, and schemas as perceptive filters for salient features of situations and events. The bulk of teaching students salient patterns of information occurs within specific courses or degree programs: for example, teaching students to understand the relevance of a patient presenting with a seemingly disconnected laundry list of symptoms or the meaning of seemingly random temperature spikes within a combustion chamber. With regard to teaching students schemas as perceptive filters, this is perhaps where the greatest depth of learning occurs. However, a robust EAC program allows for perhaps the greatest breadth of learning by providing additional information to help focus attention or to provide clues from a broader perspective. For example, engineering students grasp the moral relevance of features of their products or services that may cause physical harms; given the requirements of their programs and field, an emphasis on physical safety is key. However, such students are less likely to be attentive to the cognitive strain incurred by the use of a product and how it increases levels of stress or fatigue in the user than those who also study psychology. An EAC program increases what is salient to the student from variety of fields, and such increased understanding, attention, and awareness allows students to comprehend additional possibilities for action and how they can respond.

In the cognitive domain, moral sensitivity harnesses knowledge and concepts, both theoretical and normative, to increase the identification and understanding of appropriate action within situations. One of the specializations of particular degree programs is that they provide detailed knowledge of laws, professional codes, and expectations for job performance within the field. An EAC program allows for expanded access to greater knowledge of laws, professional codes, systems of thought, theories, and concepts. For example, *The Immortal Life of Henrietta Lacks* allows us to examine the effect of racism on scientific research, which is not typically a focus of medical education or training. Similarly, students within the social sciences rarely have the opportunity to learn details about cell propagation and the need to access tissue for medical research. And while all students engaged in research must have training through IRBs and learn about the Belmont Report's (1979) guidelines for research on human subjects, an EAC program can expand student understanding by examining concepts about the legal concept of ownership of tissue and body parts within the system of law. The ability to determine the morally appropriate course of action in a given situation must rely on this kind of in-depth knowledge, and EAC programs are uniquely poised to develop the cognitive underpinnings of moral sensitivity.[6]

---

[6]For detailed discussion about how to assess the development of the cognitive underpinnings of moral sensitivity see Thoma (2015).

In the normative domain, moral sensitivity harnesses the norms of particular communities and inculcates value. Most fields have their own individual codes of ethics that establish normative practices and values within the field such the promotion of patient rights or the value of privacy. In addition, most degree programs nationwide offer upper division courses focusing on such specialized codes and establishing expectations for young professionals on how to conduct themselves relative to their clients and obligations towards other professionals. In addition to these values of professionalism, EAC programs also inculcate value by encouraging students to think about their ethical responsibilities as professionals writ large, in providing a service to society, and in protecting individuals and members of society from harm given their professional expertise. The inculcation of value focuses attention on particular aspects of situations, and provides motivation for action by identifying priorities of value. One of the most important values students learn within EAC programs is that to be a professional is to have special ethical obligations, requiring one to be a leader within the field in maintaining those ethical obligations, and always on the lookout for moral issues regarding the actions of other professionals.

The third primary benefit of an EAC program is that it provides the opportunity to practice and develop moral sensitivity over the span of several years. Various EAC programs combine lower and upper division courses, which incrementally develops the underlying capacities for moral sensitivity and helps to integrate them into a complex expertise. Although skills make use of various sorts of knowledge, the learning of skills is unlike the learning of knowledge, which involves memorization, recall, and comprehension of statistics, factors, and facts. Like any expertise, moral sensitivity requires trial and error for its finesse, a range of opportunities to practice it, and an extended period of time in which to practice. In this respect, the longitudinal development of moral sensitivity[7] is more like improvisation in which students must build on and use their experience when faced with complex moral issues, questions, and dilemmas in novel situations. EAC programs provide this extended period of experience and practice through the various courses, workshops, public lectures, and extra-curricular activities such as service learning or internships in which students engage in real-world moral situations in real time.

The fourth primary benefit of an EAC program is the opportunity for directed training of moral sensitivity under the expert and watchful eye of professors and other professionals. As seasoned professionals know, there is no simple script or professional code to follow that does justice to the nuance and moral complexity of situations. At times, it seems that more than one option would be morally appropriate. At other times, it seems that no option would be morally acceptable yet one must act in the least morally egregious way. Moral sensitivity allows us to interact with and respond seamlessly and appropriately to others on many levels and across

[7]For discussion of how moral sensitivity can be assessed longitudinally over the span of a student's education see Jameson et al. (2015).

many contexts, sometimes acting within more than one role at a time, juggling competing obligations, and knowing deftly when to prioritize one over another. EAC programs create unique conditions for developing moral sensitivity under the guidance of professors and professionals through direct coursework and training, apprenticeship, service-learning, residency, and most importantly, personal relationships. In an interesting research finding as part of the Personal and Social Responsibility Index conducted by the American Association of Colleges & Universities, Dey et al. (2010) note that "students cite faculty, especially, as an important influence on their development of integrity and ethics. For instance, students who interacted with faculty outside of class were more likely than their peers to report that their personal and academic integrity had increased while in college" (23). Because students have the opportunity to engage their professors over the course of several years, they can discuss a variety of scenarios, both hypothetical and actual, that they encounter during the course of their education. Through intensive conversations with faculty, students can analyze the situations in which they find themselves. Faculty can help students think through problems and cases and offer points of correction to help students understand when they could have made better choices, analyzed the situation differently, or attended to something morally salient in the situation that they neglected. The mentorship provided by faculty and professionals within a robust EAC program is perhaps one of the most beneficial for the development of moral sensitivity.

EAC programs are extremely valuable because they provide a number of unique benefits for the development of moral sensitivity such as diverse resources from a variety of fields, a focus on domains of underlying capacities, longitudinal development, and mentorship. However, not all universities have the staffing and resources to create such an extensive college and university-wide program. For such institutions, another means for developing moral sensitivity is the creation of a stand-alone course.

Moral Sensitivity and Professional Ethics Courses

Although it is not optimal, educators can still meet their obligation of developing the moral sensitivity of young professionals within single courses given a specific design and focus. In this section, I describe how one can begin to develop moral sensitivity in the span of a single semester within a specially designed Professional Ethics course. Unlike many upper-division courses within degree programs that focus on professional codes of ethics for their own field, a lower-division course in Professional Ethics can capture some of the benefits offered by an EAC program by (1) providing a similar multi-field approach and (2) initiating the development of domains of underlying capacities for moral sensitivity.

Similarly to an EAC program, there is great value in adopting a multi-field approach within a Professional Ethics course because one can capitalize on the perspectives and resources of a variety of fields. Within a semester, one can examine specific moral issues and dilemmas, the conflict of moral obligations, and professional codes of ethics for fields as diverse as engineering, business, journalism, law, nursing, social work, medical research, counseling, police, and the military. Students within the course hail from those fields and many others as well (e.g., economics, political science, psychology, or history). As professionals-in-training, they absorb the culture of their fields of study. Although the students do not have the same level of knowledge or expertise as professionals within their field of study, they bring the knowledge of the theories, concepts, and systems of thought as well as the goals and values of the field, and thereby serve as representatives for their respective fields. Classroom discussions and activities can capitalize on this diversity of perspectives. Students share information they learned in their other courses as well as experiences they have had in internships or relevant employment opportunities. When examining case studies, students can work together to identify moral issues, discuss the approach and analysis from the perspective of professionals from a variety of fields, and work collaboratively to determine what the most appropriate course of action would be in the case, all things considered and from a multiplicity of areas of expertise. Although quite different from an EAC Program, a Professional Ethics course can approximate, to some degree, the benefit of examining moral issues and dilemmas from perspectives across the curriculum.

A Professional Ethics course can also initiate the development of domains of underlying capacities for moral sensitivity. No single course can suffice to develop moral sensitivity given that it is a complex expertise that develops over time, through experience, and given extensive practice. However, such a course can be used to jumpstart the process which can be capitalized on within a professional degree program. As discussed previously, within the normative domain, moral sensitivity harnesses the norms of particular communities as the grounds for value and motivation. An economics major at a large state institution includes among its learning outcomes "learning to think like an economist." Apart from the fact that this is a woefully immeasurable learning outcome, it is easy to recognize that part of what we do—in point of fact—in specific degree programs is to teach students to "think like an X." As professionals-in-training learning specific roles, students absorb the culture of their respective fields. The norms within each field provide guidance on appropriate actions within the context of a role, how to evaluate competing obligations, and the prioritization of values. However, the norms inherent within each field also serve as blinders, constraining how students evaluate moral situations, what is salient to them, what actions come to mind, and which values they consider to be sacrosanct. The function of blinders on horse bridles is to limit the scope of vision, reduce distracting stimuli, and encourage focus on what lies immediately ahead. Blinders are valuable in that they constrain our focus, yet they can be too limiting—particularly for students and professionals-in-training with limited knowledge and experience.

Consequently, a course such as Professional Ethics, full of an array of students with such field-specific blinders, provides a valuable counter-balance to constrictive training that sometimes occurs as a process of professionalization. Classroom discussions and activities allow students to challenge each other's assumptions—

each other's blinders, as it were—in thinking through basic normative assumptions, why some values are prioritized over others in specific fields, what a professional's duties are within a specific field, and the extent of a professional's obligation to one's clients. Over the years, I have observed countless numbers of students who are mystified as to the assumptions made by their fellow classmates on a given case, yet who find themselves on the defensive when their classmates challenge the assumptions they themselves brought into the course and consider obvious. Such conversations serve as correctives to the contracted norms of individual fields and push students to adopt a broader normative perspective. To aid this expansion, I introduce Kantian Deontology and Utilitarian Consequentialism to show students that there are systematic normative theories that undergird some of their assumptions and to demonstrate that the beliefs of students with competing normative intuitions are not groundless. Part of initiating the development of moral sensitivity is the expansion of the normative domain.

One can also expand the cognitive domain through a Professional Ethics course. Clearly, introducing students to even two representative moral theories provides them with supplementary knowledge and concepts. In addition, students are introduced to a new concept each week as part of growing a 'conceptual toolkit' to use when confronted with situations or evaluating a case study. Concepts such as the agent/patient distinction, bodily integrity, a fiduciary relationship, and the analysis of trust, privacy, deception, loyalty, etc. promote a deeper consideration and understanding of the dynamics within situations. Further, the inclusion of case studies allows for the expansion of the cognitive domain following the revised version of Bloom's Taxonomy of Learning (Anderson et al. 2001). In class discussions, students must remember (Knowledge) the concepts, demonstrate their understanding (Comprehension) of them in building their position, apply (Application) the concepts to the case study as a novel situation, evaluate (Synthesis) the situation using them to identify salient and key features, and create (Evaluation) an argument about what morally appropriate actions the professions in the case should do using the concepts as part of the justification. Regular practice in analyzing cases using the growing conceptual toolkit culminates in a classroom debate in which students must collaborate on a policy proposal and defend it from the opposing team. The course is designed to provide incremental cognitive development as well as the constant use of knowledge and concepts that are key for sensitivity to the moral features of situations.

Perhaps surprisingly, one can initiate development of the perceptual domain through a Professional Ethics course. Schemas provide perceptive filters through which we identify morally salient features of situations and events, and one of the ways to develop the perceptual domain is to alter students' perspectival scope. At the beginning of the course, the emphasis is on considering moral situations from a first-personal perspective with obligations to particular clients. I push students to consider each case as though they were the professionals in that situation. What would they feel? What would they identify? What would cause alarm? What actions could they take? Included in this beginning phase is explicit discussion of professional-client relationships, where again, the emphasis is consideration of a limited number of parties from a first-person perspective. How long have they known the client? Are there any conflicting loyalties between the client and the client's family (perhaps whom are also clients)? As we progress through the course, the perspectival scope broadens from a first-personal perspective of their relationships and obligations to other professionals, to a consideration of their relation and obligation to the field, and finally, to a third-personal perspective of the relation and obligation of the field to society. With each expansive shift to another perspective, students gain new schemas as perceptive filters that focus their attention on different aspects, details, persons, relations, and institutions. In what is perhaps the most dramatic section of the course, we examine a variety of models of professional relationships and detail the assumptions within each for perspectival scope, types of relationships, and obligations and duties. We then discuss how each of the models has functioned within their own conversations about the cases throughout the course. Much like Dorothy in the Wizard of Oz, students are always visibly surprised when 'the curtain is pulled' and they catch a glimpse of the wizardly schemas used in their own thinking. This section of the course begins the development of moral sensitivity by overtly altering the students' models, schemas, and perspectival scope.

Lastly, a Professional Ethics course also can initiate the development of capacities in the emotional/affective domain underlying moral sensitivity. It is one thing for students to study theory and hypothetical cases, and quite another for them to consider historic cases that involved real persons. Typically, historic cases are well-known because they involved tragedy, suffering, loss of life, or egregious abuses of justice on a large scale. Many times, students have learned about the historic events in another class; for example, the medical experimentation in Tuskegee, Alabama can figure into an African American studies course, or the explosion of the Challenger shuttle may be mentioned as part of an astronomy survey course. However, learning about a large-scale tragedy, its causes, and the aftermath in such classes is quite different from the focus in an applied ethics course. Conversations about the details of the situation and the ethical questions and challenges for the persons involved therein strike to the heart. Both empathy for individuals harmed and compassion for those who suffer, whether as impacted agents or woeful perpetrators, are common effects when discussing historical and current real-world cases.

But affect need not always be positively valenced to be beneficial educationally: sometimes, the impetus for the development of moral sensitivity comes not from empathy and compassion, but from shock, disgust, and horror at human depravity, irrationality, and poor judgment. While discussion of historic and current real-world cases can certainly spark a range of affective responses, the visual power of film is the most effective. For example, the Frontline documentary *Remember My Lai* details the horrific massacre performed by American soldiers in Vietnam from historical records, newspapers and video, and lengthy interviews with soldiers. While the tale is harrowing enough, the documentary also includes multiple photographs (taken surreptitiously by an Army photographer with his personal camera) that capture the horrors of that day. At the end of the film, students are so affected and overwhelmed, they sit like stones. In addition to the brute effect of affect, films can also be used to carry students through the miniscule yet compounding decisions and actions—as they occur in what approximates real-time for the viewer—that result in a moral dilemma. As a viewer, seeing a scene from the perspective of a character (and having to interpret and identify the morally salient aspects of it for oneself) places one in the midst of the quandary and breeds greater understanding for the complexity of the moral challenges many professionals will face.

Conclusion

The notion of moral sensitivity has a long and pedigreed history, and has long been recognized as essential for moral action. Although no single course can develop moral sensitivity, a course such as the one described above provides some diversity of perspective from across the curriculum and supports or initiates the domains of capacities that underlie moral sensitivity. Used in conjunction with upper-division courses within specialized degree programs, students can continue to develop their moral sensitivity while gaining a greater depth of knowledge in their fields. And what makes for the best of all options: a robust EAC program commencing with such a Professional Ethics course. Given that moral sensitivity is a skill needed for ethical life in general and for professional practice in particular, we should focus our attention on it so that we can develop it more successfully. The educators' charge is to develop moral sensitivity in young professionals.

References

American Association of Colleges and Universities. (2014). *Character traits associated with the five dimensions of personal and social responsibility: Core commitments: Educating students for personal and social responsibility*. www.aacu.org/sites/default/files/CharacterTraitsofthe5Dimensions.pdf. Accessed July 14, 2014.

Anderson, L., Krathwohl, D., Airasian, P., Cruikshank, K., Mayer, R., Pintrich, P., et al. (2001). *A taxonomy for learning, teaching, and assessing: A revision of Bloom's taxonomy of educational objectives*. New York, NY: Pearson, Allyn & Bacon.

Aristotle. (1985). *Nicomachean Ethics*, trans. By Terence Irwin. Indianapolis, IN: Hackett.

Baker, R. (2018). Union college. In E. E. Englehardt & M. S. Pritchard (Eds.), *Ethics across the curriculum: Pedagogical perspectives* (this volume). New York, NY: Springer.

Bebeau, M., Rest, J., & Narvaez, D. (1999). Beyond the promise: A perspective on research in moral education. *Educational Researcher, 28,* 18–26.

Blum, D. (2004). *Love at Goon Park.* New York, NY: Berkley Publishing.

Blum, L. (1991). Moral perception and particularity. *Ethics, 101,* 701–725.

Coplan, A., & Goldie, P. (Eds.). (2014). *Empathy: Philosophical and psychological perspectives*. Oxford: Oxford University Press.

Davis, M. (2018). Illinois Institute of Technology. In E. E. Englehardt & M. S. Pritchard (Eds.), *Ethics across the curriculum: Pedagogical perspectives* (this volume). New York, NY: Springer.

Department of Health, Education, and Welfare. (1979). *The Belmont Report*. https://www.hhs.gov/ohrp/regulations-and-policy/belmont-report/.
Dey, E., Antonaros, M., Ott, M., Barnhardt, C., & Holsapple, M. (2010). *Developing a moral compass: What is the campus climate or ethic and academic integrity?*. Washinton, D.C.: Association of American Colleges and Universities.
Donovan, A. (2018). Dartmouth College. In E. E. Englehardt & M. S. Pritchard (Eds.), *Ethics across the curriculum: Pedagogical perspectives* (this volume). New York, NY: Springer.
Englehardt, E. E. (2018). Utah Valley University. In E. E. Englehardt & M. S. Pritchard (Eds.), *Ethics across the curriculum: Pedagogical perspectives* (this volume). New York, NY: Springer.
Frayn, M. (1998). *Copenhagen*. (play).
Frey, W., & Cruz, J. (2018). University of Puerto Rico, Mayaguez. In E. E. Englehardt & M. S. Pritchard (Eds.), *Ethics across the curriculum: Pedagogical perspectives* (this volume). New York, NY: Springer.
Frontline. *Remember My Lai* (documentary).
Herkert, J., & Ellison, K. (2018). Arizona State University. In E. E. Englehardt & M. S. Pritchard (Eds.), *Ethics across the curriculum: Pedagogical perspectives* (this volume). New York, NY: Springer.
Hume, D. (1751/1957). *Enquiry concerning the principles of morals*. C. W. Hendell (Ed.). Indianapolis, IN: Bobbs-Merrill Company, Inc.
Hume, D. (1888/1978). *Treatise of human nature* (2nd ed.), Selby-Bigge & P. H. Nidditch (Eds.). Oxford: Clarendon Press (Oxford University Press).
Hutcheson, F. (1993). Inaugural lecture on the social nature of man. In T. Mautner (Ed.), *Francis Hutcheson: Two texts on human nature* (pp. 124–147). Cambridge, UK: Cambridge University Press.
Jameson, M. M., Fuhrman, H., & Mower, D. S. (2015). A portfolio approach to integrated moral sensitivity. In D. S. Mower, P. Vandenberg, & W. L. Robison (Eds.), *Developing moral sensitivity* (pp. 246–263). New York, NY: Routledge.
Jordan, J. (2007). Taking the first step toward a moral action: A review of moral sensitivity measurement across domains. *Journal of Genetic Psychology, 168,* 323–359.
Kaufman, M. (2000). *The Laramie Project* (play).
Maxwell, B. (2008). *Professional ethics education: Studies in compassionate empathy*. Dordrecht: Springer.
Mower, D. S., Vandenberg, P., & Robison, W. L. (Eds.). (2015). *Developing moral sensitivity*. New York, NY: Routledge.
Newton, L. (2018). Fairfield University. In E. E. Englehardt & M. S. Pritchard (Eds.), *Ethics across the curriculum: Pedagogical perspectives* (this volume). New York, NY: Springer.
Rest, J. (1982). A psychologist looks at the teaching of ethics. *Hastings Center Report, 12,* 29–36.
Rest, J. (1986). *Moral development: Advances in research and theory*. New York: Praeger.
Rest, J. (1994). Background: Theory and research. In J. R. Rest & D. Narvaez (Eds.), Moral development in the professions: Psychology and applied ethics. Mahwah, NJ: Lawrence Erlbaum Associates.
Robison, W. L. (2018). Rochester Institute of Technology. In E. E. Englehardt & M. S. Pritchard (Eds.), *Ethics across the curriculum: Pedagogical perspectives* (this volume). New York, NY: Springer.
Skloot, R. (2011). *The immortal life of Henrietta lacks*. New York, NY: Broadway Books.
Sloan, D. (1980). The teaching of ethics in the American undergraduate curriculum, 1876–1976. In D. Callahan & S. Bok (Eds.), *Ethics teaching in higher education* (pp. 1–57). The Hastings Center, Hastings-on-Hudson, New York, NY: Plenum Press.
Smith, A. (1759/2000). *The theory of moral sentiments*. Amherst, NY: Prometheus.
Smith, J. L., Ickes, W., Hall, J. A., & Hodges, S. D. (Eds.). (2011). *Managing interpersonal sensitivity: Knowing when-and when not-to understand others*. New York: Nova Science Publishers.

Steuber, K. (2006). *Rediscovering empathy: Agency, folk psychology, and the human sciences.* Cambridge, MA: MIT Press.
Swaner, L. (2004). *Educating for personal & social responsibility: A planning project for the Association of American Colleges and Universities.* A Review of the Literature. Association of American Colleges and Universities, September 13, 4. www.aacu.org/core_commitments/documents/review_of_lit_000.pdf. Accessed July 23, 2014.
Swaner, L. (2005). Educating for personal and social responsibility: A review of the literature. *Liberal Education, 91,* 14–21.
The Hastings Center. (1980). Hastings Center Project on the teaching of ethics: Summary recommendations. In D. Callahan & S. Bok (Eds.), *Ethics teaching in higher education* (pp. 299–302). The Hastings Center, Hastings-on-Hudson, New York, NY: Plenum Press.
Thoma, S. J. (2015). A Neo-Kohlbergian tale of two sensitivities. In D. S. Mower, P. Vandenberg, & W. L. Robison (Eds.), *Developing moral sensitivity* (pp. 227–245). New York, NY: Routledge.
Weaver, K., Morse, J., & Mitcham, C. (2008). Ethical sensitivity in professional practice: Concept analysis. *Journal of Advanced Nursing, 62,* 607–618.
Woodson, S., & Zhu, Q. (2018). Colorado School of Mines. In E. E. Englehardt & M. S. Pritchard (Eds.), *Ethics across the curriculum: Pedagogical perspectives* (this volume). New York, NY: Springer.

Aiming Professional Ethics Courses Toward Identity Development

Glen Miller

Abstract The many elements of professional ethics programs can be oriented by the goal of identity development situated within a teleological virtue ethics structure. This structure supports the integration of an individual's values, acts, and goals and the framework (including ethical codes, laws, common practices, and the social good) of their profession. Professional ethics understood in this way extends beyond the normal focus on propositional and practical knowledge to include other important aspects of professional activity. The kinds of activities that are of particular interest in this analysis are those that fit under Alasdair MacIntyre's concept of practice that take place within distinct moral spaces. By combining the idea of practice and distinct moral spaces, professional ethics can be expanded to draw awareness to characteristic virtues dominant in different ethical fields, offering critical distance and promoting agent self-awareness. One's identity arises through answering what Charles Taylor called "qualitative questions," used to define one's self, which depends on what one has already done and what one aims to do, guided by what one holds to be significant. Thoughtful answers require mental deliberation and discourse; their articulation and the development of a coherent moral identity that combines personal and professional intentions, actions, and goals are closely correlated with exemplary professional behavior, according to research done by social psychologists. According to this argument, one's unique identity is expressed in the imaginative composition of words, virtues developed, and practices to which they are applied, over the course of one's life.

Keywords Professional ethics · Practice · Ethical fields · Characteristic virtues Identity · Virtue ethics · Social psychology

---

G. Miller (✉)
Department of Philosophy, Texas A&M University, 4237 TAMU,
College Station 77843-4237, TX, USA
e-mail: glenmiller@tamu.edu

E. E. Englehardt and M. S. Pritchard (eds.), *Ethics Across the Curriculum—Pedagogical Perspectives*, https://doi.org/10.1007/978-3-319-78939-2_6

Introduction

The extended duration of professional careers, which often span decades, imparts a challenge to many kinds of education, but especially for professional ethics courses. For most kinds of education, the longer the gap between classroom and application, the worse the outcome. This concern exists—but is mitigated—for courses that seek to impart propositional knowledge (know-that) and practical knowledge (know-how). Acquisition of propositional and practical knowledge is often impersonal and frequently delivered through technology, making it less expensive; measurement is also usually easier, either by self-assessment or according to the multitude of external standards and commonly accepted best practices presently available. In cases of shortcomings, deficiencies in propositional knowledge can often be quickly remedied, and specific exercises can often improve practical expertise when it is inadequate.

The same problems diminish the efficacy of professional ethics courses offered to undergraduate and graduate students, but in these cases, the longitudinal dimension generates other kinds of challenges. In addition to propositional knowledge of the standards and ethical demands of a profession and the practical knowledge of how to behave in accordance with them, ethical action also depends on prudence and the faculty of the will. Prudence, or practical judgment, is the proper assessment of moral features in a particular situation and recognition of the appropriate action. Its cultivation requires self-knowledge and discernment. Moreover, in many situations encountered by professionals, relevant propositional knowledge on which prudence depends is often unstated and sometimes not ready-at-hand. The will is necessary to convert moral decisions into action; its individual nature does not admit to disinterested, objective scientific evaluation. Neither a lack of prudence nor weakness of the will admit to crash courses or can be achieved by attaining some ISO standard.

A second set of concerns further complicates the connection between the professional ethics classroom and later ethical behavior. For many professionals, job opportunities and responsibilities often evolve over the course of a career. Ethical issues for business leaders, engineers, and technology experts change as they ascend in their organizations. Initial responsibilities are usually technical and practical, with objectives are provided by others; ethical concerns are usually simple and under the individual's control. As professionals advance, they gain executive and organizational responsibilities. Ethical questions that accompany such responsibilities are broader—goals must be defined, rather than just achieved, and their achievement often depends on collective action outside of the control of any one individual—and a creative imagining of what is possible should precede the selection of a particular course of action. A young computing professional may need to adhere to professional standards when testing a new application and give proper credit for intellectual property used, for example, whereas a senior employee may be responsible for determining which social groups should be considered in requirements definition of a product and for creating just working conditions that enable workers from different backgrounds to flourish. Professionals who frequently operate more independently, such as doctors and lawyers, often gain executive responsibilities in their businesses and professional societies. Meanwhile, societal, economic, and cultural values, which provide the background in which professional activities occur, also can change, adding another degree of separation from what the mid- or late-career professional learned in a university course. Continuing ethics education, not always required or even readily available, rarely addresses these complicated issues: its scope is commonly limited to laws and legal precedents, codes of ethics, and business rules. In an ironically cruel twist, continuing ethics education for professionals usually aims to impart propositional knowledge or to inform students about how to use formal mechanisms to address illegal or otherwise prohibited actions, with apparent obliviousness to the increased degrees of freedom that professionals must navigate, and as any previous work in ethics from their university experience, which should help in such matters, fades in their memories.

These challenges are especially problematic given the special nature of professional ethics education, which includes a commitment to professional ideals. This expectation exceeds goals held by the majority of ethics courses—awareness of ethical concerns, knowledge of social expectations, and skill at moral reasoning—that have much in common with many other kinds of university courses. Most ethics courses are concerned primarily, if not exclusively, with the intellect and judgment, areas for which continuing education ethics refreshers can help. Commitment, however, is categorically different: it means that professionals are more likely to act in a way aligned with the ideals and rules of the profession, and with their spirit, than they would have had they not taken the course.[1] Awareness, knowledge, and skill of moral reasoning are necessary but subordinate to, and, in a sense, preparatory for, commitment, which is evaluated based on behavior and requires development of not only the intellect and the faculty of judgment, but also the will, self-knowledge, and imagination.

The preceding assessment, developed predominantly through personal experience, reflection, and observation of friends and colleagues, aligns with data collected in the field of social psychology. Laurence Kohlberg's influential model of moral progression through a series of stages, which assumed judgment to be central (Kohlberg 1969), was questioned by Augusto Blasi, who noted that no strong link between moral reasoning and action had been shown (Blasi 1980). Blasi's line of criticism was advanced by James Rest, who showed that only about ten percent of the variance in moral action could be traced back to data measured in the Defining Issues Test (DIT).[2] Awareness, propositional knowledge, and practice in moral reasoning, while necessary, are not sufficient for moral action.

[1]I touch on action outside of established norms in the last section of the paper. For now, it suffices to say that professional ethics courses should lead the agent to have a better considered explanation of why they acted in such a way than if they had not taken the course, and that these situations are not common.

[2]My brief summary of behavioural psychology is based on Thoma and Bebeau (2013).

Given the numerous challenges detailed above, exacerbated by the shrinking currency that professors and other authorities have in shaping the behavior of others, especially in the present period in which anti-expert sentiment is strong, the situation facing professional ethics educators seems daunting. How can professional ethics courses improve commitment to ethical ideals that result in moral action when it is difficult or impossible to know the kinds of situations students will encounter, the strengths and weaknesses they will have gained, and the social pressures that will encourage or discourage certain behaviors, especially in light of the fact that students' wills are not always malleable and while respecting freedoms expected in a liberal democracy?

I propose that the concept of identity can orient the combination of intellectual and moral aspects of professional ethics and provide a durable foundation on which its practitioners can make decisions far into the future. With this orientation, imparting knowledge about historical and common cases encountered by professionals, unacceptable actions, and the importance of preventing immoral behavior, illegal actions, and harmful consequences remains an essential part of professional ethics development, but one that contributes to a larger goal: they provide a ground for and catalyze the development of each individual's identity. Professional ethics regains its aspirational and reflective character, making it more than dogmatic instruction: it is the challenge of crafting a meaningful life that appropriately forms and is formed by commitments and identifications. Seen in this way, the ultimate task of ethics is to develop an integral view of one's self, which includes professional and personal dimensions, that is robust enough to develop over time, that is aware of professional expectations, including technical and ethical practices, that is inclined toward continued growth, and that results in thoughtful action.

The connection between professional ethics and identity can be developed in three steps. First, ethics must be understood as more than just a decision-making procedure: it must retain a teleological dimension that connects individual agents and acts with professional activities and social goods. Second, within this comprehensive understanding of ethics, smaller spaces in which dispositions, habits, actions, and goals are formed and form professionals in modern societies can be delimited as discrete ethical fields. The idea integrates Alasdair MacIntyre's concept of practice and Charles Taylor's explication of frameworks of moral space. Within each ethical field, *characteristic virtues* that are widely held, necessary, or advantageous can be identified. Third, identity arises through one's individual articulation of the good in discourse and mental deliberations as well as in the development and integration of virtues developed and expressed through certain practices directed toward specific goods. Through the writing of one's autobiography, as it were, consisting of words and deeds, an individual can exercise informed freedom and develop an intuitive sense of the "good" that can guide actions, both those that are consciously deliberated and those that are not. The resulting expansive understanding of professional ethics acknowledges the intersection of personal and professional desires, opportunities, and responsibilities that arise over the course of their careers, and it is oriented toward developing the convictions, reflectiveness, and imagination to pursue a coherent and integral vision of the good.

Professions, Agents, Acts, and *Telos*

My argument toward an identity-oriented professional ethics starts from the traditional breadth of ethics, namely as the study of free human behavior. This starting point is the opposite of the present fashion of treating professional ethics as a standalone area of inquiry, as though it were independent of other moral concerns. In its modern segregated state, ethics is often understood to be the study of decision-making principles and evaluation of isolated acts, detached from the acts that led to the present circumstances and from their effect on the agent. In this form, the moral act is characterized as a disinterested judgment, which, for better or worse, aligns more closely with the conditions of the classroom and debate than with ethical practice. Lost in this framing of the ethical problem is the relationship between the agent and the act, with the corresponding separation of the *telos* of the act from the *telos* of the agent. A consequence of this division is an incoherent separation between the social good promised by a profession and the agents and their acts that constitute it.

The connection between the nature of the agent and his or her acts has been a staple of ethical consideration since at least the time of Aristotle. This connection is explained most clearly for the "moral virtues" or "virtues of character," which are obtained by an agent through acting in a virtuous way. To become courageous, one must act courageously (Aristotle 2002, 1103a15–1103b25). The close reciprocal connection between agent and act is pithily expressed by Etienne Gilson as "a being's habits determine the manner in which it realizes its own definition"[3] (Gilson 1994, 256). Our actions form and reform us, especially when taken repeatedly, much like human bodies adapt to varying kinds of physical activity, making new actions possible or rendering previous acts impossible. To put it another way, I have become who I am by what I have done, which is expressed linguistically as verbs are transformed into nouns, e.g., one who repeatedly steals is called a thief, and, more constructively, a person who has appropriate training and practices medicine or engineering is properly predicated as a doctor or an engineer, respectively. The linguistic expression of this relationship is even clearer in the Greek of Aristotle's day: *ethos* (habit) and *êthos* (character) are cognate (Aristotle 2002, 1103a19–23).

An even more intimate connection between agent and act can be found in the intellectual virtues, identified by Aristotle as the complement to the moral virtues already discussed. Acquired by instruction and experience, intellectual virtues are of special importance in professional ethics, even though they are normally treated superficially or ignored in its discourse, where they are commonly reduced to factual knowledge. One of the distinguishing marks of a profession is that its members possess mastery of some specialized body of knowledge and the accompanying skills that can be employed to bring about some social good.

---

[3]In this paper, I understand a being's "definition" as something that the individual can construct, i.e., that it, like one's physique, can be determined to an extent, though not entirely. A similar line of reasoning works for those who adopt an anthropology with a more extensive definition of human essence, such as Thomas Aquinas.

A "good" professional has not only developed her natural intellectual capacities so that she possesses the mastery necessary to competently complete professional tasks; she also is able to gain mastery needed to complete novel tasks and is aware of the limitations of her expertise. The latter is more than propositional or practical knowledge: this development aligns with an ambiguous and complex virtue called "professional judgment," closely correlated with the classic virtues of prudence (*phronesis*) and technical judgment (*techne*), that is discussed infrequently but often assumed to develop "naturally" over the course of a career. As with the other intellectual virtues, the transformation that takes place with professional judgment is one of the agent, which may or may not lead to a different external state of affairs. Given the complexity and difficulty of many professional activities, which are only possible after extensive education and development, it seems clear that, for intellectual virtues as much as the virtues of character, the agent and the act have a reciprocal relationship, where one determines the other, and vice versa.

The virtue-oriented approach described above maintains the connection between agent and act that is necessary for the connection between agents and their ends, which ultimately enables the connection between agents, their acts, and goals, on the one hand, and the social goods promised by the profession, on the other. To the degree that professions act in the world, they do so through their members, the agents who define and bring into existence agreed upon ends through their acts. This connection between the individuals and their professions provides some ethical direction, but the degree varies from profession to profession.

As Carl Mitcham has argued, the ultimate goal or end of the acts taken in some professions are clearly defined: the end of medical profession is health, and the end of the legal profession is procedural justice (Mitcham 2009).[4] The ends of some other professions are less clearly defined: it is not immediately obvious what the end of engineering is. Is it a contribution to the material lot of humanity, to its convenience, comfort, and craving for novelty, to developing and maintaining the technical infrastructure that makes modern life possible, or some combination of the above? A contested or poorly defined *telos* means that no single set of standards exists by which the good of the profession and of individual professionals and their acts can be assessed, although even in this situation the rough definition has some value. Note that even the minimalist version of professional ethics, which is primarily preventive and prohibitive injunctions, are teleological, though in a more restrictive negative sense: they emphasize the avoidance of certain undesirable ends, e.g., for medicine, to do no harm, for engineering, to protect public health, safety, and welfare.

In this section, I have argued that professional ethics should use the framework of teleological virtue ethics so that it can support an integral understanding of agent, act, and (individual) *telos* that pairs with the aims of the profession. The focus has been on the agent and the community, on individual coherence informed by a corporate *telos*. The next step is to enhance this inherited conceptual framework by focusing on the nature of professional acts.

---

[4]As Mitcham points out, clarity of ends does not necessarily lead to their attainment.

Practices, Ethical Fields, and Characteristic Virtues

The general framework of agents and their aims, considered individually and collectively, described in the previous section can be improved by analyzing the different spaces that individuals occupy, the practices in which they engage in these spaces, and how they affect each other. In much of the discourse on professional ethics, acts are treated as atomistic judgments, subject to adjudication by the Kantian disinterested observer or through an impersonal moral arithmetic favored by utilitarians. To be sure, some professional decisions do hinge on whether individual autonomy and dignity should outweigh "the greatest good for the greatest number" or how to act in accordance with the four mid-level principles identified by Tom L. Beauchamp and James F. Childress (2009). Yet a critical loss occurs if the study of the free behavior of a professional is treated as an occasional evaluation of specific decisions—its measurement should encompass one's professional existence, including acts, competencies, and relationships that have been developed over the course of a career, seen in the context of one's other experiences, activities, and goals.

This expanded scope is superior to one limited to solitary decisions for two reasons. The first reason is that the acts of professionals today have little in common with the simplistic acts usually used to illustrate ethical concepts: as an extreme example, consider how far the trolley problem diverges from ethical issues faced by professionals. Philippa Foot's trolley problem, intended to illuminate distinctions between direct and oblique intention and positive and negative duties and rights, catalyzes discussion of causation and responsibility, action and passivity; distribution of disproportionate benefits and harms on select individuals; and the non-obviousness and possible counterintuitiveness of many ethical questions (Foot 1978). Yet these credits are more than offset by the associated debits when applied to professional ethics. The trolley problem severs the temporal and social continuity that is present in professional work and is an important feature in many of its ethical issues. The ahistorical framing of the problem omits concern for how actions in one situation can alter the dispositions of the agent. It implies that ethics should be thought of as a study of free action in dramatic events, rather than as a determination to be consistently honest, courageous, and thoughtful of the concerns of all stakeholders, a determination that can be developed over time. The ahistorical framing makes the ethical agent appear stuck in a fixed position, left with two problematic choices, whereas a professional's agency can reshape the conditions in which one works: professional dilemmas may often be avoided with anticipation, imagination, and effort. From a social perspective, the trolley problem protagonist is wholly disinterested, free of relationships with the other characters. Important professional concerns about excessive self-interest or conflicts of interest cannot be mapped to elements in the trolley problem. Moreover, many professional ethics issues arise in team settings, where coordination and collaboration may be of utmost importance in achieving an appropriate outcome.

Perhaps most importantly, to many non-philosophers the trolley problem depicts ethics as a game of reasoning for the cleverest, susceptible to G. E. M. Anscombe's critique of 1950s Oxford philosophy. It is hard to blame students who, when asked questions like "what you ought to do if you had to move forward and stepping with your right foot meant killing twenty-five fine young men while stepping with your left foot would kill fifty drooling old ones," answer as she does, "obviously the right thing to do would be to jump and polish off the lot" (Anscombe 2005, 163).

The second reason to not treat ethics as isolated decision-making is that doing so ignores the interrelation between behavior in different moral spaces: one's ethical comportment develops over time, and professional activity is often, but not always, influenced by other activities and forces. An understanding of the nature of professional acts can be deepened by utilizing Alasdair MacIntyre's idea of a practice, and the interrelation between spaces the idea of ethical fields.

Many professional acts, as well as many other formative actions and experiences, take place within practices, which MacIntyre defines as

> any coherent and complex form of socially established cooperative human activity through which goods internal to that form of activity are realized in the course of trying to achieve those standards of excellence which are appropriate to, and partially definitive of, that form of activity, with the result that human powers to achieve excellence, and human conceptions of the ends and goods involved, are systematically extended. (MacIntyre 1984, 187)

MacIntyre's concept captures the teleological nature of human action developed in the preceding section, both as it exists in the individual and as it is expressed in the collective action of those in the profession. It incorporates the importance of intellectual virtues, especially as they develop over time, that are characteristic of particular activities. It also highlights the social benefit that accrues when skills and talents that culminate in the profession are maintained and improved as is possible over extended periods of time. For example, a medical doctor with thirty years of experience and continuing education usually has better judgment and more knowledge than he possessed at the start of his career, and doctors today have more tools at their disposal than their predecessors did, due in part to advances in medical science and technologies, but also due to their predecessors' curation of the profession. Note that all of the preceding is ignored if professional ethics is considered atomistic, disinterested judgment.

A practice has an existence that is separate, in a sense, from what is done by individual agents working toward it: the social organization of practitioners and the historical development of means to achieve particular goods provide a relatively stable framework in which professional activities take place. The frameworks are structured by the discourse of practitioners, especially their vocabulary, and their formal and informal expectations, rewards, and punishments. Each profession has its own framework, as do departments, colleges, universities, corporations, political organizations, and other voluntary organizations: most people navigate several frameworks simultaneously. Explicit professional frameworks are found in documents such as codes of conduct, governing laws, and educational expectations. Implicit delineation happens in a myriad of ways, from the retelling of classic cases that shaped the field to informal mentoring to narratives conveyed at dinner—or perhaps more so, over after dinner drinks—at conferences, to the narratives about the founders or exemplars of professions.

With the recognition of multiple frameworks, various kinds of practice, and an expansive scope of ethical concerns that includes all facets of human life, a generic treatment of acts and virtues can be supplemented by the concept of *ethical fields*, which has both explanatory and formative power for professional ethics instruction and for virtue ethics as a whole. Ethical fields are spaces with moral, social, and physical dimensions that individuals can elect to enter where particular dispositions and actions, determined by the goals of the practice, are held and exercised, respectively. In contrast to Aristotelian virtues that are desirable for all humans by virtue of their animating soul, such as prudence and justice, certain virtues are more important in some fields than in others. The division between fields becomes sharper when practices and individuals are rigidly segmented, which occurs when knowledge and labor are specialized, people are relatively mobile, and social circles have little overlap. These conditions, common in modern developed economies, were not present to the same extent in the Greek *polis* or medieval city.

Within each ethical field, *characteristic virtues* can be identified. They are usually widely held by agents in the field and contribute to its successful navigation. These virtues can arise naturally, based on appropriateness to practices of the field, or through convention, based on social agreement and consensus. The intellectual virtue of scientific reasoning using physical laws is essential to engineering; the same intellectual virtue applied to the human body is essential for the practice of medicine. Yet other virtues can be prized in an ethical field that are not immediately related to practices. For a number of reasons, including longstanding gender stereotypes, many positions are associated with virtues such as assertiveness, independence, and emotional detachment that can be vices and detract from the attainment of the excellences or goods sought.

Field segmentation offers a framework that can explain how behavior in one field can be affected by activities in another and how some individuals exhibit strikingly different characteristics in various spaces. Habits, understood as repeated acts as well the disposition that leads to and is derived from these acts (Hurka 2006), are usually developed in specific ethical fields; sometimes they bleed into other spaces, but not always. On the one hand, virtues such as honesty, courage, and determination developed through formative activities and commitments in one field often are expressed in other fields over the remainder of one's life. On the other hand, some individuals hold long-term sustainable commitment to ideals such as concern for clients, truthfulness, and loyalty to the profession, at the same time as they ignore parental responsibilities, are undependable friends, and do not keep other promises. In her well-known report on Adolf Eichmann's trial, Hannah Arendt noted that his bureaucratic competence was paired with an utterly incomprehensible lack of concern for what he was doing to Jewish people, which, in her analysis, somehow never seeped into his personal relationships (Arendt 1963). The concept of ethical fields allows more to be said about such inconsistency, rather than labeling problematic acts as vicious or simply saying that individuals lack the virtue of prudence or wisdom.

While Arendt highlighted the possibility of separation between ethical fields, most of the time they are interrelated. Dispositions and acts formed in one field are imprinted on the individual and shape behavior in others. This interaction occurs historically and concurrently. Thinking historically and simplifying somewhat, an individual can be thought of as an aggregation of practices and the sum products of the virtues and vices developed in the ethical fields already inhabited. Previous inclinations and habits are applied to new practices, especially when a complex new ethical field must be navigated. Aristotle recognized the pivotal nature of prior moral development, stating that it makes "all the difference" how one is habituated from a young age (Aristotle 2002, 1103b20), though perhaps he erred by understating the redirecting power of crises, even later in life, and by omitting the idea of rehabilitation commonly held in contemporary society.[5] Concurrently, most people are attracted toward coherence between the different ethical fields they inhabit: dissonance seems to suggest a lack of integrity, an idea that will be developed further in the next section.

Recognizing the ethical fields one inhabits and their characteristic virtues promotes three aspects of self-awareness. One, an individual becomes aware of the virtues and vices developed in previous experiences, which allows one to notice gaps between what he or she possesses and what is necessary to succeed in certain practices. Two, recognition draws attention to the ethical fields in which one currently works so that the individual understands how he or she is being shaped by his or her actions in each. Armed with this knowledge, an individual may identify other fields not currently inhabited that would cultivate certain virtues and decide to enter them. One may also seek out uncharted areas of fields already inhabited to do the same. Three, one becomes cognizant of the possible need to keep or create separation between fields, especially where vices developed in one may be particularly harmful in another.

Turning to the specific aims of this paper, one can predict that an individual entering a new practice in a new ethical field will be inclined to apply virtues that were recently developed and are closely related to the new field. The moral and intellectual virtues that new doctors apply to their practices are shaped by their education, including their internships, which is their temporally and topically proximate ethical field. A ruthless and impersonal educational system would be expected to create ruthless and impersonal doctors. More generally, virtues associated with critical evaluation of ideas are more likely to arise in educational settings in which dissent is encouraged, rather than one that has a strong hierarchical character. Because of their foundational role for professions, the characteristic, virtues prized in educational spaces deserve close attention.

[5]Even if Aristotle's claim is taken literally, youthful development should be thought of as setting the floor and ceiling of one's development, or else the role of virtuous friends (Book VIII of *Nicomachean Ethics*) would make no sense.

The way an individual navigates an ethical field also depends on relationships with one's colleagues and the common practices in a profession. Aristotle notes the importance of such relationships, which are important for but do not determine ethical development, in his treatment of friendships (Aristotle 2002, 1161b11–15). No doubt many professional relationships exist for their usefulness and the pleasure they give to both parties, but professions are enriched when relationships are between those who elevate each other's excellences, moral and otherwise.[6] These relationships can redirect competitiveness, common in many professions, from envy and jealousy toward camaraderie. Common practices in a profession, especially those that are not wholly of technical character, tend to shape their agents and encourage certain attitudes.

Mentoring relationships deserve special attention because they provide an initial orientation to an ethical field and accelerate one's understanding of it. Mentors guide the novice through terminology, rewards and punishments, and opportunities and dangers that are present in the professional space. They can share their understanding of the space that they have gained through experience, supplementing objective formal information with their subjective understanding, and they can make the expansive amount of propositional and practical knowledge, which seems to have increased exponentially in the "Information Age," manageable for the aspiring professional. The personalized nature of these relationships also means that questions normally outside of the narrow scope of professional ethics, including the balance between "life and work," can be addressed, which is critical if different ethical fields do in fact affect each other.

The term "field" relates to identity in two important ways. In this section, its sense as a bounded space with certain contours, largely shaped historically by social forces that shape the identity of professional practitioners, has been foregrounded. The other sense of a field, as an expansive range, implies a freedom the practitioner possesses. The "free play" given the individual and its importance in the development of a unique identity is given primacy in the next section.

Identity and Self-authorship

The practices and the fields of virtue that individuals can develop are finite: given limited time and knowledge, each individual faces the critical tasks of developing one's professional capabilities and integrating them with other obligations, desires, and opportunities. Working in the backdrop created by the *telos* of the profession and its linguistic and social framework, individuals craft their identity through their navigation of professional and personal degrees of freedom. The sense of identity to

---

[6]"But the complete sort of friendship is that between people who are good and are alike in virtue, since they wish for good things for one another in the same way insofar as they are good, and they are good in themselves." (Aristotle 2002, 1156b7–10).

which I refer is the process of individuation that arises from one's unique answers to what Charles Taylor calls "qualitative questions" (1989). An individual answers these questions in word, either in conversation or in mental deliberation, and in act, in the practices and virtues they have sought and choose to seek. The process of answering these questions can be thought of as writing one's autobiography, a unique story that consists of one's selection of the ethical fields he or she chooses to inhabit, what is said and done in each, and how they will be integrated. The process of authorship lends coherence to these decisions and is a source of moral strength.

It is important to note that professional activities constitute only some of many practices undertaken by any one agent, which are chosen out of a myriad of possibilities available to individuals in modern liberal society. To determine which practices should be developed and how they should be integrated is to formulate an answer to Taylor's "qualitative questions," which include

> questions about how I am going to live my life which touch on the issue of what kind of life is worth living, or what kind of life would fulfill the promise implicit in my particular talents, or the demands incumbent on someone with my endowment, or of what constitutes a rich, meaningful life—as against one concerned with secondary matters or trivia. (Taylor 1989, 14)

Taylor's qualitative questions orient decisions about practices in a similar way as ethics broadly understood provides the context in which professional ethical concerns must be deliberated. In order to answer these qualitative questions, competing interests must be evaluated and a single course of action must be selected out of the many possible. As boundaries between work, social, and personal time have eroded, these decisions become more complex and obscured than they had been. Put simply, answering these questions is the task of determining one's ultimate good, the target at which an individual aims, which determines one's identity.

The answers to qualitative questions are Janus-faced, i.e., they are grounded on what has already been written and what has already been done, but they also indicate a trajectory. As Taylor argues, questions of "who am I?" always take the form of an "'and then': there was A (what I am), and then I do B (what I project to become)" (Taylor 1989, 47). While these determinations are challenging for young professionals, who have less experience and knowledge of themselves and the world, small changes in bearing will have a large ultimate impact over the course of a career; mid-career changes are possible but usually come with significant costs; for the older professional, the professional trajectory has largely been traveled; and at the moment of death, the question of who I am is entirely retrospective: there is nothing more that I can do, no further projection possible. Moreover, answering these questions takes on an iterative character: an initial answer is necessary to determine a provisional target at which one aims and to think through what it entails, and subsequent answers are informed by earlier efforts.

One might expect that the process of thoughtfully integrating personal desires and professional actions and goals, which requires reflection on fundamental principles and concerns and an effort to bring them into coherence, would have a formative effect. Data gathered by social psychologists supports this expectation: being able to articulate one's idea of the good correlates with higher levels of ethical commitment. In research building off their initial study of dental professionals, James Rule and Muriel Bebeau found a striking contrast between exemplary professionals in medicine, law, and the military and those convicted of not following minimal ethical standards. They found that only about five percent of those referred for remedial ethics training could articulate professional duties and responsibilities, whereas exemplars "were able to articulate the public duties of their profession, integrate them with their personal value frameworks, and regularly and consistently engage in socially responsible actions" (Bebeau 2014, 126). This correlation held even as measures of ethical sensitivity, moral reasoning, and competence in implementing ethical solutions varied widely between the two groups. Bebeau concludes that data strongly suggests that "development of a moral identity that is consistent with the norms and values of the profession is *the* driving force that gives rise to the development of other abilities that account for responsible professional conduct" (Bebeau 2014, 126, italics added).

An explanation for the moral power that arises with identity development is found in Taylor's explication of the emergence of a "self." To be deserving of the term, he argues, one must be reasonably aware, or working toward an understanding, of the customs, traditions, and history that shape his or her frameworks and be cognizant of his or her idea of the good. The power of "self"-definition, then, is two-fold: it orients, and it provides a basis for resistance to outside forces. As Taylor says,

> My identity is defined by the commitments and identifications which provide the frame or horizon within which I can try to determine from case to case what is good, or valuable, or what ought to be done, or what I endorse or oppose. In other words, it is the horizon within which I am capable of taking a stand….Put counterfactually, they are saying that were they to lose this commitment or identification, they would be at sea, as it were; they wouldn't know anymore, for a range of questions, what the significance of things was for them. (Taylor 1989, 26)

At best, being "at sea" means arbitrary or purposeless actions; at worst, it makes one overly susceptible to external influences from other people or from institutional practices. Either way, knowing what has significance guides one toward acts in accordance with it and allows one to recognize acts that are contrary to one's ideals in personal and professional situations.

A coherent articulation of one's "good" does not stop at verbalization; an individual's identity depends as much, if not more, on one's acts and on the combination of the virtues that one has developed and the practices to which one has contributed. As mentioned earlier, virtues take time to develop, and human ontological and epistemological finitude limits one's involvement to a few practices: exclusive choices are necessary. The physical and material development and expressions of identity must take into account one's talents, capabilities, and limitations as well as the interplay between different practices and commitments over the long term. Alignment of actions and words yields integrity, and one strengthens the other.

Articulations are always partial and virtues for each field can be developed further, but taken together, they contribute to two important characteristics of a professional. The first is a disposition toward what is good, which is important when making judgments where consequences are to a degree unpredictable; the second is a developed intuition toward ethical action. Charles Harris (2008) notes that dispositions are important in risky situations, where action is complex and collective and uncertainty is elevated.[7] As Michael Davis argues, the engineering profession demonstrates a widely held disposition: in spite of, or perhaps because of, some high profile disasters, such as *Challenger*, *Columbia*, and the Macondo blowout, its professionals regularly demonstrate commitment to taking responsibility that extends beyond their role in complex sociotechnical projects, which is in the interest of the public and may be one of the reasons the profession is respected as much as it is (Davis 2012). An intuition of the good also seems necessary: Jon A. Schmidt notes that engineers rarely refer to their codes of ethics—an observation that can likely be extended to medical doctors, lawyers, and other professionals—yet engineering practice usually aligns with them (Schmidt 2014). While some ethical expectations are codified in technical standards and others unconsciously embraced through commonly accepted practices, these guides almost always underdetermine action, which suggests that the proper disposition toward and intuitive sense of professional good, developed at least to some extent, is at work.

Articulation of one's identity, an intuitive sense of the right decision, proper dispositions, and attention to one's virtues and vices yield a foundation needed for reasoned innovation, the form of self-authorship whereby the individual most clearly distinguishes his or her self. At this point, the agent can leverage his or her specific strengths toward a particular end, conscientiously step outside of generally accepted norms in a field and reconsider its characteristic virtues, and even work to alter practices of a profession. To put it another way, it is at this point that the professional can engage his or her imagination, practical and moral, to act freely. It is important to note that more experience is not always better: a younger professional may be more willing to take risks and defend these positions in the face of resistance. Relatedly, experience or maturity does not necessarily lead to this kind of free action. Not all individuals have the same potential for such acts, nor do all those with potential develop it, and, for many, the longer one works in a space, the harder it is to maintain critical distance from one's inclinations, common practices, and colleagues. A professional acting this way, with capabilities appropriately developed, has informed freedom; professional ethics programs should be designed to make this end a possibility.

---

[7]Harris identifies sensitivity to risk, awareness of the social context, attunement to nature, and commitment to the public good as dispositions important for engineers.

Conclusion

I have sought to show how the concept of identity, understood as the iterative process of planning and action that forms a teleological quest, can orient professional ethics instruction. The concept is robust enough to support the diverse important elements that constitute professional ethics. One, the foundation allows propositional and practical knowledge associated with the codes of ethics, laws, and accepted practices of a profession to be integrated into an individual's identity, instead of always remaining as external dictates. Two, it captures the teleological aspect of professions, their *raison d'être*, and connects it to the acts of the professional. Three, it encompasses aspects of the moral act that have been shown by social psychologists to be determinant in professional ethical behavior: professional judgment, which is reflective and requires an understanding of one's fields of virtue as well as their limitations; commitment, the firmness of will that is articulated, as it were, in word and act, that can be developed through practices; and a moral imagination that enables creative solutions to ethical problems. Taken together, identity incorporates socially and professionally inherited values while providing some distance from these norms. This distance respects the freedom that individuals in liberal democracies should expect, so that they are more than just moral automatons, and permits a critical view and reform of these norms. Through self-determination of practices and virtues that are developed over a lifetime, the individual crafts his or her own end, cognizant of the moral space in which he or she operates. This crafting creates a more durable ethical position.

Professional ethics programs interested in identity formation should target a number of pedagogical objectives. The first set of two objectives is *foundational*. One, students should know the frameworks of their profession, which includes relevant codes of ethics, laws, and practices, along with formative cases and people who have shaped it. These frameworks are a readily available map for students to use as they navigate their professional journeys. Two, students should have an introductory understanding of the characteristic intellectual and moral virtues associated with the practices of their profession. At the least, individuals should know the minimal set of virtues necessary for competence; ideally, they will know what is necessary for professional excellence. As part of this process, individuals should be given some direction to consider how professional friendships, especially mentoring relationships, can contribute to achieving this objective.

The second set of objectives are called integrative because they require an individual to critically incorporate the professional framework, developed by the foundational objectives, into the professional's comprehensive identity. The first objective in this set is to have students compare the various fields of virtues that they possess to what is needed for various levels and kinds of success in their professional domains to identify strengths and shortcomings. The second is to consider how the virtues they plan to develop for certain ethical fields and the characteristic virtues in each field will shape them, to ensure that these efforts align with the identity they desire. The third objective is to have students articulate their intended autobiography, their unique combination of virtues and the practices to which they will be applied, that explains how professional principles and actions contribute to their goals as well as to the welfare of the profession and society. The fourth is to have students develop a sense of self that is informed by but retains critical distance from generally accepted norms, which can be promoted by asking students to determine ethical boundaries that, if pressured to cross, would force them to speak up or even leave their organization. This last objective can be put another way, namely, to determine how certain actions would stress, or perhaps fracture, the coherence of the individual's sense of self.

The two sets of objectives are measured in different ways in classroom settings. The propositional and practical knowledge involved in the fundamental objectives can be tested using multiple choice and short answer questions. Using such modes of evaluation is at best a challenge, though, when one thinks about how ethical issues arise for professionals: they need to have an internalized understanding of the whole of the ethical fields in which they operate, especially since they rarely reference the codes themselves, and ethical decisions are rarely are binary or multiple choice. These characteristics suggest that the integrative objectives should be assessed by essays, presentations, and discussion, in which students are encouraged to translate information about their professions, including its *telos*, into the "first person," by comparing it with their goals and ethical principles, to find points of coherence and contradiction. Imagination and self-reflection are the key elements of these works.

Finally, professional ethics educators should consider the role of the modern research university, especially those that are public, as an important early step in the ethical development of many professionals. In its present form, the university combines research and teaching to develop an educated populace able to handle the responsibilities of democratic rule.[8] The university provides the setting for moral development with limited long-term economic consequences, the upside to the "academic exercises" pejorative. Educators should ask themselves whether they have given students not just the propositional and practical knowledge they need to understand social and professional expectations but also the intellectual space that is necessary for one to integrate personal and professional responsibilities and opportunities in his or her own way. Have their students considered the practices of their professions and the virtues they have and want to develop over the course of their lives? Put another way, the success of professional ethics educators can be measured by the degree to which their students are better able to craft coherent identities, cognizant of weaknesses and open to refinement, that can withstand undesirable external forces so that the individual is able to conscientiously chart his or her future in the workplace and in the community. Ideally, students are prepared to be active participants in the profession, aware that they are being shaped by it and

[8]For more on objectives of the modern university, fashioned and defended by Wilhelm von Humboldt, which unites teaching and research for citizens' development and equality, see Fuller (2009).

intentionally shaping it. Such an expectation goes far beyond the understanding of ethics as adherence to business rules or preventive or prohibitive injunctions, but this additional step promotes the conscientious maturation that enables one to exercise his or her agency oriented by the good of the individual, the profession, and society.

References

Anscombe, G. E. M. (2005). Does Oxford moral philosophy corrupt the youth? In M. Geach and L. Gormally (Eds.), *Human life, action, and ethics: essays by G.E.M. Anscombe*, (pp. 161–167). Charlottesville, VA: Imprint Academic.

Arendt, H. (1963). Eichmann in Jerusalem. *New Yorker*. February 16.

Aristotle. (2002). *Nicomachean ethics* (J. Sachs Trans.) Newburyport, MA: Focus Publishing.

Beauchamp, T. L., & Childress, J. F. (2009). *Principles of biomedical ethics* (6th ed.). New York, NY: Oxford University Press.

Bebeau, M. J. (2014). An evidence-based guide for ethics and instruction. *Journal of Microbiology & Biology Education, 15*(12), 124–129. https://doi.org/10.1128/jmbe.v15i2.872.

Blasi, A. (1980). Bridging moral cognition and moral action: A critical review of the literature. *Psychological Bulletin, 88*(1), 1–45.

Davis, M. (2012). Ain't no one here but us social forces. *Science and Engineering Ethics, 18,* 13–34. https://doi.org/10.1007/s11948-010-9225-3.

Foot, P. (1978). The problem of abortion and the doctrine of double effect. In *Virtues and vices and other essays in moral philosophy* (pp. 19–32). Berkeley: University of California Press.

Fuller, S. (2009). *The sociology of intellectual life*. London: Sage.

Gilson, E. (1994). *The christian philosophy of St. Thomas Aquinas*. Notre Dame, IN: University of Notre Dame Press.

Harris, C. E., Jr. (2008). The good engineer: Giving virtue its due in engineering ethics. *Science and Engineering Ethics, 14,* 153–164. https://doi.org/10.1007/s11948-008-9068-3.

Hurka, T. (2006). Virtuous dispositions, virtuous acts. *Analysis, 66*(1), 69–76. https://doi.org/10.1111/j.1467-8284.2006.00591.x.

Kohlberg, L. (1969). Stage and sequence: The cognitive developmental approach to socialization. In D. A. Goslin (Ed.), *Handbook of socialization theory and research* (pp. 347–480). Chicago, IL: Rand McNally.

MacIntyre, A. (1984). *After virtue: A study in moral theory* (2nd ed.). Notre Dame, IN: University of Notre Dame Press.

Mitcham, C. (2009). A philosophical weakness of engineering. *The Monist, 92*(3), 339–356.

Schmidt, J. A. (2014). Changing the paradigm for engineering ethics. *Science and Engineering Ethics, 20,* 985–1010. https://doi.org/10.1007/s11948-013-9491-y.

Taylor, C. (1989). *Sources of the self: The making of the modern identity*. Cambridge, MA: Harvard University Press.

Thoma, S. J., & Bebeau, M. J. (2013). Moral Motivation and the Four Component Model. In K. Heinrichs, F. Oser, & T. Lovat (Eds.), *Handbook of moral motivation: Theories, models, applications* (pp. 49–68). Rotterdam: The Netherlands, Sense Publishers.

The Role of Teaching Ethics *in Teaching Ethics Across the Curriculum*

Alan Tomhave and Mark Vopat

Abstract In this article we, the current editors of *Teaching Ethics*, give a basic overview of the goals, purpose, and contribution of the journal *Teaching Ethics*. This proceeds in three sections. First, we look at the stated purpose of the journal and the *Society for Ethics Across the Curriculum*. Second, we briefly discuss six areas of clear benefit offered by *Teaching Ethics*. Finally, we briefly discuss our view of the journal and the importance of the diversity of disciplines that are represented in the journal.

Keywords Diversity of disciplines · Society for ethics across the curriculum
Ethics pedagogy · Mission · Invitational · Submissions

Introduction

Teaching Ethics is the journal published by the *Society for Ethics Across the Curriculum* (SEAC). SEAC was started in 1992 and guided by the following: "The purpose of the Society for Ethics Across the Curriculum is to stimulate scholarship on ethics and the teaching of ethics in all academic disciplines and to afford an opportunity for the exchange of research."[1] In 2001 SEAC started the journal *Teaching Ethics* as an outlet for stimulating scholarship and exchanging research on ethics and the teaching of ethics. In what follows we, the current editors of *Teaching Ethics*, will do three things. First, we will look at and discuss the goals and mission of *Teaching Ethics*. Second, we will look briefly at six areas where

[1]This comes from the website of the *Society for Ethics Across the Curriculum*, http://www.rit.edu/~w-ethics/seac/, last accessed June 14, 2017.

A. Tomhave (✉) · M. Vopat
Youngstown State University, Youngstown, OH, USA
e-mail: aetomhave@ysu.edu

M. Vopat
e-mail: mvopat@ysu.edu

E. E. Englehardt and M. S. Pritchard (eds.), *Ethics Across the Curriculum—Pedagogical Perspectives*, https://doi.org/10.1007/978-3-319-78939-2_7

articles in *Teaching Ethics* have and continue to make contributions to ethics pedagogy. Finally, we will briefly look at the future of the journal. Our goals in writing this chapter are two-fold. First, we wish to discuss the important role of the journal in ethics pedagogy. Second, we hope that the chapter is largely invitational. That is, we hope to make it clear that we take an approach that is welcoming to submissions in all aspects of teaching ethics from all disciplines.

The journal *Teaching Ethics* is guided by the following:

> ***Teaching Ethics*** is dedicated to ethical issues across the curriculum with particular attention to pedagogical methodology and practice in both academic inquiry and professional practice. The journal's editorial focus is on ethics as a dimension of all academic inquiry rather than as an isolated philosophical discipline. Its primary mission is to provide a peer-reviewed forum for academic dialogue in ethics instruction across disciplines such as business, medicine, trades, technology, law, and other areas of liberal education.[2]

For those who have never been to a SEAC conference, one of the most noticeable differences between a SEAC conference and other conferences in Philosophy is the higher presence of multiple disciplines as both attendees and presenters. Many of these presentations are eventually submitted for publication in *Teaching Ethics*. This has translated into a journal that is diverse in perspective and welcomes contributions from any discipline. It is this multidisciplinary diversity that makes *Teaching Ethics* an ideal clearinghouse and resource for those teaching ethics and for those working to institute an ethics across the curriculum program.

We would like to highlight one aspect of the mission of the journal. As seen above, "The journal's editorial focus is on ethics as a dimension of all academic inquiry rather than as an isolated philosophical discipline." One of the striking features of studying ethics is that it is applicable to all areas of life in a way that most other disciplines are not. All actions are evaluable for ethical questions. This means that all disciplines have a role to play in ethics education. The best way to ensure this is to simply make ethics "a dimension of all academic inquiry." The journal contributes to the research and inquiry for both academic purposes and professional ethical practice. It can only do this with a respect for a multidisciplinary approach to ethics. Hence, we hope it is clear that the focus of the journal is not of Ethics as a stand-alone discipline, but of "ethics as a dimension of all academic inquiry." We will now highlight six areas where *Teaching Ethics* has made, and continues to make, contributions to the pedagogy of Ethics.

---

[2]Though this can be found in numerous places, including every issue of the journal, a natural place is the website for the current publisher of the journal, the Philosophy Documentation Center, https://www.pdcnet.org/tej, last accessed June 14, 2017.

Resource Areas in Teaching Ethics

In every issue of *Teaching Ethics* readers will find articles on a diverse set of topics. Some articles may be directly about specific moral issues. Other articles may consider questions of the best pedagogical strategies to teach certain topics. Still others might discuss attempts to integrate Ethics education into curriculums, both limited in scope and reaching across curriculums. Here we will highlight six areas that we think are both of special interest to readers, and ones that can be found in just about any issue one picks up.[3]

Role of Theory in Ethics Education

One of the ongoing dates in the teaching of ethics, and in particular, teaching ethics across the curriculum is the role ethical theory ought to play in ethics education. Should students be well versed in ethical theory before examining the issues found in their discipline? Is a grounding in ethical theory necessary for instilling an appreciation of the ethical issues found in things like business or engineering? Is ethical theory really necessary for ethical decision making? While it might be assumed that it is the professionally trained philosophers who advocate for a theory based approach, there is in fact wide disagreement by both ethicists and non-ethicists alike. Some non-philosophers engaging in ethical instruction argue for the inclusion of theory, while philosophers who have spent their careers working on theory argue the contrary. *Teaching Ethics* has provided, and continues to provide, a forum for debating the merits of both approaches.[4]

Bridging the Gap Between Professional Ethicists and Professionals in Other Disciplines

Another important function of the journal is to act as a bridge between professional ethicists and professionals from other disciplines. In some respects this issue is a

[3]This list is by no means exhaustive. Many of these issues may already be discussed by other chapters in this volume.

[4]This debate has included the following articles in *Teaching Ethics*: Henderson (2002), Harris (2009a, b, 2011), Davis (2009, 2010) and Gert (2010). This debate has also spread beyond *Teaching Ethics*, with the former editors of *Teaching Ethics* publishing a contribution to the discussion in *International Journal of Applied Philosophy*: Englehardt and Pritchard (2013). The debate has continued in many other journals as well. Further, the discussion continues with recent scholarly work, including a paper at the annual meeting of the Association for Practical and Professional Ethics (APPE) in Dallas in February, 2017, with a presentation by Valerye Milleson, titled "On the Obligation to Use Moral Theory in Teaching Professional Ethics.".

subset of the theory versus non-theory debate to teaching ethics. Some professional ethicists hold that the teaching of ethics requires a firm foundation in ethical theory. While it may be viewed as a form of professional snobbery, it is a snobbery that has been implicitly fed by non-professionals. Many disciplines implicitly or explicitly support the idea that ethics should be left to the philosophers, or worse, that ethics isn't particularly relevant to a certain discipline. In the case of the latter view, there are those that hold that ethics really amounts to a kind of legalistic following of a code of ethics. It is within this space that *Teaching Ethics* attempts to bridge this divide by demonstrating how professional ethicists and professionals in other disciplines can learn from one another. The non-professional ethicist can benefit by seeing how ethicists elucidate the ethical complexities that are inherent in many disciplines and professions—complexities that are not always captured by professional codes of ethics. On the other hand, professional ethicists can gain a better understanding of the real-world complexities and issues that may not be apparent to those looking at a profession or discipline from the outside. Thinking of "ethics as a dimension of all academic inquiry" requires an awareness of the nuance and complexity of both ethics and the issues at play in professional practice. The full understanding that is sought is only possible by bridging the gap between ethicists and other professionals. *Teaching Ethics* helps bridge the gap, playing an important role in bringing professionals of all kinds together.

Specific Topics Within Applied and Professional Ethics

Ethics, particularly applied ethics, is a constantly changing area of inquiry. Just as the bridging of the gap that was just discussed is a subset of the theory versus non-theory debate, these specific topics are partially an offshoot of the bridging of the gap between professional ethicists and professionals in other disciplines. New issues in applied and professional ethics emerge as new industries and technologies are developed. In other cases, new issues develop as society's normative ethical views evolve. What once may have been considered an acceptable business practice may no longer be viewed as such in light of a new understanding of normative concepts like social responsibility, rights, or equality. Similarly, our ethical understanding or moral development does not always keep pace with the exponential rate of innovation in the hard sciences, engineering, and computer science. Artificial intelligence, the automation of industry and the military, big data, and data mining all raise questions of distributive justice, privacy, and moral responsibility. This in turn raises issues in ethics instruction: what do we teach, and how do we teach it, become even more pressing questions. Keeping abreast of the various topics with applied and professional ethics is an important purpose of the journal.

These first three resource areas have proceeded from the theory versus non-theory debate to the specific topics. While these resources have been treated as distinct areas of focus of the journal, such a distinction is artificial. It should be clear that these first three areas are not entirely separate from each other. To help illustrate this, let us consider them briefly in reverse. As new issues arise that are in need of ethical study, either due to changing views or emerging technologies, or some other source, non-ethicists and ethicists come together to consider the different issues. Thus, the new issues also help bridge the gap between ethicists and non-ethicists. This, in turn, naturally raises the question of whether or not theory is something that needs to be taught in ethics courses. Thus, though these three areas can be discussed separately for our purposes here, they are really tightly connected. Let us now turn to our final three areas of clear benefit from *Teaching Ethics*.

Promoting and Establishing Ethics Across the Curriculum Programs

The journal's origin as a compliment or extension of the purposes of SEAC obviously entails that Ethics across the curriculum (EAC) is of interest to the journal and its readers. Hence, one of the issues that is frequently mentioned in *Teaching Ethics* is curriculum matters on a programmatic level. These articles come in a variety of forms. Some look at starting points and make initial suggestions. Others are retrospective in nature and look at the results of attempted EAC programs. Whichever topics specific EAC articles cover, it will be of interest to those faculty who are interested in starting or continuing and improving an EAC program at their home institution.[5]

Additionally, of special interest to many of us are the number of articles that deal with learning outcomes and assessment of EAC programs and Ethics in general. In the current higher education climate, where assessment is more important than ever, it is helpful for those of us who are not well versed in the kind of assessment that most of our institutions require, to have a place to start thinking about assessing our programs and classes. Assessment is necessary to both justify continued funding in institutions of higher learning and to prove to accrediting bodies that we are doing what we say we are doing. More importantly to many of us though, we care about ethics. A major question that many in ethics education (indeed, most educators everywhere) have is simply, am I doing a good job of teaching what I say I am teaching? Thus, *Teaching Ethics* plays an important role in helping ethics professionals engage in the assessment of their classes and programs. These assessments are vital to improving the quality of student learning, communicating with upper level administrators about programs, and meeting university accreditation requirements.[6]

[5]For example, see the following: Rhodes (2003), Davis (2006), Sia (2008) and Cruz-Cruz et al. (2010).

[6]For a selection of papers dealing with assessment, consider the following: Ozar (2001), Newton (2001), Keefer and Davis (2012) and Davis (2016).

Resources for Teaching Ethics: Book and Textbook Reviews

The journal plays an important role as a collection of reviews regarding books having to do with teaching ethics. These books might be textbooks for various ethics courses, both applied, theoretical, and professional. They might also be stand-alone books that are appropriate for ethics courses because of the subject matter of the book. Furthermore, these reviews, whether they are dealing with a standard book or a textbook, are critical reviews. In the case of textbooks, the book under review is often directly compared to other new or “classic” texts on the same topic. This approach to reviews is particularly helpful to the first time instructor of a particular course, who may have been drafted to teach a course in an area outside of their particular area of expertise. Further, anyone who has ever switched from one textbook to another knows the volume of options is staggering. Taking the time to review each of them is sometimes not possible. One of our goals as the current editors of the journal is to establish a reputation for being the first place that instructors look when thinking about what books they might want to use when teaching Ethics.

Teaching Ethics

Although the stated purpose of the journal is focused on ethical issues across the curriculum, it does at times address ethical issue about teaching, though not necessarily teaching ethics. There are ethical issues related to teaching that may not fall within the usual questions of theory or applied areas of ethics that students will encounter when, for example, voting or later in their careers. These are ethical issues that arise when teaching. The diversity of the authors represented in *Teaching Ethics* helps to raise issues that many would not think about, but that are important to consider. Just a quick perusing of the topics of the journal will lead to the finding of articles dealing with issues from the ethics of trigger warnings to questions about cheating and student perceptions of cheating. These types of articles also have the advantage of reinforcing the idea that ethics is not something you leave in the ethics classroom.[7]

[7]It is impossible to give an accurate picture of the topics that have been covered in articles published by *Teaching Ethics*. Hence, here we provide references only for the two articles mentioned in this paragraph. Wyatt (2016) and Shrader et al. (2013).

The Future of the Journal and Approach of the Editors

As mentioned above, one of the striking features of a SEAC conference is the diverse set of perspectives that are present at the conference. While it does draw its share of philosophers, it also brings together lawyers, medical professionals, engineers, and many others working in university level or pre-college teaching positions.

The diversity that exists at SEAC conferences is present in the authors of the articles published in *Teaching Ethics*. As highlighted above, the journal publishes articles on integrating Ethics into a variety of disciplines, including, but not limited to, Economics, Foreign Languages, Graphic Design, the sciences, Business, Engineering, Exercise Science, and Criminal Justice. It is this diverse set of authors and articles that allows the journal to play the role of bridging the gap between professional ethicists and other professionals. Everyone is actually welcome, both at SEAC conferences and as authors in *Teaching Ethics*.

This diversity serves three very important roles in education. The first is simply that this diversity allows for a rich exchange of both ideas and examples. Ethicists know the ethics part of the equation and other professionals know the examples and issues that are most pressing in their academic specialties. When dealing with professional ethics, both aspects are necessary.

Second, ethics education is a standard requirement for accreditation in many disciplines (e.g., the accreditation standards for both the Association to Advance Collegiate Schools of Business and the Accreditation Board for Engineering and Technology include ethics as part of the curriculum). The interaction between ethicists and other professionals provides those that do not work in ethics as a major part of their job with resources to further ethics education in their programs. Thus, the diversity at SEAC conferences and within *Teaching Ethics* serves the very practical role of helping with the accreditation of programs and improving ethics education.

The third and final role involves the actual pedagogy of ethics. Ethics classes alone are not enough to get students to begin to take ethics seriously and to work it into their thinking when it comes to professional settings. It is not enough for the philosophers teaching business ethics to tell students that they must think about the impact to communities and workers when laying off workers. In order to get students to think about these issues when making decisions as future business leaders the same message must be given by their professors in Business. The same goes for the future engineers who take engineering ethics. It is not enough for the ethicist to tell students that the best design is one that prevents user errors. Students must hear this message from engineering faculty. It is the role of *Teaching Ethics* to assist instructors in the best, most effective ways to present this material. Especially important is the role the journal plays in assisting non-professional ethicists to effectively frame and present the ethical issues of their disciplines.

It is this diverse approach to ethics instruction that makes the journal *Teaching Ethics* unique when compared to most Philosophy journals. In order for the journal to continue serving its role of bringing together all aspects of ethics education, wherever ethics comes up—which is everywhere—the journal must continue to publish articles from a variety of perspectives and disciplines. Therefore, we, as the current editors of the journal, would like to extend an invitation to submit articles on anything having to do with ethics education to *Teaching Ethics*. We are honored to be the current editors and look forward to reading the journal for many years to come.

References

Cruz-Cruz, J., Curbelo, A., & Frey, W. (2010). Doing ethics across the curriculum: The EAC toolkit. *Teaching Ethics, 10*(2), 47–69.

Davis, M. (2006). IIT's workshops for integrating ethics into technical courses: Some lessons learned. *Teaching Ethics, 6*(2), 29–42.

Davis, M. (2009). The usefulness of moral theory in practical ethics: A question of comparative cost. *Teaching Ethics, 10*(1), 69–78.

Davis, M. (2010). The usefulness of moral theory in teaching practical ethics: A reply to Gert and Harris. *Teaching Ethics, 12*(1), 51–60.

Davis, M. (2016). From practice to research: A plan for cross-course assessment of instruction in professional ethics. *Teaching Ethics, 16*(2), 251–260.

Englehardt, E., & Pritchard, M. (2013). Teaching practical ethics. *International Journal of Applied Philosophy, 27,* 161–173.

Gert, B. (2010). The usefulness of a comprehensive systematic moral theory. *Teaching Ethics, 12* (1), 25–38.

Harris, C. E. (2009a). Is moral theory useful in practical ethics? *Teaching Ethics, 10*(1), 51–68.

Harris, C. E. (2009b). Response to Michael Davis: The cost is minimal and worth it. *Teaching Ethics, 10*(1), 79–86.

Harris, C. E. (2011). A reply to Bernard Gert. *Teaching Ethics, 12*(1), 39–50.

Henderson, B. (2002). A reminder on recognizing ethical problems are practical: Distinctions in teaching theory and practice. *Teaching Ethics, 2*(2), 1–18.

Keefer, M., & Davis, M. (2012). Curricular design and assessment in professional ethics education: Some practical advice. *Teaching Ethics, 13*(1), 81–90.

Newton, L. (2001). Outcomes assessment of an ethics program: Purposes and challenges. *Teaching Ethics, 2*(1), 29–67.

Ozar, D. (2001). Learning outcomes for ethics across the curriculum programs. *Teaching Ethics, 2* (1), 1–27.

Rhodes, B. (2003). Ethics across the curriculum and the nature of morality: Some fundamental propositions. *Teaching Ethics, 3*(2), 59–65.

Shrader, C., Revenscroft, S., Kaufman, J., & West, T. (2013). Classroom cheating and student perceptions of ethical climate. *Teaching Ethics, 13*(1), 105–128.

Sia, S. (2008). Ethics across the curriculum: Some observations. *Teaching Ethics, 9*(1), 5–11.

Wyatt, W. (2016). The ethics of trigger warnings. *Teaching Ethics, 16*(1), 17–35.

Part II
Teaching Challenges

Teaching Practical Ethics

Elaine E. Englehardt and Michael S. Pritchard

Abstract For several decades, we have had an interest in introducing ethics across the curriculum at our universities, especially in areas that help students prepare for their working careers. This has involved encouraging faculty from a variety of disciplines other than philosophy to reflect with their students on the ethical problems one is most likely to face in professional and working life. Many students are already working and are anxious to discuss the ethical difficulties that occur daily in their work places. Without a strong background in philosophical ethics is it appropriate for faculty in these disciplines to take on the educational task of trying to help their students anticipate and think through possible solutions to the ethical problems they are likely to encounter? Our answer is, yes; and, as we will maintain below, we believe Thomas Reid would give the same response.

Keywords Ethical problems · Philosophical background · Thomas Reid
Curriculum · System of morals · Grand theories · Euthyphro · Perplexing problems

Introduction

A marked change in the teaching of ethics in higher education over, roughly, the last half-century is the rise of serious interest in ethics across virtually the entire curriculum. In turn, this has given rise to questions about what the aims and goals of

---

Much of this article is found in "Teaching Practical Ethics," *The International Journal of Applied Philosophy,* December 11, 2013. We thank the journal and its editor Elliott Cohen for granting permission for us to include these materials.

E. E. Englehardt (✉)
Utah Valley University (UVU), Orem, UT, USA
e-mail: Elainee@uvu.edu

M. S. Pritchard
Western Michigan University, Kalamazoo, MI, USA

E. E. Englehardt and M. S. Pritchard (eds.), *Ethics Across the Curriculum—Pedagogical Perspectives*, https://doi.org/10.1007/978-3-319-78939-2_8

such teaching should be, in what academic programs this teaching should take place, what the specific content of such programs should be, and what qualifications are needed to teach ethics in this or that area.[1] These questions have been particularly vexing for those accustomed to thinking that the academic discipline of philosophy should play a special role in the teaching of ethics in higher education. A central concern has been to determine what sorts of contributions philosophy can, or should, make to the clearly practical concerns in academic areas such as business, engineering, law, medicine, social work, and other professional programs in higher education.

One view is that academic courses or educational units in ethics should somehow address the central philosophical concerns about ethics. A worry this view prompts is that even an entire course in, say, engineering ethics cannot afford the time required adequately to present a theory of ethics like utilitarianism or Kantian ethics and still do justice to the sorts of ethical questions engineering students need to prepare themselves to address. Another worry is that, even if more efficient ways of presenting these theories were employed, they simply would not do the sort of work one might hope for—viz., help students in engineering better understand the practical ethical problems engineers face, thus increasing the chances that they will, as engineers, make better ethical decisions than they would without being introduced to these theories. A concern is that, in order both to take into account the students' lack of background in philosophical ethics and their need to understand ethical problems that arise in engineering practice, the favored ethical theories would need to be "watered down," over-simplified and, in short, distorted. So, even if the theories were then applied to ethical problems in engineering, the applications would not yield conclusions in good keeping with either the theories or the practical problems themselves.

Help from the Past

In addressing these pedagogical concerns, we will begin with some reflections on the views of 18th century Scottish philosopher Thomas Reid (1710–1796). Reid spent much of his philosophical life in Glasgow, where he succeeded Adam Smith as holder of the chair of Moral Philosophy. As such, he taught a great variety of courses, including courses on practical ethics and on what he called "the theory of morals". Although Reid devoted much of his philosophical energy to "the theory of morals" (which he took to be concerned with the powers of the mind that enable us

[1]For an early discussion of these matters, see Daniel Callahan and Sissela Bok, eds., *Ethics Teaching in Higher Education* (New York: Plenum, 1980), as well as the many monographs in specific areas of the curriculum. For more current discussions, see issues of *Teaching Ethics*, the official journal of the Society for Ethics Across the Curriculum. (See, especially, the recent exchanges by C. E. Harris, Michael Davis, and Bernard Gert beginning in the Fall 2009 issue and continuing in the Fall 2011 issue.)

to be moral agents at all and, thereby, to take morality seriously in our daily lives), he did not think that even the best theory of morals would be of much, if any, value in improving our moral judgment. Just as a theory of vision does not, by itself, improve our vision, Reid held that a theory of morals, does not, by itself, improve our moral discernment.[2] For such improvement, Reid thought, we should turn to the domain of practical ethics, rather than a "theory of morals". It is this aspect of Reid's thought that we will explore here.

When we turn to practical ethics, says Reid, we will find "systems of morals," ways of organizing our moral thinking that can aid us in recognizing our duties, their relationships to one another, and their connections with the various moral virtues. Although, for Reid, a system of morals can help us organize our moral thinking, he holds that such a system is more like a system of botany or mineralogy than a system of geometry (AP, 281). In a moment we will discuss briefly what he likely has in mind in drawing this contrast. But, here what we want to emphasize is that this exploration of Reid is not primarily an historical endeavour to understand Reid in his day. We are discussing Reid because we believe that his views can provide important insights into how, even today, we might effectively approach teaching ethics in various professional disciplines, practices, vocations and research areas.

Reid's *Active Powers of the Mind* (AP, 2010) and his recently published lecture notes, *Practical Ethics* (PE, 2007), are filled with ideas that illuminate our need to examine our moral practices, often with the aid of available systems of morals, which are readily understandable by philosophers and non-philosophers alike.[3] He insists that it is a serious mistake to think that "in order to understand his duty, a man must needs be a philosopher and a metaphysician" (AP, 283). Basic moral notions should be within the grasp of anyone willing to spend the time and energy to focus clearly on the practical issues they raise. In his *Practical Ethics,* Reid says:

> The practical Part of Ethicks is for the most part easy and level to all capacities. There is hardly any moral Duty which when properly explained and delineated does not recommend itself to the heart of every candid and unbiased man. For every Man has within him a touchstone of Morals, the dictates of his own Conscience which approves of what is Right and condemns what is wrong, when it is fairly represented and considered without prejudice. (PE, 10–11)

---

[2]This seems to overlook the possibility that a theory of vision might help us improve our vision by providing insights into how to develop visual aids such as eyeglasses, artificial lens, microscopes, and telescopes, as well as surgical procedures that improve vision. Analogously, recent explorations in psychology, sociology, and neuroscience might help us both discover shortcomings in some of our basic ways of addressing moral matters and to develop effective ways of dealing with these shortcomings. Just as we have had to learn to recognize and deal with visual "blind spots" in driving, we are now trying to learn how to recognize and deal with moral "blind spots." On the latter, see, for example: Bazerman and Tenbrunsel (2011) and Werhane et al. (2013).

[3]Reid (2010). This will be referred to in the text as 'AP'. Reid (2007). This is referred to in the text as 'PE'.

Reid's expression, 'systems of morals,' indicates that he thinks that there is more than one moral system available to us. But he holds that the same basic principles can be expected to run through all the systems. Reid compares these principles to the laws of motion–few and simple, but they regulate everywhere. It would seem that everyday experience is a key to being able to navigate well morally in this world. However, unlike the laws of motion, the limits of systems of morals are not "fixed by nature, but by the wide circle of human transactions" (AP, 281).

This "wide circle of human transactions" is navigated, not just (or even primarily) by philosophers and metaphysicians, but by people from all walks of life who, nevertheless, are quite capable of understanding the moral complexities of the circumstances in which they find themselves. Whether this capability is well developed in actual practice is another matter. Here Reid is more realistic than optimistic—although he would place some hope in various forms of educational practice.

For several decades, we have had an interest in introducing ethics across the curriculum at our universities, especially in areas that help students prepare for their working careers. This has involved encouraging faculty from a variety of disciplines other than philosophy to reflect with their students on the ethical problems one is most likely to face in professional and working life. Many students are already working and are anxious to discuss the ethical difficulties that occur daily in their work places. Without a strong background in philosophical ethics is it appropriate for faculty in these disciplines to take on the educational task of trying to help their students anticipate and think through possible solutions to the ethical problems they are likely to encounter? Our answer is, yes; and, as we will maintain below, we believe Reid would give the same response.

As we have said, for Reid, "the theory of morals" is concerned with clarifying our understanding of the active powers of mind that enable us to be moral agents. He distinguishes this from practical ethics, which focuses on useful ways of organizing our understanding of how we ought to conduct our lives. Reid regards both types of inquiry to require careful, rigorous thinking. The first, moral theory, he says, is as "difficult and perplexing" as any area of philosophy. However, practical ethics, is much more readily within our reach and, he adds, is "in most cases very plain," even though there are some "intricate and perplexed cases" (PE, 11).

Many might contend that, whatever the terrain of practical ethics may have been like in Reid's time and place, there is no shortage of "intricate and perplexed" cases that require our attention in today's professions and practices. Even so, "moral systems" remind us of matters that are rather clear and straightforward, as well as help us better frame our understanding of those issues that are more "intricate and perplexed." Although organizing our moral understanding in this way requires careful, rigorous thinking, Reid is convinced that this can be undertaken by philosophers and non-philosophers alike.

Reid's comparison of moral systems with those of botany and mineralogy, rather than with a more tightly constructed system of geometry, suggests a somewhat open-ended approach to practical ethics. In areas such as business, engineering, law, and medicine there is much to discover and clarify, just as there is in botany and mineralogy. Although we might aspire to expand our systematic understanding, progress is piecemeal; and being confident that one has had an important insight or discovery in one area need not wait on our being able to work out its possible relationships to everything else that might eventually fit into a system.

It is interesting that Reid uses botany and mineralogy as models for systems of morals, rather than geometry. Botany and mineralogy in Reid's day were emerging sciences. The taxonomic schemes that were being developed were wedded to human use—medicine, herbs, and foods were the primary focus in botany. Human uses of minerals were the primary focus in mineralogy. Furthermore, both kinds of system were incomplete—open to new entries, and even to modifications in classification. The reach and usefulness of these schemes was very much dependent on what might be discovered by empirical means. The way in which Reid depicts a geometric system is strikingly different. Ideally, it is a closed system, with its basic parts set, both in themselves and in relation to one another.[4] Discoveries come through a priori assumptions and deduction rather than empirical searching, sifting, and winnowing.

So, moral systems for Reid leave room for challenge and revision—though, likely, not total displacement. But, even at their best, they will leave some answers incomplete, even to the point of not being very helpful. Still frustration or dissatisfaction in one area need not infect other areas. Also, doing well in one area may not help much in other areas.

In contrast to Reid's approach, consider Plato's early dialogue, *The Euthyphro*. In response to Euthyphro's confident assertion that it is right for him to prosecute his own father for murder, Socrates challenges Euthyphro to explain what it is that makes all right acts right (Plato 1975, 8). He rejects Euthyphro's initial appeal to particular examples of rightness; seemingly, he thinks that we need to know what all right acts have in common before concluding that this or that act (or even this or that kind of act) is right. How do we go from 'some' to 'all' if we cannot explain why the 'some' we are offering qualify as being of the right sort or part of the 'all' for which we are searching? However, Reid would likely object that there may be no one thing that makes all right acts right—but this should not prevent us from seeing that some sorts of actions are right, or wrong. If there is one thing that all right acts have in common that makes them right, we may some day discover this, Reid could say. Meanwhile, we can make some headway with less than that.

Although moral systems may require relatively few principles, Reid concedes that they "swell to great magnitude" because the applications of these principles "extend to every part of human conduct, in every condition, every relation, and every transaction in life" (AP, 280). What these extensions are likely to involve can be anticipated to some extent, but much needs to rely on practical experience rather than philosophical or metaphysical speculation. This needs to be borne in mind in

[4]This depiction best fits the predominant Euclidean plane geometry of Reid's day. Whether Reid would have depicted the non-Euclidean forms of geometry that he was beginning to explore as closed, deductive systems is less clear.

teaching courses in practical ethics that are designed to help students anticipate the sorts of ethical issues they will face once they are in the workplace.

It should be noted that, although practical ethics should be within the grasp of all without requiring us to become philosophers or metaphysicians, Reid acknowledges that obtaining a firm, clear grasp does pose serious difficulties. After asserting that there is "no branch of Science wherein Men would be more harmonious in their opinions than in Morals were they free from all Biass and Prejudice," Reid concedes:

> But this is hardly the case with any Man. Men's private interests, their Passions, and vicious inclinations and habits, do often blind their understandings, and biass their judgments. And as Men are much disposed to take the Rules of Conduct from fashion rather than from the Dictates of reason, so with Regard to Vices which are authorized by fashion the judgments of men are apt to be blinded by the Authority of the Multitude especially when Interest or Appetite leads the same Way. It is therefore of great consequence to those who would judge right in matters relating to their own Conduct or that of others, to have the Rules of Morals fixed and settled in their Minds, before they have occasion to apply them to cases wherein they may be interested. (PE, 2007, 11)

Still, he adds optimistically, if our duty is "properly explained and delineated" to those who are "candid and unbiased," they can be expected to understand and accept its moral force (PE, 11). So, helping students understand these challenges and consider ways of dealing constructively with them in morally acceptable ways would seem to be useful objectives in courses in practical ethics.

It is important to bear in mind that Reid sees the function of fixed and settled "Rules of Morals" in particular circumstances, not as algorithmic tools, but mainly as helping us focus on what we need to take into account, rather than simply allowing our "Biass and Prejudice" to rule the day. Even with the appropriate rules firmly fixed in mind and "Biass and Prejudice" pushed aside, *judgment* is needed. The sort of fixed and settled "Rules of Morals" Reid has in mind can stand up to critical challenge (as distinct from rules based solely on "fashion") as aides to gaining clear, moral focus and in resisting the lure of "Passion," "Appetite," "Interest," and "Fashion," when they tempt us in the direction of "Biass and Prejudice".

> When the principles are applied to these [i.e., the 'wide circle of human transactions'] in detail, the detail is pleasant and profitable. It requires no profound reasoning (excepting perhaps, in a few disputable points). It admits of the most agreeable illustration from examples and authorities; it serves to exercise and thereby to strengthen moral judgment. And one who has given much attention to the duty of man, in all the various relations and circumstances of life, will probably be more enlightened in his own duty, and more able to enlighten others." (PE, 11)

Notice, however, the condition that must be met if there is to be ease of understanding that is "pleasant and profitable"—one must have "given much attention to the duty of man, in all the various relations and circumstances of life". This is hard work, even if one finds it "pleasant and profitable". This hard work can occur in courses in practical and professional ethics—even if only some of the "wide circle of human transactions" is examined.

Back to the Present

Michael Davis's views on teaching courses in practical ethics would seem to resonate well with Reid's approach to practical ethics. In a series of recent articles in the journal *Teaching Ethics,* he exchanges ideas with C. E. Harris about the relevance of traditional moral theory in teaching courses in practical ethics.[5] Davis contends that teaching grand theories, such as utilitarianism or a Kantian theory of respect for persons, are not needed in trying to help students determine morally good courses of action in their professional settings. In fact, they can get in the way. Davis concludes:

> Using moral theory in a course in business or professional ethics is like calculating logarithms from scratch when you have a reliable table available (and are not good at mathematics). You will take more time doing the calculation, have more errors (because of the complexity of the calculations), and (if, but only if, all goes well) end up with much the same result as if you had used the table. (Davis 2011, 56)

For Davis, trying to teach versions of the complicated theories of moral philosophers is problematic. Those with academic background in moral theory may be able to do quite well in teaching those theories. However, he observes, "most moral theorists have, I think, noticed how often those who know something of moral theory but are not expert get a moral theory wrong or, at least, fail to appreciate how problematic certain interpretations of it are" (Davis 2011, 56). He adds that there are no quality controls on how these theories are taught, and this can do harm to the student and the professions.

> There is no evidence that students who take even several courses in moral theory are more likely to act ethically than those who take none. And we should, I think, have substantial evidence that moral theory does benefit students enough in that way before imposing a required course on them for that reason. A requirement should rest on more than a well-meaning belief that the course will do some good. Anyway, the failures of teaching a little moral theory do not force the conclusion that what is needed is more moral theory. There is good reason to think that no amount of teaching moral theory can be justified by the better practical decision-making likely to result. (Davis 2011, 52)

Davis has long argued that students and professionals are best served in practical ethics courses by focusing on a set of practical questions, rather than on the grand theories that predominate in standard courses in philosophical ethics. He offers a list of questions, such as the following, that can be asked when examining cases calling for ethical choices in practical ethics courses. (Davis 2009, 73–74):

- Harm test—does this option do less harm than any alternative?
- Publicity test—would I want my choice of this option published in the newspaper?
- Defensibility test—could I defend my choice of this option before a Congressional committee, a committee of my peers, or my parents?

[5]These discussions can be found in *Teaching Ethics*, 10.1, Fall 2009 and 12.1, Fall 2011.

- Reversibility test—would I still think the choice of this option good if I were one of those adversely affected by it?
- Virtue test—what would I become if I choose this option often?
- Professional test—what might my profession's ethics committee say about this option?
- Colleague test—what do my colleagues say when I describe my problem and suggest this option as my solution?
- Organization test—what does the organization's ethics officer or legal counsel say about this?

Davis says that this list is not the only possible, or even the best, list; but it is one that works well with his students when engaging in moral problem solving.

In support of including some consideration of moral theories in courses and programs in particular professions, Harris (2009a) discusses the usefulness of utilitarian and respect for persons theories in framing ethical issues that arise in engineering practice. Sometimes these theories work in concert in supporting views about what should or should not be done. Sometimes they are in tension, if not outright conflict. Even so, Harris argues, they can help us better understand what is at stake morally. He illustrates this with the following example[6]:

> In 1993, it was publicly revealed that Germany's Heidelberg University conducted automobile crash tests, using more than 200 corpses, including more than eight children. The public controversy that followed included a statement from a spokesman for the Roman Catholic German Bishop's Conference, who argued that "even the dead possess human dignity." On the other side, advocates for the tests argued that relatives of the deceased had given permission and that the test data could result in the saving of many lives.
>
> The public controversy took the form of a contest between those who believed that priority should be given to respecting human dignity (including the derivative dignity that should be ascribed to corpses), and those who believed that the promise of the tests to save lives and thus promote the general good was the most important consideration. Both perspectives are partial and inadequate for appreciating the full dimensions of the case. Thus, in order to appreciate the issue in its full complexity, one must consider both perspectives and take into account the limitations of each.

So, does this example support Harris's view that moral theories have a useful place in practical ethics courses? To see why Harris believes it does, it will be helpful to consider how he begins his essay, "A Reply to Bernard Gert" (2011, 39):

> In teaching ethics, I have found it important to correctly describe a moral problem as it most naturally presents itself to a person in a situation of moral choice. We can call this attempt to correctly describe the structure of a moral problem an attempt to achieve phenomenological accuracy. Enumerating some of the ways in which moral problems can present themselves will provide the context for my understanding of the usefulness of moral theories.

---

[6]This example is discussed in *Engineering Ethics: Concepts and Cases*, an engineering ethics text we co-authored with Harris, Ray James, and the late Michael Rabins (Harris et al. 2014). The next several paragraphs are drawn from our discussion of the debate among Pritchard and Englehardt (2015).

It is Harris's view that the public discussion of the use of cadavers in crash testing vehicles indicated that both utilitarian and respect for persons theories were at least implicitly in play. That is, a phenomenologically accurate account of these responses could appropriately include some reference to these theories.

Harris concludes his reply to Davis's "The Usefulness of Moral Theory in Practical Ethics: A Question of Comparative Cost (A Response to Harris)" with a final word in favor of including some moral theory in the teaching of engineering ethics:

> [I]t is always possible to tell students and practitioners that in their real-world experience there are a few simple guidelines that they should remember. One of these is that they should be aware that, in cases of moral conflict, the issues will often take the form of an opposition between considerations of harm and benefit to the public on the one hand and considerations of the rights of individuals on the other.... These guidelines are as simple as anything Davis presents, perhaps simpler. Students and practitioners will probably not find them unduly technical or forbidding. If they do, they shouldn't". (2009b, 86)

Harris is not urging that a place should be found for a full-scale presentation of utilitarian and Kantian theories. But he does favor making explicit some of the basic ideas in those theories. As Davis himself acknowledges, his own list of questions, in fact, invites students to consider the moral issues in terms that move in the direction of theories familiar to moral philosophers. For Harris, the point of making this explicit to students is not to introduce moral theories for their own sake, but to do this in ways that will enhance their practical reflections about the cases and issues under discussion. Both Harris and Davis rely on common morality and common sense in supporting their approaches to teaching practical ethics.

The straightforwardness of Davis' questions mirrors Reid's reflections on our ability to understand moral problems and responsibilities. Despite our lack of full or precise understanding of the origins and development of our moral capacities (a problem for a "theory of morals" to address), Reid insists that this does not provide an excuse for those who might claim they do not understand duties to others and their related virtues. Once our rational capacities are reasonably well developed, Reid holds, we are capable of understanding the duties associated with the various virtues. However, even among those whose rational capacities are well developed, it is "Men's private Interests, their Passions, and vicious inclinations & habits" that are the primary offenders in provoking the bias and prejudice that so often distort moral judgment (PE, 14).

So, as inexperienced workers extend their "moral systems" into a world of new challenges, there will be moral surprises; but with alertness, seriousness of purpose, and a willingness to employ one's moral faculties thus far developed, Reid is convinced that, for the most part, they should be able to discern what is morally required of them and act accordingly. The academic study of practical ethics cannot be expected fully to prepare students for these challenges, but it can, if done well, help set a promising path.

Aims and Goals of Teaching Ethics in Higher Education

Just as practical ethics and philosophy have been going through an awkward relationship in today's educational institutions, practical ethics' relationship to disciplines other than philosophy has been strained. Noticing the proliferation of new courses in practical ethics in a variety of disciplines (not just Philosophy departments), in the late 1970s the Hastings Center, an ethics "think tank" in the New York area, formed a group of educators from across the curriculum to discuss what they thought the aims and goals of teaching ethics in higher education should be.

To the surprise of many, a consensus emerged among those who met periodically over a two-year stretch of time at the Hastings Center. A series of publications on teaching ethics was published for a wide range of areas (e.g., business, engineering, law, medicine, social work) (Callahan 1980). The five points of agreement for this group were that courses in ethics should:

- Help stimulate students' moral imagination.
- Help students recognize moral issues.
- Help students analyze key moral concepts and principles.
- Help elicit a sense of responsibility in the student.
- Help students deal constructively with moral ambiguity and disagreement.

Incorporating these objectives in courses presumes that students should be regarded as active learners who may already have considerable aptitude for undertaking the study of ethics in areas of life for which they are preparing. They will have had at least some prior, first hand experience dealing with everyday situations calling for moral discernment and judgment, and many will have already spent a fair amount of time working with others, learning about customers, making exchanges for money or service, and the like. However, Reid offers a word of warning:

> The most obvious truths are not perceived without some ripeness of judgment… Judgment even in things self-evident, requires a clear, distinct, and steady conception of the things about which we judge. (2007, 319)

Furthermore, the "conception of things" about which students will to have familiarize themselves includes an understanding of various professions (for example, engineering, law, medicine, social work) and aspects of the world of work they have not yet confronted. It is in these areas of relative inexperience that students will need to be able to discover that which can, on reflection, sometimes be seen as self-evident—and that which, even after careful reflection, cannot.

Common Sense Ethics

Reid and Davis advocate a common sense approach to practical ethics. Davis recommends that students ask themselves a set of questions that will already strike them as somewhat familiar. He explains:

> What makes these tests easier to teach than moral theory is that they are drawn directly from common sense. Students can apply them with reasonable accuracy almost as soon as they have read them because they have in fact already been applying them more or less (though generally using one to make a decision and forgetting the rest). The problem with this method, if it is a problem, is that there is no simple routine for dealing with an option that passes some tests but not others—except to develop a new option that does better.

If we wish to get clearer about matters of ethics in a given area, how should we proceed? Well-developed codes of ethics (as found in, for example, professional engineering societies) can be useful starting points, as these are developed by experienced, thoughtful practitioners. However, it is important to bear in mind that codes of ethics undergo changes. They are best viewed, not as the last word, but as the current word; and their provisions are not self-interpreting, but require good judgment.

Case studies featuring ethical problems are often an effective means for developing strong discussions about ethical questions that can arise in the professions and the work place. Codes of ethics can be included in discussions, but not as something to which one should blindly and uncritically adhere. Teachers can take the lead in developing and discussing cases with their students. These cases can be written and analyzed from their own professional experiences or from experiences others have had in the field in question. The cases can also come from news features, literature, and film. But these are only suggestions. There are many possible approaches, the main requirement of which is that a steady eye should be kept on the practical arena within ethical issues need to be addressed.

Sometimes faculty learn about ethical challenges their students face after they have left the classroom. Imagine a former student, Troy, telling his instructor about an ethical problem at his engineering firm. He is asked to support a bid his company is making for a large contract. Mark, a former classmate of Troy has, is similarly assigned to support his company's competing bid for the same contract. Mark learns of Troy's assignment, contacts him and asks if there are any job openings where Troy works. He tells Troy that he has sent him an e-mail that contains the bid information Mark's firm is submitting. He tells Troy, "You can open it or not; but please realize, I want a job with your firm. I know your firm needs this contract. My firm will do fine without it."

This is an example in which the approach to practical ethics advocated by Reid and Davis–both classroom teachers–could assist Troy in thinking through his "opportunity". A noteworthy feature of Troy's situation is that he might well find himself tempted by Mark's offer. To that extent, we might say, Troy has a motive for looking at what Mark has sent him. However, Reid might add, human beings are capable of assessing and, when appropriate, resisting such motivation. Motives,

Reid holds, might influence our actions, making our decisions easier or more difficult; but they do not necessarily cause them—we moral agents do.

In the case under discussion, let us imagine that Troy exercises his judgment in a way that reflects his personal integrity and commitment to honesty. Though not formally schooled in philosophical ethics, Troy could also explain his decision in terms of the code of ethics endorsed by his employer, an engineering firm that adheres to the responsibilities, rules and ethics needed to be a respectable professional. But even without using the professional code, Troy understands that he would morally object to a competitor using such tactics at the expense of his firm. This, Troy, concludes, would be cheating—as wrong if he resorts to this as it would be for one of his company's competitors. He places his trust the standard bid process, as it is fundamental to his notion of fairness in business. Asking Davis's questions could help Troy clarify his discomfort with Mark's email and help provide him with a thoughtful and nuanced reason for distancing himself from the information Mark is offering.

Although Davis's questions do not provide an algorithmic way of resolving ethical problems in business and the professions, asking this same set of questions, regardless of the business or profession in question, suggests that there are ethical principles and considerations that cut across the moral domain—much as Reid says about "moral systems." However, caution in generalizing from one kind of case to others is needed. There is only one principle that Reid is prepared to say applies everywhere without exception. This is what today is commonly called the "principle of universalizability". Here is how Reid characterizes it:

> In every case, we ought to act that part towards another, which we would judge to be right in him to act toward us, if we were in his circumstances and he in ours; or more generally—What we approve in others, that we ought to practice in like circumstances and what we condemn in others we ought not to do. (AP, 274)

Reid says acceptance of the principle of universalizability is necessary if we are to be moral agents at all. So, Troy must think beyond his particular case to others that are relevantly similar. However, it is not necessary for him to have a comprehensive theory of justice or fairness (let alone of morality in general) in order to make a reliable decision. Uncertainty, or even confusion, in one area does not need to hinder confident, clear-headed thinking in other areas of ethics.

We can understand that basic moral considerations are related to one another, and it may seem that coming up with a unified theory headed by a single, grounding principle would be ideal. But, as Dugald Stewart, one of Reid's students and strong followers warned, philosophers can be misled by "an excessive love of simplicity" and in their quest for one master principle may have succeeded only in sidestepping "the real complication of our active principles" (Stewart 1829, 7). From Reid and Stewart's perspectives, it is quite acceptable to have pieces missing from our moral jigsaw puzzle. In the end, Reid was quite satisfied that practical ethics can do without a single, master principle that provides unity and comprehensiveness to a moral system.

Reid may have underestimated the extent to which moral disagreement can remain even after getting past prejudice, distorting passions, personal ambition, and the like. For example, advances in medical technology have given rise to some perplexing problems in ethics related to the use of sophisticated technology, including life-saving or life-extending devices, justice issues in providing and distributing healthcare, and so on. Philosophers should not necessarily be expected to shed more light on these practical problems than other thoughtful persons. What is needed is good, hard thinking at practical levels accessible to thoughtful people from all walks of life.

Nineteenth century British philosopher, William Whewell, also resisted the lure of seeking one master principle in morality. He was a mineralogist as well as a philosopher, suggesting that he might have agreed with Reid's view that systems of morality are more like systems of botany or mineralogy than geometry. In any case, at the outset of his widely used *Elements of Morality*, Whewell says, "I am desirous that [the reader] should understand that, though I do not speak of my work as a Philosophy of Morality, I have tried to make it a work of rigorous reasoning, and therefore, so far at least, philosophical" (Whewell 1864, vol. I, vii). So, although one does not need to be a philosopher in order to think carefully and well about ethics, it does not follow that one will not have to think philosophically. But, one does not have to be a mathematician in order to think mathematically, either.

Who Can Teach Practical Ethics?

Although we know of no place where Reid explicitly addresses the question of who can, or should, teach practical ethics, we believe that today he would encourage professors in medicine, botany, engineering, and the like to take up the challenge of teaching practical ethics to their students. He would not want professors from professions other than philosophy to be discouraged from this because they lack formal background in philosophical ethics. Students need the assistance of a professor/mentor who can discuss ethical problems they are likely to encounter. Neither professors nor students need to be philosophers to comprehend nuances of these discussions—or their practical importance. Reid rightly points out that the "theory of morals" is a very difficult subject, one that requires a rather high level of philosophical acumen to undertake seriously, and also a subject about which we can expect much disagreement among philosophers. Something similar can be said about the philosophical quest for a system of morals headed by one, comprehensive master principle. However, Reid thinks it is not necessary for professors in medicine, botany, engineering, and the like to undertake these quests. Practical ethics provides plenty of pedagogical challenges of its own. These two quests, Reid, Stewart, Whewell, and Davis would agree, need not be among them.

References

Bazerman, M., & Tenbrunsel, A. (2011). *Blind spots*. Princeton, NJ: Princeton University Press.

Callahan, D. (1980). Goals in the teaching of ethics. In D. Callahan, & S. Bok (Eds.) *Ethics teaching in higher education*. A Hastings Center Monograph. New York: Plenum.

Davis, M. (2009). The usefulness of moral theory in practical ethics: A question of comparative cost. *Teaching Ethics*, *10*(1) Fall, 69–77.

Davis, M. (2011). The usefulness of moral theory in teaching practical ethics: A reply to Gert and Harris. *Teaching Ethics, 12*(1), 51–58.

Harris, C. E. (2009a). Is moral theory useful in practical ethics? *Teaching Ethics, 10*(1), 51–68.

Harris, C. E. (2009b). Response to Michael Davis: The cost is minimal and worth it. *Teaching Ethics, 10*(1), 79–86.

Harris, C. E. (2011). A reply to Bernard Gert. *Teaching Ethics, 12*(1), 41–50.

Harris, C. E., Pritchard, M. S., Rabins, M., James, R., & Englehardt, E. E. (2014). *Ethics in engineering: Concepts and cases* (5th ed.). Belmont, CA: Cengage.

Plato. (1975). *The trial and death of socrates* (trans: Grube, G. M. A.). Indianapolis, IN: Hackett Publishing.

Pritchard, M. S., & Englehardt E. E. (2015). Ethical theories and teaching engineering ethics. In S. S. Sethy (Ed.) *Contemporary ethical issues in engineering* (pp. 111–120). Hershey, PA: IGI Global Publishing.

Reid, T. (2007). *Practical ethics*. In K. Haakonsen (Ed.) *Practical ethics*. Pennsylvania: Penn State Press.

Reid, T. (2010). *Essays on the active powers man*, edited by K. Haakonsen, & J. Harris. Edinburgh: Edinburgh University Press.

Stewart, D. (1829). *The works of Dugald Stewart, Vol. 5: The philosophy of the active and moral powers of man*. Cambridge: Hilliard and Brown.

Werhane, P. H., Hartman, L. P., Archer, C., Englehardt, E. E., & Pritchard, M. S. (2013). *Obstacles to ethical decision-making*. New York: Cambridge University Press.

Whewell, W. (1864). *The elements of morality*, Vols. I and II, 4th ed. London: Cambridge University Press.

Ethics Theory and Ethics Practice

Christopher Meyers

Abstract This article's thesis is that teachers of practical and professional ethics should adopt what Nick Fotion calls a "weak" approach to ethics theory and reasoning, one that is situated within a practice-driven attitude toward ethics engagement. This approach insists that ethics reasoning should strive to find real solutions to real world problems, while also embracing that such solutions are more often than not only tentative. That is, the approach does not demand that one's reasoning model will always provide the complete and final answer to every ethics question, but instead strives for the best answer within acknowledged epistemological deficits. This humbler attitude recognizes that one's proposed solutions could very well be upended by facts not currently available or by more skilled reasoning. Adopting such an approach can also do much to alleviate what I call the "roller-coaster" effect: namely, the all-too-common experience of reading the traditional theorists, quite rightly finding that they provide powerful and compelling—even convincing—arguments, only to then have that conviction undercut by the often devastating critique presented by the text or professor. On this model, one instead embraces key insights from those theorists and works to merge them into a practical method of ethics reasoning. I close with a proposal for one such method.

Keywords Ethics reasoning · Strong moral theory · Weak moral theory
Ethics as practice

There is some irony in the increasingly widespread ethics-across-the-curriculum (EAC) programs now present in universities around the world: How can we—philosophers—urge others to engage in sophisticated ethics reasoning when we cannot even agree on how to do it?

Many of the ideas presented here, including specific language, are taken from Meyers (2011, 2016, 2018).

C. Meyers (✉)
Emeritus, Kegley Institute of Ethics, California State University, Bakersfield, USA
e-mail: cmeyers@csub.edu

E. E. Englehardt and M. S. Pritchard (eds.), *Ethics Across the Curriculum—Pedagogical Perspectives*, https://doi.org/10.1007/978-3-319-78939-2_9

Such disagreement is reflected in two millennia of fierce battles over moral theory, battles that are also directly present in the typical structure of the theory sections of most practical and professional ethics (PPE) textbooks and courses. Students are given an overview of central theories, wherein each is shown to be powerful and compelling and potentially a good tool for real ethics decision-making. And *then* the texts' authors close the section with a typically devastating critique.

This progression is also frequently mimicked in class lectures and discussions: The professor assumes the role of the theorist of the day, extolling their virtues, defending against critiques, pulling out their clear insights into the human condition —*until* she closes with her own typically devastating critique.

All of this contributes to what I call the roller-coaster effect: Students engage Aristotle and conclude he's right—until his fatal errors are revealed. They then move on to Kant and go through the process all over again. Followed by Mill, Ross, Rawls … you get the point. For many, I fear, this leads to the opposite of our goal: Instead of reinforcing the power of these theorists' ideas, the process instead contributes to a kind of nihilism: "If such brilliant philosophers cannot get it right, or even agree on such central moral notions as the relative value of motive and consequences, who am I to try to figure out the correct model?"

But they are told they must learn the theories, maybe even enough to show how a Kantian (or Millian or …) would address some complex moral problem. The smart (or cynical) ones learn enough to be able to write a paper explaining the theories, thereby pleasing the instructor and getting a preferred grade. But all that happens merely at an intellectual level; they do not *embrace* the theories themselves. This is easy enough to discover: Challenge them in an informal, non-class setting on how they would solve a tough moral quandary and see how they consistently fall back on ad hoc responses or their religious upbringing.

And why should they do otherwise? Their text—and likely their professor—has presented the theories as internally consistent and complete, wholly distinct from all others, and all one needs to solve any moral problem. And then they are shown that each has serious, indeed probably fatal errors.

Ethics theories presented in this roller-coaster fashion come across, thus, as more like exercises in puzzle-solving than as genuine methods for moral reasoning. And if it is all merely a puzzle, students do not have to consider it as directly relevant to their lives and decision making activities. They can pick one to get through the paper/test/class, but then quickly forget once the course is complete. Worse, it may motivate them to reject theory (and often also philosophy) altogether.

Furthermore, they have likely lived enough years to know that simply winging it —that is, relying on what they learned at their mothers' and fathers' knees, in their church or temple, or through their social groups—will likely result in morally acceptable life choices. So what good is moral theory when they can engage in moral relations just fine without it?

To make matters worse, some students may very well stumble upon contemporary arguments in which various anti-theorists (Clarke 1987; Dancy 2009) also reject theory. Such scholars claim no theory can be coherently defended and they are not necessary for moral reasoning.

I am unambiguously in the theory camp; without it, choices are, by definition, extemporized only, with no necessary consistency among them. Whatever choices persons might make, based on whatever grounds they wished, would be as valid as any other. Particularists might argue that such inconsistent arbitrariness simply will not happen; humans just are not hard-wired that way. Even if such optimism is warranted—a shaky proposition, at best—a particularist approach is at a loss in cases of moral uncertainty. Theory provides the general norms and procedures for managing new or particularly difficult moral situations.

Most moral theories also attempt to get at the bigger philosophical picture—metaphysics (What sorts of creatures and cognitive processes underlie ethics reasoning?), epistemology (How do we overcome bias and sensory confusion to achieve justified true beliefs?), and anthropology (How have persons in fact, historically and contemporaneously, addressed ethics questions and to what effect?). Without such theoretical grounding, choices are at best personally pragmatic or, more commonly, merely expedient.

But if we accept that theory is necessary and also recognize the wisdom and insight present in the canon, what is to be done about the roller-coaster? Its hills and dips can, I think, at least be somewhat smoothed out by adopting what Fotion (2014) calls a "weak" approach to moral theory: One should presume that traditional theories are very good as far as they go and that representatives from each of the major traditions have clearly captured key insights about what it means to be a moral person and how to best engage moral reasoning. But one should also fully embrace—and teach—the likelihood that each is at best incomplete and the solutions it generates suggestive but maybe not definitive.

This contrasts, Fotion argues, with a "strong" approach in which any given theory is taken to be representative of the final word on morality. That is, it is presented as internally consistent without flaw and as able to address *all* important questions—everything from the metaphysics of moral agency, to how to characterize moral truth, to a method for resolving all possible conflicts. The problem with insisting on such accuracy and sufficiency is that none to date meets that standard—see the above referenced "devastating critiques." Whether some theorist will ever develop such an all-inclusive theory is an open question, but in the meantime students have no clear method for working through tough ethical problems (nor, for that matter, do theoretically motivated practicing ethicists).

Think of it another way: The strong approach adopts a Cartesian model of deductive certainty—certainty with respect to the accuracy of the theory and to the accuracy of the outcome the theory generates when correctly applied to any moral problem. In short, think Kant. The weak approach, in comparison, takes a humbler, inductive approach—both the theory and the solutions it generates *seem plausible*, but given better arguments or more data, might need to be amended or corrected. Think Ross and Aristotle.

My thesis in this chapter is that we—teachers of ethics, practicing ethicists, even theory builders—should adopt Fotion's weak approach as part of ethics engagement that is decidedly practice-driven. Teachers and ethicists will, I hope, readily see that this approach better aligns with the real-world needs that underlie teaching, learning, and doing ethics. For the theorist, among other advantages is that the associated humility puts one in a better position to revise and improve one's model, once the inevitable "devastating critique" is lobbed in one's own direction.

In what follows I will explain and defend this claim: Namely, ethics teaching—at least in PPE courses and certainly in the limited modules that often make up the heart of EAC activities—should be rooted in *practice*. From there I discuss teaching strategies for employing a weak approach, even when one's text presents theory in the standard roller-coaster framework. I conclude with a brief overview of one suggested model of ethics reasoning.

Ethics as Practice

First, let me make it clear that I am not disparaging philosophical work that is strictly theoretical, devoted to making sense of moral agency (or metaphysics, epistemology, cognition, …) and to promoting conceptual clarity and argument consistency—all with no goal other than such sense-making and clarity. Doing that is part of what makes for great philosophy and for great classes. There is little more fun than getting a bunch of super-bright and committed students in a room puzzling over, for example, just what Kant meant by "an a priori synthetic proposition" (Kant 1785/1998: 59).

But such puzzle-solving, as intellectually stimulating as it might be for some students, is not among the primary goals of PPE courses, let alone of the often piece-meal, module-based interactions common to EAC programs. Rather, those goals are, or should be, much more practice-driven, helping students find their way through moral life, helping them figure out how to ethically flourish—both themselves and those their choices impact. Specifically, the goals include:

- Raising ***awareness*** among students (and, often, other faculty) that a commitment to ethics is vital to their identity—their personal identity, but even more their future professional or occupational identity. This is especially true for students going into traditional professional fields—for example, engineering, medicine, the law—since those fields are at their core normative. That is, professions are largely defined through their avowed moral commitment to competency and prioritization of client well-being (Meyers 2018), a point accrediting agencies emphasize through strong ethics requirements. These requirements are intended, in largest part, to inculcate that normativity in budding professionals, to teach them that one cannot be a true professional without those moral foundations.
- Awareness, though, is not sufficient; students also need to ***engage*** their role as moral members of a community, profession or organization. That is, they need to analyze and resolve ethical quandaries, ones they might face at any moment, as well as ones they will likely encounter in their future work roles.

Moral imagination, motivated via case studies and role-playing, is the underpinning to such engagement, as it motivates consideration of perspectives beyond one's own, which in turn can help generate solutions beyond initial gut reactions (Werhane 1999).

- Without a ***reasoning*** method, however, case study reviews cannot move beyond ad hoc reactions: "This conclusion seems right but I'm not sure why or whether I would reach the same one down the road. Nor do I have a way of coherently responding to someone who challenges it." Motivating engaged awareness is a necessary first step, but students must also be given the tools for analyzing ethics problems and coming up with best, or at least better, solutions.
  Such tools acquisition has been the intended purpose of theory sections: Students learn several and then pick the best, or at least the one they can most coherently write about, and apply it to real-world ethical dilemmas. The problem, again, is that those theories are presented as if each is internally consistent, independently sufficient, and wholly distinct from all others—a standard they cannot sustain. Add to this the reality that students[1] have not acquired the wherewithal for rationally choosing one over another and we are back to the ad hoc: "This conclusion seems right but I'm not sure why…" Or worse, the result is skeptical cynicism that the whole exercise is a mere game and students just need to figure out the rules: "I can make enough sense out of Kant to write an A paper." From there it is another easy step to a cynical take on the whole course —"Another hoop they're making me jump so I can get my degree." Not exactly the outcome we, or accrediting agencies, seek with ethics requirements.
- The final and overarching goal is ***character*** enhancement. While one hopes accrediting bodies are not so naïve as to think that we can teach students to be ethical persons—by the time we encounter them, those roots have been long-established—university-level engagement *can* remind students of those roots and draw the connections between a life of honor and particular professional or occupational choices. Metaphorically, persons are like trees: The trunk represents the basic personality and character traits acquired at a young age, while the branches are choices, each with their own inducements. Some are virtuous, consistent with a life of honor; some are vicious; and part of our job should be to help them to discern the difference and reinforce their, hopefully, pre-existing honorable intentions.

Note these goals are atypical of those taught to, and by, philosopher-ethicists, wherein the emphasis has traditionally been on problem *analysis*, via conceptual and theoretical elucidation, not on problem *solving*. Nor are they reflected in the structure of most textbooks, with their roller-coaster theory sections, followed by chapters focusing on specific problems common to different occupations and professions.

---

[1]I focus here on students in PPE courses or those getting brief exposure via a modular EAC program. In all such cases, one should assume that this is the students' first exposure to moral theory, with its often abstruse and complex verbiage.

Now, the three goals are not the *only* goals for PPE courses, especially for those targeted at non-philosophy students and that fulfill campus General Education requirements. Such classes are often the only exposure students have to these key thinkers and ideas. Thus it is appropriate to also want the students to have exposure to some of history's greatest thinkers, with their extraordinary insights on the human condition. They should also be exposed to rigorous philosophical methods, even if sometimes it is mainly to enhance students' analytic and problem-solving skills, something they cannot get from ad hoc case reviews alone. My argument here is that these goals—correct as they are—have too often come at the *expense* of the practice-driven and unnecessarily so; history, insight, and analytic skill-building are not mutually exclusive of a practice sensibility. When texts and courses are structured effectively, including with a 'weak' approach to theory, one can achieve both.

Teaching Strategies

As the lists of goals above suggest, PPE courses—and EAC programs—can and should be rich experiences, filled with intellectual challenge while also providing valuable tools for dealing with life's challenges.

One of the difficulties PPE instructors typically face is students' vacillation between moral relativism and absolutism. Most will enter a class having been well-trained to respect others' views on religion, politics, and ethics. They often thus qualify any judgments with such phrases as, "Well, that's just my opinion, …" or "Others may believe differently, but…".

It is easy enough to push them off at least naïve relativism by challenging them on such deep intuitions as the immorality of torturing children or, closer to home, tossing out the grading criteria at the end of the term so as to favor those students who provide the proper 'incentive'. Students almost universally recognize that both examples—or other similarly slam-dunk versions—are *Wrong*.

What often follows, though, is a quick flip to a facile absolutism, an insistence that the answers they have reached are *Right*. Kant, with his epitome of strong theory, becomes very attractive at this point—not because he is easy to understand, but because once one has at least a minimal grasp of his theory, applying it to ethical problems is a relatively simple task: Simply plug the proposed answer into the Categorical Imperative and out pops absolutely correct answers.

It doesn't take much work, though, to show students that ethics reasoning in the real world is not so neat. Although lying is clearly wrong, it is often *less* wrong than the moral alternative—see, for example, the infamous "soldiers at the door." Similarly, while all persons may be of equal and infinite worth, relationships and history also dictate that our duties are sometimes stronger to some than to others—under normal circumstances, if I can save only one person from being scrunched by an oncoming train and my daughter is one of them, surely I have a greater duty to pull her out of the way than a stranger.

The noted caveat, "under normal circumstances," is to my mind among the more important phrases in ethics reasoning, as it reflects its contextual nature. While, yes, lying is wrong—just as is hurting others, breaking promises, being unjust—it is so only *prima facie*. That is, and as Ross taught us, we are, plausibly, bound by a plurality of duties, ones that, everything else being equal, we should follow and promote. Such duties, though, regularly come into conflict with one another and the ethics reasoner must rely upon practical wisdom—including the incorporation of principles, consequences, relationships, concerns for justice, and impact on character—to determine which duty one should plausibly follow in that situation.

There is a lot imbedded in and implied by those two paragraphs, much of which I will address below. For now, it should be apparent that this model of ethics reasoning is built upon weak theory—see the "plausibly" qualifiers and the reliance on all of the major traditions. When employed effectively, it grants students the creative license to explore what is compelling in each of those traditions, from which they may decide that one is the most compelling, if, again, only as part of a "plausible and good enough for now" mentality. They may also decide the best approach is some melding of all. And in all cases it makes for wonderful teaching moments, as students explore not only the traditional philosophical standards for theorizing—internal consistency, conceptual accuracy, and practical value—but what makes sense *to them*, which theory or theories helps them find their place in their moral lives.

There are a variety of teaching strategies that can help them get there. A faculty-centered version is to assign the text's readings on the central theorists, but teach all of them together, as part of a unified project in ethics reasoning. While avoiding obvious logical contradictions, can one achieve convergence among the major foci of those traditions? That is, can one come up with an answer that respects persons and principles, that achieves best aggregate outcomes and that promotes one's character?

A student-centered version is the commonly employed role-playing method, in which students assume the perspective of a particular theorist, as it is provided by their text, and determine how that theorist would answer a particular problem. Not common to that method, though, is to have students go on to discuss how a competing theorist would critique that view, followed by an argument for which position *they* think is more compelling and why.

Both strategies, I find, motivate an appreciation for the core views and key insights of multiple theorists. They thus break down some of the rigidity students may fall into when representing a particular theory. It also gives students the opportunity to engage in the analytic skill-building that comes with trying to determine how one would resolve a problem from within a theoretical framework. These strategies go on, however, to insist that ethics reasoning is not *just* about solving the puzzle; rather, it is also about a making a personal commitment to seek out the best available solutions to real problems. Furthermore, these strategies highlight the pitfalls attached to insisting—in a 'strong' way—that one approach is correct, to the exclusion of all others.

Note all of this can be done even while using a text that employs the theoretical roller-coaster. The instructor just needs to be careful not to teach each theory in isolation from the others (or from their respective historical contexts, for that matter). Rather, they can be presented as part of an *ongoing conversation* in which a bunch of really smart folks have been trying to capture what is both vital and unique to ethics reasoning. These strategies reinforce the philosophers' need for analytic skill-building, while also providing exposure to some of history's greatest thinkers. That is, the approach urged here is wholly consistent with those traditional philosophical goals, while also reinforcing ethics reasoning's practice sensibilities. Their effective use, however, also requires the humility, the "still-seeking" mindset, of weak theory.

A Suggested Model of Ethics Reasoning

Those practice sensibilities suggest several things:

First and foremost, the goal in ethics reasoning is the same as with all other forms of reasoning: To get the correct, or at least most correct, answer to the problem at hand.

Second, just as we have learned in logical reasoning, an inductive approach typically aligns better with real world problems than a deductive one. Formal deductive logic is indispensable for many of the activities central to human life—mathematics, computing, theoretical physics—but it almost always falls short in matters of human relationships, including the moral life. Furthermore, because ethics reasoning is deeply rooted in available and known facts, and because the totality of needed facts will rarely be available to ethics decision-makers, one will seldom be able to say that those solutions can be affirmed with the kind of certainty that is the hallmark of deductive reasoning. This reinforces the "weak," "good enough for now," approach: Because decision-makers routinely have to act from within an epistemological deficit, they must embrace that additional facts might change their best estimates of right choices, just as do those working in the empirical sciences. Importantly, however, this does not negate the *possibility* of certainty in ethics (or science)—if one does, in fact, have all the relevant data and reasons properly, certainty *would* emerge.[2]

Third, as the second point affirms, ethics reasoning is grounded in the messiness of empirical reality; one cannot do good ethics without good facts. One cannot know which principles, consequences, virtues, or justice considerations are at stake in a problem unless one knows, *thoroughly* knows, just what is going on, including:

[2]This conclusion, in fact, provides another method for overcoming students' naïve relativism: Most believe in an omniscient (and omnibenevolent) god, one who would know everything there is to know about any given ethics problem and thus would always be reasoning deductively, producing certain answers.

- Who is being impacted and how much are they?
- What is the problem's history?
- What are viable options and what associated implications, including costs and benefits—financial, relational, emotional—are attached to those options?
- What legal factors does one have to take into consideration?
- Are there spiritual, political or educational concerns?

In my experience as a clinical ethicist, in fact, once those empirical questions are adequately resolved, the ethical conflict routinely disappears.

Fourth, conceptual clarity—or at least conceptual agreement—is critical to ethics reasoning. All participants in the conversation must agree on the meaning of key terms—general ones like autonomy or justice, and context-based ones like informed consent, legal advocacy or confidentiality. Without such shared meaning, participants will simply be talking past one another, all the while thinking they are in fact on the same page.

These meta-ethics foundations established, good ethics decision-making is like all other forms of reasoning: One needs a method for determining who and what are at stake, what are likely outcomes, and what norms are relevant to best choices. Here is one such.

A Method[3]

Start with three top-level principles:

- Moral decision makers must accept a core commitment to achieving the ethically best choice, even if it does not necessarily align with the prudentially best.
- What counts as flourishing, as Aristotle argued, is relative to the economic, historical and physical circumstances of the persons in question. A flourishing life for a 21st Century Californian is significantly different than for a 12th Century Saxon.
- The ethically preferred choice is one that achieves convergence among[4]:
  - Best outcomes, in particular those that enhance individual and group flourishing and that create aggregate benefit over harm;
  - Character promotion—again, individual and communal—through choices that serve to reinforce persons of honor, worthy of emulation; and
  - Adherence to ethical principles, especially those that promote respect for persons.

Next, work through the components of the ethical problem. One can think of these in terms of steps, but only *prima facie*: Other than the first step,

[3]Much of what follows is taken from Meyers (2018).

[4]This characterization is borrowed from a similar process developed by Dan Wueste, from the Rutland Institute of Ethics at Clemson University (Wueste 2013).

fact-gathering, one need not follow a rigid lineage; the specifics of the case will largely determine the order in which the elements need be addressed.

1. **Determine the facts**. This might seem obvious, but it is routinely among the hardest components of ethics reasoning and thus the most likely to be done badly. It is the hardest because acquiring them is often resource-, especially time-, intensive and entails creative sleuthing. Furthermore, how persons understand key facts is invariably entwined in institutional politics and in the "scripts" that guide how participants make sense of their world (Werhane 1999). Persons who are themselves embedded in those scripts will not always even *see* the relevant facts: Institutional politics and histories can, for example, affect the extent to which someone accepts an antagonistic co-worker's problem description, often sub-consciously. Hierarchies and other power structures can produce similar factual gaps.
   It is done badly because decision-making participants are often rushed or impatient to get to the (perceived) meat of the matter—value conflicts. Furthermore, what one person takes to be an obvious fact, another may consider to be ideological bias—consider how people along the political spectrum differently perceive such facts as climate change and the economic impact of immigrants.
   Getting at the facts is critical for two reasons: First, an accurate and thorough examination is the only means by which one can determine if the problem even is an ethical concern and, if so, of what type and with what stakes. Second, as noted above, perceived ethical conflicts are more often than not disagreements about what is factually at work in the case. For example, a family might *hear* that a given treatment will likely produce a successful medical outcome, when the physician *meant* that success is possible but chances are very slim. Getting all relevant parties on the same factual page frequently allows the dilemma to disappear.
   Careful empirical analyses also help students move beyond initial sympathies toward naïve relativism. The misleading temptation is to interpret *factual* variance across cultures or institutions as *value* variance. That one group might highly value, for example, loyalty and respect for elders, while another might more highly prize honesty and autonomous independence, is a contextually driven fact, not a disagreement over principle. Both see all the associated principles as morally binding, but cultural contingencies result in different contextual emphases.
   This is similar to how different cultures legally codify their ethical foundations: Californians, emphasizing beneficence and health, have legally banned tobacco smoking in all enclosed work spaces, while South Carolinians, emphasizing free choice and economic advantage, place no restrictions on private work spaces and even allow smoking in some public buildings' "designated areas." Does this mean Californians do not value liberty, or that South Carolinians reject the value of good health? Of course not, but historical and contemporary contingencies motivate differing valuations.

2. **Determine *who* will be impacted**. A sub-set of a factual analysis, the focus here is on *persons*, especially those who will be most directly affected by respective options. Again, this is trickier than it might appear: A common problem is to draw the impact circle too narrowly, excluding secondarily affected individuals.[5] In health care choices, for example, those affected in the first-tier typically include patients and families; second-tier might include close friends and immediate health care providers; while third-tier extends to such people as the patient's co-workers, other patients, extended relatives, and ancillary health care providers.
   The extent to which these persons' ethical concerns need be taken into account is, again, contextual, determined by how deep and wide the impact is. To ignore them outright, though, is to insufficiently fulfill one's ethical obligations. But, by the same token, one can expand the circle *too* far, agonizing over the impact on persons sometimes layers and layers removed from the primary stakeholders. Experience, with the associated acquisition of practical wisdom, helps one determine where to stop, so as to avoid 'paralysis by analysis'. Among the considerations such wisdom provides is an appreciation for the role of relationships among the various players: How *deeply* will someone be impacted by a choice? A co-worker might be profoundly affected by the death of a colleague, whereas a long-estranged brother is only minimally touched.
3. **Determine what *type* of conflict it is**. This is vital since different types require different responses:

   - *Moral distress*, wherein one is confident about the right choice, but is prevented from carrying it out because of any of a range of constraints, including institutional power structures, the law, and economics. Among the more frustrating of ethics problems, it may be that no true solution is possible. Instead, persons are reduced to better problem management—including assistance from a power-equalizing advocate—and to moral imagination, through which one can strive to mitigate harms through successful workarounds.
     As the reference to power asymmetry suggests, moral distress is commonly experienced by those hierarchically lower on the food chain, such as nurses, patrol police officers and infantry soldiers. Ethically committed organizations strive to create mechanisms for safe appeal to more powerful allies; organizations without such a commitment, by contrast, contribute to early burn-out and post-traumatic stress.

---

[5]One should also be careful not to restrict the evaluation to *human* persons. Many choices have the potential to impose great harm or provide great benefit to non-human animals, particular with respect to pains and pleasures. See Solomon (1999).

- *Moral ignorance*, wherein one is in all new moral territory and does not even know how to begin the evaluation. Here, the wisdom of experienced mentors will be an invaluable resource, as will input from a diverse set of advisors—essentially crowd-sourcing.
  Early in my career I encountered such a case, namely, whether a hospital should attempt to sustain and eventually deliver an early term (20-week), non-viable, fetus in a (brain) dead mother's body. While we now have considerable data on such cases, at that time there was almost no information on whether it was even possible, let alone what long-term ramifications successfully delivered children might encounter. The other members of the ethics committee and I were truly in the dark; our equivalent of crowd-sourcing—calling national colleagues and an extensive literature search—turned up only one case, whose circumstances were different enough as not to be very helpful. We thus took our best guess and decided to try to keep the mother's body 'alive' until viability. Unfortunately, the body quickly became septic and the fetus died.
- *Moral dilemma*, wherein one has the classic conflict of competing principles. Any choice one makes will violate one or more of the principles, cause harm, and make it hard to be virtuous.The most common of ethics problems—classic examples include whether to lie to the soldiers at the door so as to protect an innocent friend, whether to break a promise to meet a friend so as to assist a stranger in great need, and whether to tell a friend that her partner may be cheating—they provide the paradigm situation for careful ethics reasoning. How can one possibly decide, on other than ad hoc grounds, what the better choices are unless one carefully deliberates upon the elements of the sort described in these steps?

4. **Determine what principles are at stake in the problem**. These include general principles like non-maleficence and fidelity, along with role-specific ones motivated by the particular professional and relational obligations inherent in one's role, for example, obtaining informed consent, being a mentor for a student, and financially assisting one's child before helping strangers.
5. **Determine *the extent* to which those principles are at stake**. Might, for example, respecting a professional duty to look out for the well-being of a client conflict with a general duty of honesty (for instance by being 'expansive' on a diagnosis so a patient can get insurance coverage) and, if so, which is the more egregious violation? Which action will more likely produce an environment in which persons, individually and communally, can flourish? How does the choice impact the character of the decision-maker (and, for that matter, of the recipient)? Has a range of morally imaginative options been considered, with the goal of seeking the best, most creative solution? How likely is it that proposed solutions can be effectively implemented? What impact do organizational cultures, ideologies, and power relationships have on implementation?

This step is at the heart of ethics decision-making process and reinforces that ethics reasoning is always contextual: One cannot determine the extent to which principles are involved without giving adequate attention to the contingencies of the particular problem. It thus also reinforces how difficult ethics reasoning can be—determining and evaluating those contingencies is demanding work. The call here, thus, is for moral agents to be active, engaged, and sincere in their deliberations and to do their very best within the practical and epistemological limitations that are likely constraining ideal choices.

6. **Determine what obligations emerge as a corollary to the decision**. That is, once one resolves, to the best of their ability, the best available choice, does this produce, for example, a new obligation to repair associated harm? Given that even the ideal solution to an ethical dilemma entails the violation of at least one other principle, there will inevitably be remainder, i.e., the good that was attached to the violated principle(s) has been left hanging. If, in the "promise to meet a friend" example, one decides one's greater duty is to stop and help the stranger, one probably has a duty to make it up to the neglected friend.
7. **Determine what needs to done to see the choice through to conclusion.** For most persons, in most situations, the mere making of a choice is concomitant with its fulfillment. For those in advisory roles (e.g., ethics consultants), this step is often among the most difficult to complete, simply because one is not the primary agent. All persons in the process must thus embrace the process and the choice and also commit to full transparency—to how the decision was made, how it was fulfilled and what were its ramifications.

Note how the steps also combine to help one achieve the noted convergence among results, character, and principles. They therefore pull together key insights from Kant, Mill, Ross, and Aristotle. They also emphasize that relationships impact how principles are at stake, thereby pulling in a key aspect of contemporary "ethics of care" approaches.

Conclusion

Consistent with the recommended "weak" meta-ethical starting point, following this method will rarely produce certainty in one's reasoning. For that, one would need to be omniscient, always capable of rational deliberation with no corruption from bias, ideology or conceptual scheme, and also pure in intention. In other words, one would have to be something other than *human*. The informational deficits, biases, and moral failings all of us mere mortals must overcome in ethics deliberation means we simply do the best we can. If that deliberation is undertaken with a true commitment to getting it as right as possible, and if there are no major gaps or errors in the facts one has gathered, then one can confidently conclude that the resulting options fall within a narrow range of acceptable choices. The better the deliberation, the narrower the range—including, possibly, the one, universally correct choice, though one may not recognize that success at the time, since one has not yet experienced the emergent consequences.

For those with Kantian—"gotta have certainty"—inclinations, the good news is this approach does not preclude the possibility of correct moral choices; indeed, it insists on them. The bad news is that one may never know whether one's *right* reasoning also produced the *good*.

Many students will wholeheartedly embrace this ambiguity, especially those disturbed by the arrogance of a strong theoretical approaches. The challenge for the instructor is to prevent contextual *ambiguity* from sliding into *relativism*. As the deliberation gets tough, students will be tempted to fall back unto relativism ("Well, that's just my opinion."). Our job is to remind them of their earlier acceptance of such core intuitions as the immorality of torturing children or violating the syllabus and then to guide them through continued reasoning, to directly show them that better deliberation produces better moral choices. From there we can engage their moral imagination: What parts of the deliberation seem right, what parts need correction, and what does that process tell us about the nature of ethics reasoning?

A last point: A complete and sophisticated engagement with a reasoning method of this sort is far beyond what one can hope to achieve in the limited, module-style interactions of many EAC projects. For that, one might focus on three key elements:

1. The rejection of naïve relativism;
2. The importance of context;
3. The convergence among respect for principles, best aggregate outcomes, and promotion of personal character.

If students walk away from a 90-min session on ethics reasoning with an appreciation for those elements—and the associated realization that such reasoning is *dang hard stuff*—we will have done much to reinforce their role as ethics agents —in school, in work, and in life.

References

Clark, S. G. (1987). Anti-theory in ethics. *American Philosophical Quarterly, 24*(3), 237–244.

Dancy, J. (2009). Moral Particularism. In *The stanford encyclopedia of philosophy* (Spring 2009 Edition), edited by E. N. Zalta. http://plato.stanford.edu/archives/spr2009/entries/moral-particularism/. Accessed March 2017.

Fotion, N. (2014). *Theory vs. anti-theory in ethics: A misconceived conflict.* New York: Oxford University Press.

Kant, I. (1785/1998). *Groundwork for the metaphysics of morals* (trans: Gregor, M.). Cambridge, UK: Cambridge University Press.

Meyers, C. (2011). Reappreciating W. D. Ross: Naturalizing prima facie duties and a proposed method. *Journal of Mass Media Ethics*, *26*(4), 316–331.

Meyers, C. (2016). Universals without absolutes: A theory of media ethics. *Journal of Media Ethics, 31*(4), 198–214.
Meyers, C. (2018). *The professional ethics toolkit*. Oxford, UK: Wiley [forthcoming].
Solomon, R. C. (1999). Peter Singer's expanding circle: Compassion and the liberation of ethics. In D. Jamieson (Ed.), *Singer and his critics* (pp. 64–84). Oxford: Blackwell.
Werhane, P. (1999). *Moral imagination and management decision making*. New York: Oxford University Press.
Wueste, D. (2013). *Ethical decision making: A toolbox approach*. Publication of the Rutland Institute for Ethics.

Developing Habits of Moral Reflection: Dewey, Moral Inquiry, and Practical Ethics

Alan A. Preti

Introduction

I begin with a quotation from a 1980 study of the state of ethics instruction in American higher education. In their summary recommendations, the authors maintain that:

> a "higher education" that does not foster, support, and implement an examination of the moral life will fail its own purposes, the needs of its students, and the welfare of society. The university offers a unique context for a careful examination of moral claims and moral purposes. We ask only that such an examination be made formal and explicit, and that sufficient imagination, energy, and resources be invested in the teaching of ethics that its importance will become manifest, both within and outside the university. (Callahan and Bok 1980, 300)

Undertaken by the Hastings Center, a bioethics research institute founded in 1969 in Hastings-on-Hudson, New York, the study came at a seminal moment, as interest in applied ethics and the teaching of ethics in higher education had increased rapidly over the course of the preceding decade.[1] It is the first study of its

The impetus for this chapter came from participation in "Moral Psychology and Education: Putting the Humanities to Work," a National Endowment for the Humanities Summer Institute held at Grand Valley State University, May 30–June 24, 2016. I am grateful to the NEH and to the Institute directors for this opportunity. I also wish to acknowledge two Institute faculty members who inspired me to explore Dewey's moral philosophy in greater depth, Stephen Fesmire and Mark Johnson.

[1]The Hastings Center itself was born of this growing interest, specifically regarding advances in medical technologies and practice, and the resulting policy vacuums.

A. A. Preti (✉)
Rosemont College, Philadelphia, PA, USA
e-mail: apreti@rosemont.edu

E. E. Englehardt and M. S. Pritchard (eds.), *Ethics Across the Curriculum—Pedagogical Perspectives*, https://doi.org/10.1007/978-3-319-78939-2_10

kind of which I am aware to have systematically explored this burgeoning development, together with a host of questions confronted by teachers, administrators, and interested observers concerning its significance.

In the nearly four decades that have passed since the publication of the report, the country has witnessed a sea change in how institutions of higher learning have approached the subject of ethics and ethics instruction. Whether we call it an "ethics boom" (or "booms") (Davis 1999, Chap. 1), a "turn to ethics" (Garber et al. 2000), or even a "return to ethics" (Kiss and Euben 2010, 3), it is clear that the Hastings Center's clarion call for serious reflection on the role of ethics in higher education has been heeded. Required courses in applied or practical ethics[2] have become common in a variety of programs at the undergraduate and graduate levels throughout the U.S., and the number of programs designed to confer degrees in applied or practical ethics (both generally and in specific fields) has increased rapidly in recent years; the same is true of the establishment of college or university-housed ethics centers and the number of professional ethics organizations, ethics publications, and ethics training programs. Ethics across the Curriculum initiatives have of course been a part of this trend as well.[3]

I imagine the original contributors to the Hastings Center study would be somewhat pleased with the current state of affairs. But, to set the stage for the current chapter, let us go directly to the heart of the matter and ask: What is the point of teaching ethics? What are we, as ethics educators, really trying to do? What would we like students completing one or more of our courses to demonstrate—a more attuned moral sensitivity, skill in moral reasoning, a commitment to addressing an issue of moral import? I suppose it fair to say that there is no single aim, no unique outcome that would find unequivocal support among those who teach ethics as the *summum bonum* of their practice. The sheer variety of settings and contexts for the teaching of ethics in higher education complicates any attempt to isolate an overarching goal that would satisfy everyone. And yet, I imagine most would concur that skill in addressing the ethical problems or dilemmas that will arise throughout students' personal and professional lives is of paramount importance; this may not be *the* ultimate aim of ethics instruction, but it is likely near the top of the list for most teachers. If what is meant by 'addressing ethical problems' is the ability to identify what is morally at stake in a given problematic situation and to come to a reasoned decision about what is morally required in order to best resolve it, then the importance of additional aims becomes evident: moral awareness or sensitivity; an understanding of and appreciation for fundamental moral concepts, principles, or theories; and skill in moral reasoning or practical judgment. Such a cluster of related aims seems to have carried the day, and whatever other

[2]While some would distinguish between 'applied' and 'practical' ethics, I will use the terms interchangeably, as the overarching argument will not thereby be affected.

[3]I am not suggesting that the Hastings Center study had a direct impact on these developments; regardless of how many educators and administrators were familiar with the study, ethics was certainly "in the air" at the time.

disagreements over goals and objectives there may be, I will assume the issue more or less settled, at least for my present purposes.[4]

With these preliminaries in mind, my aim in this chapter is to provide an account of how John Dewey's moral philosophy has anticipated both the aims and, in some cases, the methodology of contemporary applied or practical ethics instruction. It is a matter of some irony that Dewey's pragmatist ethics remains relatively unexplored as a resource for ethics instructors, particularly those who teach practical ethics. Apart from the fact that questions of morality and value were perennially on Dewey's mind and infused his work on education, logic, socio-political philosophy, aesthetics and other areas throughout his long life, the irony lies in the fact that Dewey's specific ideas on the nature and function of moral inquiry are especially commensurate with the aims and methods of practical ethics instruction. While it may not be quite accurate to identify Dewey as the founder of applied ethics as the term is generally understood today, his conception of moral philosophy as a method for dealing with the ethically problematic situations that are part and parcel of human experience (professional and otherwise) certainly prefigured the distinct approach to ethics that would eventually come to be identified as applied or practical ethics.[5]

Contemporary practical ethics education, I suggest, bears the stamp of Dewey's moral philosophy; however, the philosopher himself is rarely acknowledged in this regard. It is common, for example, to find in introductory applied ethics textbooks an initial chapter summarizing several traditional ethical theories before moving on to the specific issues with which the text is concerned. Students may thus get a dose (however minimal) of Aristotle, Kant, Mill, Rawls, and perhaps a nod to Noddings, among several "lesser" players, but I have yet to come across an introductory ethics text that includes more than a passing mention of Dewey as a contributor to the

---

[4]The rise of outcomes assessment throughout higher education has precipitated a significant amount of reflection on the goals and objectives of teaching ethics. For an account of common goals in Ethics across the Curriculum initiatives, see Elliott, this volume. See also Newton (2001) and Ozar (2001) for a discussion of both EAC goals and outcomes assessment. Keefer and Davis (2012) discuss outcomes assessment specifically in connection with the teaching of professional ethics.

[5]Dewey's and James Tuft's *Ethics*, originally published in 1908, was perhaps the first "applied ethics" textbook published in the United States. Extremely popular, it was adopted almost immediately following publication by a number of colleges and went through twenty-five printings before a second edition was published in 1932 (Sloan 1980, 23). In the preface, the authors remark that the aim of the second part, "Theory of the Moral Life," was to "show the development of theories out of the problems and experience of every-day conduct, and to suggest how these theories may be fruitfully applied in practical exigencies" (1908, iv). Part III, "The World of Action," continues the stated objective with an application of the previously mentioned theories to a number of social problems.

"great conversation" in ethics.[6] And yet, his contribution is central to the self-understanding of practical ethics, by which I mean what those who teach and do research in the field believe themselves to be doing, and how they go about doing it. To anticipate, contemporary accounts of moral decision-making and moral imagination include elements that are fundamental to Dewey's conception of moral philosophy. In what follows, I shall make explicit how this is the case. By making this connection explicit, I seek not only to give Dewey his due, but to suggest that ethics instructors would do well to add to their playbook strategies for developing habits of moral reflection that have their basis in Dewey's view of ethics as moral inquiry.

Ethics as Moral Inquiry

The general features of Dewey's pragmatism are familiar in outline, if not in detail. As one of the central figures of the pragmatist movement that flourished in the United States during the first quarter of the 20th century, Dewey shared with his colleagues a rejection of certain epistemological assumptions about the nature of knowledge, truth, and objectivity that formed part of a thorough reconception of the aims of philosophy and philosophical method. Traditional philosophical theorizing had not been particularly faithful to the complexity and richness of human experience, resulting in empty abstractions and metaphysical systems irrelevant to the pressing needs of society, however useful they may have been in times past. The layperson's charge that philosophy is of little or no practical value was quite to the point. If philosophy was to remain relevant and viable, it would have to concern itself with addressing the problems of human affairs, and this meant conceiving of philosophy primarily as practice.[7]

To say that philosophy is concerned with practice is to acknowledge that as a form of reflective inquiry it originates in experience, is informed by experience, and aims to clarify, inform, and ultimately improve experience. Following fellow pragmatist C. S. Peirce, Dewey found the paradigmatic model for inquiry in scientific method, which as a process begins in a condition of doubt and uncertainty and terminates with the restoration of the stability that prevailed prior to the discord. In broad outlines: our initial awareness of a problem prompts a desire to identify the

[6]This is not meant to imply, of course, that there are no such texts. Still, I believe that my admittedly inexhaustive survey was wide enough to be representative.

[7]The classic statement of this conception of philosophy is Dewey's *Reconstruction in Philosophy* (originally published in 1920). See in particular the introduction to the enlarged edition of 1948, and Chap. 1. For a valuable introduction to Dewey's thought through the lens of philosophical reconstruction, see Fesmire (2015).

specific nature of the problem; the problem having been defined, tentative hypotheses are proposed as possible solutions, implications of the hypotheses are worked out, and the most promising hypothesis is tested in experiment or direct action. The restoration of equilibrium or harmony is a confirmation of the hypothesis; the problematic situation is resolved. It is in such a sense that experience is said to be the origin and terminus of all inquiry.[8]

Because Dewey did not conceive of moral problems as different in kind from scientific (or any other type of) problems, he found in the method of inquiry an ideal tool for addressing issues concerning human conduct, especially given its successes in advancing the natural sciences: "we are only pleading for the adoption in moral reflection of the logic that has proved to make for security, stringency and fertility in passing judgments upon physical phenomena" (1948, 165). With regard to human choice and action, as long as entrenched and unreflective habits succeed in addressing current perceived needs and interests, there is no felt necessity for deliberating about the relative merits of a particular act or line of action. When a person's habitual responses are not adequate to changing and novel circumstances, however, the imbalance produces a felt disjunction or perplexity, signifying the initial awareness of a problematic situation. This is the crucial first step of the process of moral inquiry through which an ethical problem is defined and the resulting tension eventually resolved with the action most appropriate for the situation.

For Dewey, then, ethics is to be understood as a process of inquiry that seeks to discover the most satisfactory plan of action in the particular, concrete situations in which humans find themselves. Such an account acknowledges that the moral life is multidimensional and messy, and consequently ill-served by the attempt to proffer a single and universal normative principle or set of rules to guide human action within a uniquely moral domain separate in quality from, and requiring a distinctive mode of analysis than, the diverse dimensions of human experience. The presumption of having identified a priori a universal standard of conduct—whether that of divine decree, utility, the Categorical Imperative, what the virtuous person would do, and so on—is that goods and ends are fixed and all that need be done is discover what the proposed principle or rule requires. Such a project has been built upon a narrow circumscription of moral experience that ignores the genuine uncertainty of morally problematic situations and the possibility that in any such situation there

[8]Dewey's account of inquiry, developed over a number of years throughout a number of his writings, is fully elaborated in his *Logic: The Theory of Inquiry*, in which he defines inquiry as "the controlled or directed transformation of an indeterminate situation into one that is so determinate in its constituent distinctions and relations as to convert the elements of the original situation into a unified whole" (1938, 104–105).

can be (and often are) plural, competing goods.[9] Ultimately, only with a "transfer of the burden of the moral life from following rules or pursuing fixed ends over to the detection of the ills that need remedy in a special case and the formation of plans and methods for dealing with them" can moral practice be ameliorated (1948, 165).

All this is not to deny the usefulness of moral principles or rules of conduct; having emerged over the course of history from reflection on human needs and aspirations, they are indispensable tools for guiding moral inquiry. Each furnishes us with a point of view from which to appreciate what is at stake, consider competing demands, foresee how various courses of action will affect the interests of all concerned, and determine which action best suits the occasion. Moral principles illuminate rather than dictate. Only the moral zealot can lay claim to having discovered the correct way to organize moral reflection and action.

Dewey's ethics, then, is at once situational, pluralistic, and experimental. As situational, it recognizes the uniqueness of each particular occasion, even while acknowledging that there are general features shared by all situations, thus facilitating inquiry; as pluralistic, it recognizes that there is no single, fixed end or good toward which all action is to be directed, and that each situation has its own particular good which can be discovered through inquiry; as experimental, it is tentative, allowing for potential solutions to be explored in the imagination before being tested in practice. This last point is a central feature of moral inquiry, and will be discussed in a later section; for now, I turn to a consideration of how Dewey's view of moral inquiry has been implicitly adopted by practical ethics educators in the form of moral decision-making models.

Moral Decision-Making Models

As noted in the introduction, skill in addressing moral problems or dilemmas is a primary goal of practical ethics instruction, where such a skill is understood as consisting of a cluster of related capacities including moral sensitivity or awareness,

---

[9]"Whatever may be the differences which separate moral theories, all postulate one single principle as an explanation of the moral life. Under such conditions, it is not possible to have either uncertainty or conflict: *morally* speaking, the conflict is only specious and apparent" (1998 [1930], 315 [original emphasis]). That is, the seeming conflict is between knowing (or at least believing oneself to know) what morality demands, and failing to act accordingly; this is manifested in the tension, say, in choosing between good and evil, appetite and reason, the vicious and the virtuous, etc. As noted above, however, for Dewey conflict is inherent in the morally problematic situation; it is precisely the uncertainty of the situation itself—often a manifestation of competing goods—that moral theorizing, on his view, had not recognized. It should be noted that exceptions to Dewey's generalization that all moral theories posit a single normative principle can be found in the Scottish Enlightenment philosophers Thomas Reid, Adam Smith, and Dugald Stewart. The 19th century British intuitionist William Whewell and utilitarian Henry Sidgwick can also be cited as counterexamples. I am grateful to the editors of this volume for drawing my attention to this point.

an understanding of fundamental ethical concepts, principles, or theories, and the ability to exercise practical judgment in the making of moral decisions. Consistent with this aim, an increasingly common feature of applied or practical ethics textbooks is the so-called moral decision-making model, whose steps are designed to facilitate the development of these and related capacities. While such capacities can certainly be developed through other pedagogical methods (e.g. case studies, fictive narrative, role playing, etc.), moral decision-making models serve to capture the elements of ethical theory most relevant to addressing moral issues in various contexts and integrate them into the paradigmatic problem solving method in a way that would seem to be particularly effective.[10] Assuming students will have internalized the process, they will be well-placed to adjudicating ethical issues throughout their personal and professional lives.

Below are three examples of moral decision-making models; the first from a textbook on business ethics, the second from the website of an academic ethics center, the third developed originally for an Ethics across the Curriculum faculty workshop delivered in 1991 at the Illinois Institute of Technology.

(1) **An Ethical Decision-Making Process**[11]

1. **Determine the facts**. Gather all of the relevant facts. It is critical at this stage that we do not unintentionally bias our later decision by gathering only the facts in support of one particular outcome.
2. **Identify the ethical issues involved**. What is the ethical dimension? What is the ethical issue? Often we do not even notice the ethical dilemma. Avoid normative myopia.
3. **Identify stakeholders**. Who will be affected by this decision? What are their relationships, to me, and what is their power over my decisions and results? Who has a stake in the outcome? Do not limit your inquiry only to those stakeholders to whom you believe you owe a duty; sometimes a duty becomes clear only once the impact on a stakeholder is assessed. For instance, you might not necessarily first consider your competitors as stakeholders; however, once you understand the impact of your decision on those competitors, an ethical duty may arise.

[10]I suspect that interest in moral decision-making models has been influenced in part by a growing dissatisfaction with approaches to teaching practical ethics that emphasize ethical theory. For a discussion of the relative merits of moral theory and moral decision-making procedures or tests for teaching practical ethics, see Harris (2009a, b) and Davis (2009). This initial debate in the journal *Teaching Ethics* 10 (1) led to a spirited exchange among Davis, Harris, and Gert in a subsequent issue; see *Teaching Ethics* 12 (1) (2011). See Englehardt and Pritchard (2013) for additional reflections on the matter.

[11]Hartman et al. (2018, 90–91). As the title of the book (*Business Ethics: Decision-Making for Personal Integrity and Social Responsibility*) suggests, the authors' emphasis throughout is precisely on decision-making in the business context. In true Dewey an spirit, they assert that "a process of rational decision making, a process that involves careful thought and deliberation, can and will result in behavior that is more reasonable, accountable, and ethical" (11).

4. **Consider the available alternatives**. Exercise "moral imagination." Are there creative ways to solve conflicts? Explore not only the obvious choices, but also those that are less obvious and require some creative thinking "outside the box."
5. **Compare and weigh the alternatives**. Take the point of view of other people involved. How is each stakeholder affected by my decision? Compare and weigh the alternatives; ethical theories and traditions can help here.

a. Consequences

   i. Beneficial and harmful consequences

b. Duties, rights, principles

   i. What does the law say?
   ii. Are there professional duties involved?
   iii. Which principles are most obligatory?
   iv. Are people being treated fairly, with respect for their autonomy and equality?

c. Implications for personal integrity and character

   i. What type of person am I becoming through this decision?
   ii. What are my own principles and purposes?
   iii. Can I live with public disclosure of this decision?

6. **Make a decision**. Is this a point-in-time decision, or something that will be carried out over time? What is your plan, and how are you going to implement it? What will you do if something unexpected happens as a result?
7. **Monitor and learn**. Have you built in mechanisms for assessment of your decision and possible modifications? Make sure that you learn from each decision and move forward with that increased knowledge; you may face similar decisions in the future or find it necessary to make changes to your current situation. Do policies or procedures need to be revised as a result of this situation or its resolution?

(2) **A Framework for Ethical Decision Making**[12]

Recognize an Ethical Issue

1. Could this decision or situation be damaging to someone or to some group? Does this decision involve a choice between a good and bad alternative, or perhaps between two "goods" or between two "bads"?
2. Is this issue about more than just what is legal or what is most efficient? If so, how?

[12]Developed by faculty at Santa Clara University for the Markkula Center for Applied Ethics. Original URL: www.scu.edu/ethics/ethics-resources/ethical-decision-making/a-framework-for-ethical-decision-making/. Downloaded May 17, 2017.

Get the Facts

3. What are the relevant facts of the case? What facts are not known? Can I learn more about the situation? Do I know enough to make a decision?
4. What individuals or groups have an important stake in the outcome? Are some concerns more important? Why?
5. What are the options for acting? Have all the relevant persons and groups been consulted? Have I identified any creative options?

Evaluate Alternative Actions

6. Evaluate the options by asking the following questions:
   Which option will produce the most good and do the least harm? (The Utilitarian Approach)
   Which option best respects the dignity of those who will be affected? (The Kantian Approach)
   Which option best respects the rights of all who have a stake? (The Rights Approach)
   Which option treats people equally or proportionately? (The Justice Approach)
   Which option best serves the community as a whole, not just some members? (The Common Good Approach)
   Which option leads me to act as the sort of person I want to be? (The Virtue Approach)

Make a Decision and Test It

7. Considering all these approaches, which option best addresses the situation?
8. If I told someone I respect—or told a television audience—which option I have chosen, what would they say?

Act and Reflect on the Outcome

9. How can my decision be implemented with the greatest care and attention to the concerns of all stakeholders?
10. How did my decision turn out and what have I learned from this specific situation?

(3) **Format for Ethical Decision Making**[13]

1. **State problem** (e.g. "Do I have a conflict of interest?" Or even "This makes me uncomfortable").
2. **Check facts** (some problems disappear upon closer examination of the situation; others change radically).
3. **State specifications** (limits and objectives)—laws, professional code, and corporate rules to be satisfied, cost constraints (e.g. under $200), children to feed, place in life plan (e.g. save company).

[13]Michael Davis, Center for Study of Ethics in the Professions, Illinois Institute of Technology, Chicago ©2008. Original URL: http://ethics.iit.edu/eb/Format.%20for%20Ethical%20Decision%20Making.pdf. Downloaded May 17, 2017.

4. **Develop list of options** (be imaginative, try to avoid "dilemma"—not "yes" or "no" but who to go to, what to say).
5. **Test options**, using such tests as the following:

   - **Harm test**—does this option do less harm than any alternative?
   - **Publicity test**—would I want my choice of this option published in the newspaper?
   - **Defensibility test**—could I defend my choice of this option before a Congressional committee, a committee of my peers, or my parents?
   - **Reversibility test**—would I still think the choice of this option good if I were one of those adversely affected by it?
   - **Virtue test**—what would I become if I chose this option often?
   - **Professional test**—what might my profession's ethics committee say about this option?
   - **Colleague test**—what ***do*** my colleagues say when I describe my problem and suggest this option as my solution?
   - **Organization test**—what ***does*** the organization's ethics officer or legal counsel say about this option?

6. **Make a tentative choice** based on steps 1–5. Did you solve the problem with which you began?
7. **Make final choice (after reviewing steps 1–6), act, and then ask:** What could make it less likely that you would have to make such a decision again?

The first thing to note is that despite the differences among these examples, each in its own way illustrates the general pattern of inquiry highlighted in the previous section: becoming aware of a problem; locating and defining the problem; proposing tentative hypotheses; reasoning out the consequences of the hypotheses; selecting the most promising hypothesis and putting it to the test. That there is no precise correspondence between the specific series of steps in each example and the general pattern of inquiry is beside the point, which is that each captures quite clearly Dewey's view of the aim of ethics as a process of moral problem solving while integrating moral principles or concepts as tools for facilitating the process. The moral decision-making procedure is, in effect, a contemporary manifestation of the core of Deweyan moral theory.

Second, moral decision-making models are not intended as inflexible decision-procedures, with each step to be executed consecutively in rote fashion; rather, they distill the stages of a fluid deliberative process in which stages can overlap or merge into one another with no clearly defined boundary. Depending on the circumstances, awareness of the problem may precede the identification of relevant facts or it may emerge as facts are gathered; consideration of the alternatives for action may lead to a revisiting of the facts or to a deeper understanding of what is at stake, and so on. Additionally, questions serve as prompts for students to think creatively about alternative solutions and how they might be implemented conscientiously, with post-decision reflection clarifying further the extent to which the action was called for under the circumstances, how attentively it was carried out, and how well it addressed the problem. Granted, unless students are guided through the process thoughtfully and repeatedly, such models can become sterile heuristics; nevertheless, their potential for assisting in the development of skill in moral problem solving should be evident.

Lastly, the elements of fluidity and creativity alluded to above point to a central feature of the deliberative process, the exercise of the moral imagination. The first model above identifies moral imagination specifically with step 4 ("Consider the available alternatives. Exercise 'moral imagination'"), and the second and third models mention creativity or imagination in step 5 ("have I identified any creative options?") and step 4 ("be imaginative…") respectively, although I suspect the authors would not deny the importance of moral imagination in connection with the other steps as well. It would perhaps be more accurate to say that the moral imagination is active throughout much of the deliberative process (better: the most fruitful deliberative process concerning morally problematic situations will be one in which the moral imagination is skillfully employed at appropriate stages). In the following section, I examine the notion of the moral imagination with a view to showing how it too, in addition to the moral decision-making model itself, has been anticipated by Dewey's account of moral inquiry.

Moral Imagination and Dramatic Rehearsal

The concept of moral imagination[14] has become established currency among scholars in a variety of fields, including developmental psychology, organizational management, leadership studies, and business ethics, among others. The business ethics literature in particular has in recent years seen a burgeoning of research on moral imagination, chiefly as it bears on decision-making in the business or leadership context (Ciulla 2004; Godwin 2012; Hargrave 2009; McVea 2009; Moberg and Caldwell 2007; Moberg and Seabright 2000; Vidaver-Cohen 1998; Werhane 1999, 2002, 2008; Werhane et al. 2013). In most of this work, attention is given generally toward a descriptive account of moral imagination and how it might be implemented in practice, the implication being that its habitual exercise by those in positions of leadership will contribute positively to an ethical business culture.[15]

---

[14]While the first appearance in print of the words 'moral imagination' has been attributed to Edmund Burke (Kirk 1981, 37), the notion of an imaginative exercise having a specific moral dimension or revealing morally salient features of experience can be traced to the Scottish sentimentalist philosophers, in particular Adam Smith and his account of sympathy (what we would call empathy) and the related conception of the impartial spectator. Smith's analysis has had a significant influence on current accounts of moral imagination, emphasizing as it does the role of empathy in appreciating another's perspective and making moral judgments (Werhane 1999, 93–96).

[15]While none of the authors cited above addresses potential pedagogical implications of their accounts, I suspect all would agree that business ethics courses would benefit by the inclusion of instructional methods designed to stimulate moral imagination.

Moral imagination can be characterized as a cognitive capacity whose exercise facilitates awareness of the morally significant features of a situation and of how the consequences of a decision made within that context will bear on all those affected. Moral imagination thus involves "articulating and examining alternatives, weighing them and their probable implications, considering their effects on one's other plans and interests, and considering their possible effects on the interests and feelings of others" (Jacobs 1991, 25). Similarly, Mark Johnson defines moral imagination as "an ability to imaginatively discern various possibilities for acting within a given situation and to envision the potential help and harm that are likely to result from a given action" (1993, 202). Patricia Werhane's account, which has been particularly influential in business ethics, characterizes moral imagination as "the ability to discover, evaluate and act upon possibilities not merely determined by a particular circumstance, or limited by a set of operating mental models, or merely framed by a set of rules" (1999, 93). Moral imagination, on this view, helps draw attention to the conceptual framework or mindset which serves as the default mechanism for interpreting experience and informing the decision-making process. In the absence of an active moral imagination, important data and other considerations (and thus alternative perspectives) are bound to be left out of the picture, resulting in thought and action that follows a narrow script and which ultimately compromises moral sensitivity and decision-making.[16]

While Dewey himself does not use the term 'moral imagination,' his account of how deliberation—what he calls *dramatic rehearsal*—functions in inquiry anticipates current views on moral imagination and its role in decision-making.[17] Let us return to Dewey's model of moral inquiry. A problem having been defined adequately, success in moral inquiry will depend upon skillful and creative deliberation in which the imagination is tapped for discovering alternatives for action. For Dewey,

> Deliberation is actually an imaginative rehearsal of various courses of conduct. We give way, *in our mind*, to some impulse; we try, *in our mind*, some plan. Following its career through various steps, we find ourselves in imagination in the presence of the consequences that would follow: and as we then like or approve, or dislike and disapprove, these consequences, we find the original impulse good or bad (2008, 275 [original emphasis]).

As we imaginatively and creatively work through the possibilities for action, we participate in an experiential simulation that engages the cognitive, affective, and

[16]Werhane's account sheds light on standard examples of moral failures in the organizational context such as the Ford Pinto case, the Challenger disaster, and the WorldCom scandal, among others. The common denominator in these and similar cases is the fact that the relevant decision-makers were unable to extract themselves from their individual mindset in order to evaluate its shortcomings and consider alternative frames of reference, and thus, other viable decisions. The missing ingredient was moral imagination. See Werhane (1999).

[17]See Fesmire (2003, Chaps. 4 and 5) and Pappas (2008, Chap. 5) for helpful accounts of how Dewey conceives the function of the imagination in the context of moral deliberation.

conative elements of our moral psychology. Such an exercise has a decidedly qualitative dimension; we *feel* the way we would feel if we were to act on a possible alternative, we *feel* how we imagine those affected by a particular course of action would feel. The ability to empathically project ourselves into the experience of those impacted by our decision is crucial, as it serves to overcome the ingrained habits of short-sighted self-interest that all too often guide human thought and behavior; for Dewey, empathy is "the animating mold of moral judgment … the tool, *par excellence*, for resolving complex situations" (2008, 270 [original emphasis]). As necessary as it is for moral judgment, however, empathy is not sufficient; it is but one element in the search for reasons that ground our discovery of what the situation requires (2008, 269–270).

Dramatic rehearsal is at once creative, prospective and open-ended; along the way there will be fits and starts, dead ends, and new avenues emerging in unforeseen ways: "the imagining of plans carried out furnishes an opportunity for many impulses which at first are not in evidence at all, to get under way.

Many and varied direct sensings, appreciations, take place" (2008, 275). Such intuitive apprehensions are not themselves indicative of what is morally appropriate for dealing with the situation; rather, as is the case with our intuitive responses to others' actions, each "direct sensing" is a spur to further reflective deliberation: "spontaneous 'intuitions' of value have to be entertained subject to correction, to confirmation and revision, by personal observation of consequences and cross-questioning of their quality and scope" (2008, 272). Analysis and discrimination are required to check our intuitions and revise our initial judgments, if called for. Even if deliberation ultimately leads to our finding the "original impulse good or bad," as Dewey says in the quotation above, we cannot be assured of its fitness until put to the final test in experience. As noted in the earlier discussion on inquiry, only implementation in action can serve either to confirm or disconfirm the hypothesis.

In sum, Dewey's account of deliberation is meant to draw attention to the way in which people actually deliberate, and to indicate its potential for amplifying our sensitivity to moral problems and developing our ability to address them—in short, for becoming morally conscientious individuals. Acknowledging the imaginative foundation and interrelatedness of empathy, creativity, intuition, and analysis and reflection, dramatic rehearsal captures the complexity of our moral psychology, thus giving the lie to the traditional dichotomy between reason and emotion, with the former playing the role of final arbiter in the making of moral judgments. Deliberation is not a mathematical calculus or procedure in which costs and benefits or the demands of duty are tallied in a cool and detached manner; it is an imaginative exploration which opens us to possibilities not available when "applying ethics" is equated with the rigidity of a utility calculus, a categorical imperative, or some other principle or set of rules. If engaged in sincerely and habitually, dramatic rehearsal has the potential for not only helping individuals develop facility in addressing moral problems, but, as I suggest below, in contributing to their own moral development.

Habits of Moral Reflection and Moral Growth

The preceding section has drawn attention to the role of the imagination in moral decision-making as found both in recent accounts and Dewey's notion of dramatic rehearsal. As noted earlier, most work on moral imagination has focused on this specific dimension. Yet there is a feature of moral imagination that cuts deeper than such accounts acknowledge, and that is its role in the adoption of a new perspective of the relation between oneself and others, and how this perspective impacts one's self-image as a moral agent. Through the exercise of the moral imagination, we come to recognize the limits we have placed on our moral convictions and commitments, and to appreciate how they—and, in turn, we ourselves—might be transformed. In this section, I highlight briefly this deeper significance of dramatic rehearsal in Dewey's thought.

Dewey's account of habit helps connect the exercise of imagination in moral deliberation with the growth of a moral self.[18] For Dewey, the self is constituted by the interaction (what he refers to in his later works as 'transaction') between the organism and the natural and social environment. Through the transactional process, the natural impulses of the organism are channeled into specific behaviors and lines of behavior that eventually become settled dispositions or habits; these can range from simple instinctive reactions to complex patterns of thought. Dewey is less concerned with the external manifestation of habits than with the amalgam of desires, intentions, and choices that constitute volitional acts. For Dewey, it is these latter that are of direct bearing on the nature of the self, in that they serve as the causal impetus for the repeated actions that reflect an abiding and stable character:

> [H]abit reaches even more significantly down into the very structure of the self; it signifies a building up and solidifying of certain desires; an increased sensitiveness and responsiveness to certain stimuli, a confirmed or impaired capacity to attend to and think about certain things (2008, 171).

Sedimented habits and dispositions thus serve to shape character and identity; indeed, it would not be misleading to claim that at bottom our habits *are* our character: "[C]onduct and character are strictly correlative. Continuity, consistency, throughout a series of acts is the expression of the unity of attitudes and habits" (2008, 172). The person is as the person does.

On Dewey's view, moral deliberation has a decided effect on character by shaping, changing, and reinforcing habits conducive to moral growth. Clearly, the greater the proficiency at dramatic rehearsal, the greater the sensitivity to the moral features of situations, to various available alternatives for action, to others' needs and interests, how best to navigate the uncertainty of the matter, and so on. This is not a particularly novel revelation; practice may not make perfect, but it does make for improvement. What is perhaps not as appreciated is the extent to which facility in imaginative deliberation can serve to attune individuals to their own character,

---

[18]Dewey's rich account of habit is developed in *Human Nature and Conduct* (1950 [1922]).

and to the habits of thinking, appraising, and acting that bear on its development.[19] For Dewey, moral decisions are not one-off events that are isolated from our past and future; how we deliberate is informed by past habits of thought and reflection, and our current habits inform and shape our capacity for doing so in future situations. The quality of our deliberation thus reflects our character just as much as do overt actions; to improve our deliberative capability is to take part in our moral growth.

An appreciation for the transformative capacity of habits of deliberation opens us up to the challenge of refashioning the self through new paths of thought and action conducive to a richer and more complete life. Such challenges are not merely opportunities; in a very real sense they make demands on the individual to move beyond present self-identity and take an active role in the process of moral development. As Dewey puts it:

> The growing, enlarging, liberated self … goes forth to meet new demands and occasions, and readapts and remakes itself in the process. The necessity for choice between the interests of the old and of the forming, moving self, is recurrent. For everywhere there is an opportunity and a need to go beyond what one has been, beyond "himself" if the self is identified with the body of desires, affections, and habits which has been potent in the past. Indeed, we may say that the good person is precisely the one who is most conscious of the alternative, and is the most concerned to find openings for the newly forming or growing self; since no matter how "good" he has been, he becomes "bad" … as soon as he fails to respond to the demand for growth (2008, 307).

Once aware of the opportunity, the individual incurs a responsibility to transcend the former self and take an active role in the process of moral growth. This will entail the development of second-order desires to eliminate habits now recognized as detrimental to growth, as well as new desires and choices reflecting the concern for self-cultivation. Such a process, for Dewey, has no terminus, consistent with his view that there is no such thing as a substantial self, let alone an ideal virtuous self serving as the ultimate goal of moral agency: "[W]e set up this and that end to be reached, but *the* end is growth itself" (2008, 306 [original emphasis]). For Dewey, the responsibility to assume the mantle of moral growth represents the ultimate freedom for human beings. The gravity of the moral life lies precisely in equating this freedom with an injunction to grow as a moral agent.

---

[19]This point, although not generally emphasized, is not entirely absent from the literature on moral imagination. Business ethicists Dennis Moberg and Mark Seabright, for example, highlight the role played by "possible moral selves," which are individuals' notions of who they could be as moral persons. In this connection, moral imagination serves to produce a certain picture of one's ideal self, which in turn creates intentions that nudge one toward its realization (Moberg and Seabright 2000, 868). Because Dewey rejects the notion of an 'ideal self,' he would put the matter somewhat differently, emphasizing the impact of the moral imagination on the formation of new desires and choices that contribute to a continually developing moral self.

Conclusion

I have argued that Dewey's theory of moral inquiry is implicit in the aims and methodology of contemporary practical ethics education, specifically in the form of moral decision-making models and accounts of the moral imagination. Both, I have shown, are informed by Dewey's pragmatist view of ethics as problem solving and his account of the role of dramatic rehearsal in addressing moral problems. I have also noted that for Dewey, the moral imagination has a more significant function than is typically recognized in the literature and by ethics instructors, namely, its role in the development of a moral self. While this latter point may not be seen by most instructors as part of their charge, this need not trouble them. It is true that character development may not be an explicit goal of teaching ethics in higher education (excepting, perhaps, some religious institutions). But to accept as a central goal of teaching ethics the ability to address ethical issues affirms implicitly the value of making better moral decisions, and by extension, of being better persons. It would be curious indeed if there were no connection between the two. Ethics instructors can rest happy knowing simply that approaching their practice through the lens of problem solving and the use of decision-making models may have a direct effect on students' ability to make reasoned moral decisions, and an indirect effect on expanding their moral horizons so as to include a desire for growing as a moral agent. Of course, the assumption here, as always, is that students recognize the value of our course content and pedagogical practices. The old saw about leading a horse to water is apt—we can only do what we can; the rest is up to the students. Teaching ethics as moral inquiry is a good place to start.

References

Callahan, D., & Bok, S. (Eds.). (1980). *Ethics teaching in higher education*. New York: The Hastings Center.

Ciulla, J. B. (2004). Moral imagination. In G. Goethals, G. J. Sorenson, & J. MacGregor Burns (Eds.) *Encyclopedia of leadership* (pp. 1019–1021). Thousand Oaks, CA: Sage Publications.

Davis, M. (1999). *Ethics and the university*. London: Routledge.

Davis, M. (2009). The usefulness of moral theory in practical ethics: A question of comparative cost. *Teaching Ethics, 10*(1), 69–78.

Davis, M. (2011). The usefulness of moral theory in teaching practical ethics: A reply to Gert and Harris. *Teaching Ethics, 12*(1), 51–60.

Dewey, J. (1938). *Logic: The theory of inquiry*. New York: Henry Holt and Co.

Dewey, J. (1948). *Reconstruction in philosophy* (enlarged ed.). Boston: Beacon Press.

Dewey, J. (1950 [1922]). *Human nature and conduct: An introduction to social psychology*. New York: The Modern Library.

Dewey, J. (1998 [1930]). Three independent factors in morals. In L. Hickman, & T. Alexander (Eds.) *The essential dewey, volume 2: Ethics, logic, psychology* (pp. 315–320). Bloomington: Indiana University Press.

Dewey, J. (2008 [1932]). *The collected works of John Dewey: The later works, 1925–1953, vol. 7: 1932*, edited by J. A. Boydston. Carbondale: Southern Illinois University Press.

Dewey, J., & Tufts, J. (1908). *Ethics*. New York: Henry Holt and Co.

Englehardt, E., & Pritchard, M. (2013). Teaching practical ethics. *The International Journal of Applied Philosophy, 27*(2), 161–173.

Fesmire, S. (2003). *John Dewey and moral imagination: Pragmatism in ethics*. Bloomington: University of Indiana Press.

Fesmire, S. (2015). *Dewey*. New York: Routledge.

Garber, M., Hanssen, B., & Walkowitz, R. L. (Eds.). (2000). *The turn to ethics*. New York: Routledge.

Gert, B. (2011). The usefulness of a comprehensive systematic moral theory. *Teaching Ethics, 12*(1), 25–38.

Godwin, L. N. (2012). Examining the impact of moral imagination on organizational decision making (dissertation abstract). *Business and Society, 20*(10), 1–25.

Hargrave, T. (2009). Moral imagination, collective action, and the achievement of moral outcomes. *Business Ethics Quarterly, 19*(1), 87–104.

Harris, C. E. (2009a). Is moral theory useful in practical ethics? *Teaching Ethics, 10*(1), 51–68.

Harris, C. E. (2009b). Response to Michael Davis: The cost is minimal and worth it. *Teaching Ethics, 10*(1), 79–86.

Harris, C. E. (2011a). A reply to Bernard Gert. *Teaching Ethics, 12*(1), 39–49.

Harris, C. E. (2011b). Response to Michael Davis: The usefulness of moral theory in teaching practical ethics. *Teaching Ethics, 12*(1), 39–49.

Hartman, L. P., DesJardins, J., & MacDonald, C. (2018). *Business ethics: Decision-making for personal integrity and social responsibility* (4th ed.). New York: McGraw-Hill.

Jacobs, J. (1991). Moral imagination, objectivity, and practical wisdom. *International Philosophical Quarterly, 31*(1), 23–37.

Johnson, M. (1993). *Moral imagination*. Chicago: University of Chicago Press.

Keefer, M., & Davis, M. (2012). Curricular design and assessment in professional ethics. *Teaching Ethics, 13*(1), 81–90.

Kirk, R. (1981). The moral imagination. *Literature and Belief, 1*, 37–49.

Kiss, Elizabeth, & Euben, P. (Eds.). (2010). *Debating moral education: Rethinking the role of the modern university*. Durham, NC: Duke University Press.

McVea, J. (2009). A field study of entrepreneurial decision-making and moral imagination. *Journal of Business Venturing, 24*(5), 491–504.

Moberg, D., & Caldwell, D. (2007). An exploratory investigation of the effect of ethical culture in activating moral imagination. *Journal of Business Ethics, 73*, 193–204.

Moberg, D., & Seabright, M. (2000). The development of moral imagination. *Business Ethics Quarterly, 10*(4), 845–884.

Newton, L. (2001). Outcomes assessment of an ethics program. *Teaching Ethics, 2*(1), 29–67.

Ozar, D. (2001). Learning outcomes for ethics across the curriculum programs. *Teaching Ethics, 2*(1), 1–27.

Pappas, G. F. (2008). *John Dewey's ethics: Democracy and experience*. Bloomington: University of Indiana Press.

Sloan, D. (1980). The teaching of ethics in the American undergraduate curriculum, 1876–1976. In D. Callahan & S. Bok (Eds.), *Ethics teaching in higher education* (pp. 1–57). New York: The Hastings Center.

Vidaver-Cohen, D. (1998). Moral imagination in organizational problem solving: An institutional perspective. *Business Ethics Quarterly*. The Ruffin Series, Special Issue no. 1, 123–147.

Werhane, P. H. (1999). *Moral imagination and management decision making*. New York: Oxford University Press.

Werhane, P. H. (2002). Moral imagination and systems thinking. *Journal of Business Ethics, 38*, 33–42.

Werhane, P. H. (2008). Mental models, moral imagination and system thinking in the age of globalization. *Journal of Business Ethics, 78*, 463–474.

Werhane, P. H., Hartman, L., Archer, C., Englehardt, E., & Pritchard, M. (2013). *Obstacles to ethical decision-making: Mental models, milgram, and the problem of obedience*. Cambridge: Cambridge University Press.

The Occupational Imperative: Engaging the Professions in Teaching Ethics

Lisa H. Newton

Abstract The teaching of ethics, this paper suggests, is shaped by economic and other influences from the world outside the university gates, and by the expectations we as a society hold for those who have mastered the subject. For much of American history, university graduates were expected to settle into an elite group of civic leaders, requiring for that role a rich understanding of the classical sources of ethical discussion. In the more fluid society that followed, they were expected to use the critical thinking that they had developed in ethics class to keep the society rational in a sea of conflicting claims. Now, in the anomic society of unfamiliar economic and political threats, the best way to teach ethics is through the moral sensibility that underlies it, by sending the students off campus altogether and exposing them, through internships and other experience, to the operations of the world they will soon have to join.

Keywords Ethics · Principles · Traditions · Critical thinking · Post-truth

Introduction

There are 32,000 undergraduate students at your university, roughly evenly distributed among the four undergraduate years. Your job is to teach Ethics, to beginning students. When you meet your first section of the required Freshman Philosophy 101 (or Humanities, or Introduction to Religious Studies, or some variant on Human Values), you are looking at a random group from the entering class of 8000 students. They do not know you, or each other, or for the most part, why they are there, except they must study to get good grades to get a good job and make money, viz. enough money to pay off their college education. Your job is to teach them ethics. How might you do that most effectively, efficiently and humanely?

L. H. Newton (✉)
Fairfield University, Fairfield, USA
e-mail: LHNewton@fairfield.edu

E. E. Englehardt and M. S. Pritchard (eds.), *Ethics Across the Curriculum—Pedagogical Perspectives*, https://doi.org/10.1007/978-3-319-78939-2_11

Conversations for the Reflective

There used to be a standard approach to the problem. When I sat in that classroom as an undergraduate, the purpose of college was to initiate the student into the ongoing discussions of the ages, in all fields. Ethics has been subject matter for 2500 years at least, so we started 2500 years ago with a Socratic Dialogue (the *Crito,* as I recall). We went on to sections from Aristotle's *Nicomachean Ethics*, then something from Augustine's *Confessions* and Thomas Aquinas, then Hobbes, Locke, Hume, Kant, Mill, Dewey, and Sartre (whom we did not understand). The epitome of a college education was the University of Chicago Great Books curriculum, developed by Robert Hutchins: learning ethics (and history, literature, religion and philosophy) meant reading and discussing all the great contributions to the Western tradition. Globally, the curriculum was limited: We ignored China, India, and Africa; it was a while before we learned that Europe, aided by the United States, was not actually the only center of world learning. But it was a really good start.

What was the point of the exercise? Human thought, the theory went, most especially human thought in the pursuit of the good life, was essentially an intelligent conversation, carried out among the thinkers of all periods. We who engaged in that conversation, which was centered on the question attributed to Socrates—"How, then, ought we to live?"—became just the latest participants, and would join the line of philosophers we studied in order to hand the conversation on, improved by our contributions and supplied with footnotes on contemporary scholars, to the next generation. That was the purpose of the teaching of ethics, and philosophy in general, and the Humanities, and the Arts and Sciences as a whole. That was why we were there. (Except for some of the budding professionals, pre-meds, engineers and the like, college was not about making a living after college. Those who were fascinated by the commercial aspect of life went to trade schools, we understood, or into apprenticeships. They did not go to college.)

Of course we were an "elite." Not everyone could go to college, taking a four-year moratorium on facing the problems of the real world, including starting a career and a family. Not everyone had the leisure to engage in these conversations. Not everyone had the leisure to think deeply at all, occupied as they were in making a living and staying on the right side of the law. We knew that we were privileged to receive this initiation into the wisdom handed down to us. We were sternly commanded by all the authorities—parents, grandparents, teachers at all levels—to recognize the responsibility that went with the privilege: to preserve the goodness of the civilization we inherited, to live moderately according to its precepts, to teach it to all who would listen, to govern wisely as the opportunity arose, and to contribute our wealth (which we would undoubtedly enjoy, unless we became college professors in our turn) to the institutions that enshrined it. We understood the command, and whether or not we said a lot about it, we generally accepted it.

Analytical Skills for the Active

Later, when I stood in front of that class as a Philosophy instructor, the purpose of college had changed. No longer was the wisdom of the ages the ruling value, a conversation to be engaged and perpetuated. Now the purpose of the course was to equip students with the tools they would need to become active and effective citizens in the world beyond the ivy-covered walls, especially the tool known as "critical thinking." They had to learn to question accepted authority, challenge the boss, mistrust what they read in books and newspapers, compare diverse opinions and weigh them against accepted principles, and collect empirical evidence to resolve questions of fact. For the basics of this task, I could find no better textbook than Plato's *Euthyphro*: in the ordinary circumstances of life (Socrates and Euthyphro's encounter occurred in a waiting room), students must learn to confront claims of right with two immediate questions—What do you mean? And How do you know? And if the one claiming right cannot answer those questions, the claim has no validity: Socrates' means of establishing that were as devastating as they were charming.

We'll come back to that. For that's not how I started the class. Students by this time were unwilling to settle back, as we did in our youth, and breathe the limpid air of ancient Athens. They wanted a contemporary setting. They were the sons and daughters of the suburbs of Connecticut and neighboring states, commuter country. So I put them in a parking garage. Here's the case:

> **The Impaired Driver**
>
> You have stayed about an hour longer than you intended to at a very pleasant party with your old college friends. While you were getting your law degree and starting practice, your roommate Marty made it big on Wall Street. He hosted the party in his huge Riverside Drive apartment. All the old college ties were there—great memories, beer, booze, marijuana . . . hadn't seen *that* in awhile. Good stuff, too.
>
> Realizing you're late, you race to the parking garage, elevator to the third floor, hop in your SUV, and tear around the turn toward the exit. Smash! Car parked in just the wrong place. You hit it dead center. You back up, get out, note that there is extensive damage to the other car—both doors on the driver's side badly dented—but none to yours. What should you do?
>
> You know damn well what to do. There's clearly damage, lots of it, so you have to take out your cell phone, call the police, and wait there till they come. Watching you propping yourself up against your SUV, they'll insist on the inconvenience of a breathalyzer test. When they get the results of that, they'll give you a chauffeured ride to the precinct station and insist further on a urine test. When they get the results of that, you may get to know the folks in the precinct very well before you see the sky again. You may very well—probably will—lose your license to operate a motor vehicle. The fines will be substantial; you may lose your SUV. You may even go to jail. The damage to your reputation, and to your position in your law practice, will probably be irreparable; depending on the state, they may yank your license to practice law. That's a lot to think about. Meanwhile, you are the only occupant of this parking garage at this hour. You could just drive back to Connecticut and not say anything to anyone.

All right, it's not a very sophisticated case. And some ex-police officers who showed up in one of my evening classes pointed out that most parking garages now have surveillance cameras that will see the accident, read your license plate, report it, and have police on your Connecticut doorstep early the next morning. But it gave me a chance to introduce some of the terminology I wanted them to use, as follows:

How Do We Make Decisions in These Cases?

1. What course of action will cause the greatest good to the greatest number, minimizing pain to all parties and maximizing happiness? We call this kind of thinking **consequentialist**, or **teleological** (from the Greek word for "end" or "goal"), since it judges the moral quality of the action by its consequences or by the end it achieves. In classic **Utilitarianism**, as set forth by **Jeremy Bentham** (*Principles of Morals and Legislation*, 1823) and **John Stuart Mill** (*Utilitarianism*, 1863) the only consequences that matter are **happiness and unhappiness**, pleasure and pain, for everyone affected by the act. Measuring pleasure and pain for all parties, including your family, the owner of the other car, even the world at large, it looks like your best course is to take off for Connecticut without doing anything at all. Drive slowly so you don't get stopped. After all, the pain felt by the car owner upon finding his damaged car is nothing compared to the pain that you and your family would feel if you lost your ability to earn a living, let alone if you went to jail. Besides, his insurance will probably cover the whole bill.
2. Yes, but think of it this way. That law is there for a purpose. What you are supposed to do, as a citizen, right now, is call the police. That's your **duty.** You've enjoyed all the benefits of citizenship, now it's time to honor your part of the bargain. **What if everyone who got into an accident just took off? Would the world be a better place?** Could you approve of a law that said, when you find you've caused damage to life or limb or property, if it isn't convenient to stay around, just take off? If you can't, and you probably can't, then you have no right to make an exception of yourself in this case. That rule is the substance of **Immanuel Kant's Categorical Imperative**, which he set forth in his *Groundwork of the Metaphysics of Morals* (1785): Act so that you can simultaneously will that the maxim of your action (the reasoning that led you to do it) should become universal law. In heading back to Connecticut without calling the police, you set yourself above the law and contribute to a lawless society. Kantian reasoning is called **non-consequentialist**, or **deontological** (from the Greek word for "duty"), since it looks not at the consequences of the action but at the law or duty that governs it.
3. Here's another way to think about what you're doing, or about to do. When you get home, suppose you find your father, or the rector of your church, or your older brother, or anyone you trust, love, and admire, sitting in the kitchen. Somehow he knows what happened in that garage, and he asks you to explain just what you did, confronted with that difficult situation, and your reasoning to your decision. **Why did you do what you did?** Well, how would you explain it? How would you justify taking off like that, when you were clearly in the wrong? If that scenario doesn't suggest an approach to the problem, put a reporter from *The New York Times* (or *The Wall Street Journal,* if you're that type) sitting beside him in another kitchen chair. The reporter is going to describe the whole situation, including your reasons for acting as you did, **in the newspaper tomorrow, on the front page**. What kind of person would you look like in that story? Is that the kind of person you want to be? There are certain traits that we value in ourselves and others, traits like honesty, integrity, and courage, that we call **virtues**.

> Morality is not just about consequences, nor is it just about laws and duties—often it's about the sort of person you are, your very being, so we call the reasoning that draws on these considerations **virtue-based** or **ontological**, from the Greek word for "being." **Aristotle** (4th century BC) based his Ethics upon ontological reasoning; we've never really lost track of it.
>
> These are agonizing decisions, and they govern life—the future life of the person who has to make them, and the way history will judge her or him. More complex decisions are addressed in the discipline of ethics, and the rest of this chapter will consider more complicated dilemmas; but we must not forget that the fundamental moral quantities are honesty, integrity, and courage, those that the impaired driver must call upon right at the moment he finds himself alone in that garage with a smashed car in front of him.

It's the same dead white men in the analysis, but in a form that I hoped could emerge from a classroom discussion.

The rest of my introduction to Ethics did indeed go to more complicated cases, introducing along the way some accepted principles that would take them beyond personal dilemmas to the problems of the legislature and the courtroom. For these larger cases, I adopted the "Belmont Principles," first enunciated in an Appendix to the Report of the National Commission for the Protection of Human Subjects of Biomedical and Behavioral Research. That commission, empowered by Congress to write a set of rules for the practice of human subjects research (after a series of scandals about the use and abuse of human subjects by medical and psychological professionals had made newspaper headlines and thoroughly frightened the American people), had begun its deliberations with a modest disavowal of any ultimate principles at all. The decision was entirely practical: its diverse membership spanned several ethical orientations (at least the three mentioned above) and a variety of faith traditions, and there was no single principle or set of same that all could agree with. So they proceeded by the "case method," familiar to any who have dealt with our Common Law tradition: decisions as to permissibility or otherwise would be made in a few clear cases, then the discussion of the next case would be based on those decisions by way of analogy, then further decisions would rely on or distinguish themselves from those precedents. No "principles" were acknowledged as governing through all this. Then, with the need to report back to Congress looming at the end of the cases, the Commission hired Tom Beauchamp, a philosopher from Georgetown University, to go over the case discussions as the Commission had recorded them, and formulate some summary principles from their conclusions. Just from the reasoning offered in the discussions to support the decisions, Beauchamp concluded that in their decisions, the Commission had in fact appealed to three (or four) essential principles: Do No Harm (sometimes completed with And Also Do Good), Do Justice, and Respect Personal Autonomy. The first (and second) principles centered on the value of human welfare, and required that the subjects of experiments not be injured, and that the general public be helped, by any contemplated research; the third required that human subjects not be chosen primarily from vulnerable or disadvantaged groups; and the fourth required that human subjects always have a right to withdraw from research that had become overly burdensome.

I hoped that the tests included in these principles, generalized to the society at large, would serve to clarify ethical reasoning in the students' public lives. The point of the cases I introduced, and the ways the principles could be applied, centered on their future lives as active citizens of a democracy, not above-the-fray sages in converse with Plato. I asked them to think as legislators—tailoring their thought and action to the public good in all cases.

I will go on to suggest an additional method for teaching ethics to the present generation. But first we should note, that the importance of teaching critical thinking has never been greater. A new "philosophy" has emerged, not from some fetid cellars of academia, but from the highest levels of government, and it is deeply concerning, some might say terrifying.

The purpose of the university, according to tradition, is "the pursuit, preservation, and transmission of the truth." Easy to understand, easy to remember. But what if the very notion of "truth" has been problematized beyond recognition? A recent bestseller, Daniel Levitin's *Weaponized Lies*, opens by pointing out that the "Oxford Dictionary's Word of the Year for 2016 was *post-truth*, which they define as an adjective 'relating to or denoting circumstances in which objective facts are less influential in shaping public opinion than appeals to emotion or personal belief.'" Daniel J. Levitin, *Weaponized Lies: How to Think Critically in the Post-Truth Era*, New York: Dutton (Random House); 2017. Published 2016 as *A Field Guide to Lies.*

The problem that the colleges must address, to preserve their souls and the reason for their existence, is the appearance of a generation of students who may not be convinced that there is any "real" truth at all; they may believe that there is the academy's truth, and then there is alternative truth (or alternative facts), and you can choose between them.

Our literature has portrayed societies where truth was not honored: they are foreshadowed in Plato's *Republic*, in Book I, where Thrasymachus insists that "justice" is nothing but the interest of the stronger party, and in Book VIII, in the democratic man who acknowledges no objective good in life, and in George Orwell's *1984*, where a dictatorial Big Brother announces as Truth whatever suits the present interest of the state. Still, in Plato, the stronger party had at least a consistent rule for his determinations, and the democratic man really hurt only himself; in Orwell, as in *Republic* Book I, Truth was indeed solid and consistent, it just was the firm interest of the dictator. But the colleges are seeing something else—not a selfish demand that "truth" identify with my present interests, but a denial that there is any truth at all, leaving the term open to my present whims, advantageous to me or not. In a time when powerful national leaders dismiss any facts they dislike as "fake news," university professors, once bastions of the respectable Establishment, suddenly must be counter-cultural. Part of learning Ethics turns out to be learning respect for facts and clear thinking in a world which may respect neither.

Engagement with the World of Work

As my career continued, the students changed, almost imperceptibly at first: now they lived in parent-enabled anxiety about how they will ever make the money to succeed well enough to justify the insanely high college bills burdening their parents, or worse, feeding into the student loans that will burden them as soon as they graduate, and last for most of the rest of their lives. Students had always grumbled about the Ethics course as a waste of time; but this new generation is acutely conscious of the fact that they have no time to waste—the bills are coming due immediately, and every course must justify itself as relevant to their future as a bill-payer. (Why do they bother with college at all? That's a live question. To this date the evidence is still that the holder of a bachelor's degree will out-earn one with no or partial college.)

Watching this new generation in action in the halls of my university, I was struck by gradual changes in behavior that map onto the anxiety they expressed. They were always on their cell phones. That's a truism of contemporary life, but its significance for reflection on ethics only gradually dawned on me: The moratorium was gone for good, because the distance was gone. They weren't always talking to their friends, I came to understand; much of the time they were talking to their parents, often twice or three times a day, most often about conditions of their lives and the implications for their success in college: how are the grades? How'd you do on that last test? Have you talked to the loan officer about extending the loan (some loans were still negotiated on campus)? The collapse of distance, the end to the semi-isolated campus, had meant that the students never got a break from the focus on financial success. Trying to engage them in the quest for the world of Ideas, as shown in the Allegory of the Cave, was a waste of time. All I could do was show them that the quest was there, that many of our forebears had considered it important, indeed ultimately important—and promise that it would still be there if they ever got their heads above water long enough to look at the sky.

Where are the students now? The beauties of communion with the ages, or of sharp analytical engagement with current issues, no longer justify the ethics class, not in their minds, and, more importantly, not in the minds of the parents and loan officers. The change in the economic climate has focused our students on money, but it has changed them in other ways, too. Part of the growing dislike of the classroom is, as above, their disregard for the professor's claims to truth; what if there's a truth I feel more at home with? This disregard, combined with their economic anxiety, leads to a conviction that all of classroom learning is irrelevant: work, solid work, beckons, as spiritual fulfillment as well as financial viability, and they want to get to it. Can we teach them ethics through work?

We can certainly try. Perhaps the major trend in university education now is not new at all; it goes back to John Dewey, who seemed to share our students' dislike for classrooms. It was called "learning by doing" when I was in elementary school, and now goes under the moniker of "experiential learning." The accepted description was given by David Kolb in 1984, as a cycle: the learners start in abstract conceptualization, by raising a question emerging from knowledge they already have, or accepting a problem to solve; they then move to active experimentation, developing skills and organizing ideas for the task; they then engage in concrete experience, working in the world to address the problem; then, that finished, they conclude with reflective observation, organizing what has been learned and integrating it with the frame of knowledge with which they started. David Kolb, *Experiential Learning: Experience as the Source of Learning and Development*, Englewood Cliffs, NJ: Prentice Hall, 1984. (An awkward formulation, perhaps, but that is the proposed model and everyone seems to accept it.) The attempt to integrate work experience into university education is also useful in satisfying employers' demands that students arrive at their doors with work experience in their dossiers.

An "experiential learning" assignment bears a ghostly similarity to a traditional apprenticeship, the process by which all the non-elites who didn't go to college learned the trade in which they would make their living for the rest of their lives. An apprentice learned good work habits (showing up on time, for instance), learned the use of the tools of the trade under the watchful eye of a master of the trade, learned how to produce what his trade was expected to produce, learned how to judge the quality of that product, for on that quality he would be judged as acceptable as a journeyman or not. It was important to get it right, on penalty of failure to advance in his trade, jeopardizing a life of profitable and respected work. Along the way, he (it was usually a he) learned how to explain what he was doing, to justify his work to the master, and to become a master and teacher in his turn. Some of the elements of an apprenticeship show up in experiential learning as now understood: understanding the goals of the work, careful supervision and correction by an authority, intense effort on the part of the apprentice to understand what was demanded and to get it right, and the transformation of the callow beginner into a confident performer. Of course, this is happening in the university, which changes things. In the natural/laboratory sciences, the apprenticeship model can be closely followed, only with more textbook material. In business, internships in companies that are part of the industry the student wishes to join as a graduate may follow the apprenticeship pattern very closely, and if the internship works out, the student may well be employed by the company. But in history, psychology, philosophy—and ethics—the link between current real-world experience and lifetime professional engagement is tenuous.

First of all, it may be difficult to formulate precisely how a particular task, or any task at all, will improve a student's insights in ethics; the claim that any task could do that may in fact be a contradiction in terms. So we tend to back off "ethics" and send students off in hopes that the task will deepen the student's moral sensitivity, or compassion, qualities that move below the surface of ethical thought (although they make ethics relevant to the human condition). What we are doing, in fact, is providing the students with case studies in which they have played a part, so that when they return to the classroom to consider the standard case studies in the field, they will do so with an emotional grasp of the importance of thinking through the implications and consequences of action taken in the case.

We send students off to work with local corporations, hospitals, native Americans, homeless shelters, refugee resettlement organizations, and food banks, expecting that students' contact with the world in which the ethical problems arise will prepare them for actual work with those institutions, and that their contact with the disadvantaged will show them the importance of insisting, after they graduate and take the places for which their privileged lives have prepared them, on consideration of the rights and interests of those enjoying less privilege than themselves. It's a good idea, and the students say they have profited from the experience. Those of us who have tried this know the comedy-of-errors pitfalls that we can land in if we try to supervise these projects ourselves, and the battles with university administrations that will follow if we try to obtain adequate supervision for them. For these assignments are, in the end, terribly expensive: the locations to which the students are assigned may ask for fees, sometimes extra insurance must be purchased, a full-time non-teaching project coordinator is essential if the programs are to be run smoothly, and transportation often must be provided. This is the way to extend the students' experience in ethical reasoning, but very few small colleges have the resources to run the program well.

Meanwhile, the students preparing for the traditional professions (that require advanced degrees, so mandate a spell in graduate school) may profit less from these "work experiences." They can still be engaged with the dilemmas of the professions that we have been teaching to medical students (and practicing physicians) for many years. What are they going to be when they start paying the bills on their own? They will be something identifiable in the society—physicians, nurses, other health care professionals, lawyers, engineers, accountants, corporate officers of some other sort in the fields of business, varieties of service professionals, maybe farmers (not likely). But whatever they're going to be, they will be in fields that have their own ethical codes and standards, and against those, a wide selection of ethical dilemmas—scenarios where good and bad meet and balance, and hard thinking is going to be necessary to find ways past them. Then possibly the best idea is to encourage the students to formulate the identities they will have after college, start thinking about them now, then confront dilemmas typical of the fields they have chosen.

I've had some luck asking students to consider what their chosen identity might entail in the way of commitments to rules and principles of ethics. If the appropriate ethic is simply the inner order of the profession—the morality without which the profession cannot be practiced—the students can usefully access the literature in which this order is teased out and articulated. Then the discussion can proceed to some of the dilemmas that have confronted and continue to confront practicing professionals in the chosen fields (the fields identified by the students as their future). Generally I have started with some of the well-known cases in the fields—Baby Doe for the pre-meds, the Pinto case for the budding automobile tycoons—then go on to newer cases, even now emerging in the pages of *The New York Times*.

Conclusion, of a Sort

I end with no real solutions to the problems of teaching ethics in the current climate. I know we must hold on to the principles of our profession: to teach students the truth, starting with the conviction that there is such a thing as truth, and to instruct them in the vocabulary and forms of reasoning of traditional ethics, just so they'll be able to participate in conversations on the subject. I believe that the experiences of "experiential learning" units are valuable, for reasons that have nothing to do with the teaching of ethics and everything to do with the breadth and depth of their humanity. I sense that the university now is very different from the university where I began my career 50 years ago, and I am plagued by a feeling of inadequacy to address its current distress, financial and moral. Yet students will learn, as they always have, probably as much from taking their cues from us as role models as from reading the books we give them. May we prove worthy of the role.

Appendix

Materials for handling cases of some complexity, taught through case studies: some examples. Note: these passages are largely taken from my *Ethical Decision Making: Introduction to Cases and Concepts in Ethics*, New York: Springer, 2013.

ORDER: Confronting Complexity

A problem of moral concern, rules or character, is not the same as an ethical dilemma. In a moral problem, we know what is right, but we may have very good reasons not to want to do it, or we may be puzzled about the right way to say "No" or the best means to obtain the best outcome. (Additionally, we may be tempted to preserve our level of comfort by doing nothing at all!) In an ethical dilemma, we really do not know the right thing to do. Consider the following case:

> Peter and Dora Vlasovic, 51 years and 43 years of age respectively, are at a loss as to what to do about Dora's 67–year old mother, who lives with them. She is suffering from Alzheimer's disease, and while her periods of confusion are not yet continual, she is becoming too unreliable to be left alone. Both husband and wife work outside the house, and cannot stay with Mother during the day. They started looking into appropriate Nursing Homes, but Mother, who was a schoolteacher and fiercely independent all her life, has put her foot down: no Homes. "Look, you know how valuable my mind and my dignity have been to me," she finally said to them. "I simply cannot endure the thought of ending my days tied in a chair drooling on my lap. See that pillow on my bed? When I can't function any more, my life is over, as far as I'm concerned, and I want you just to put that pillow over my face and sit on it for about twenty minutes. Just call the doctor in the morning and say I died in my sleep. I won't contradict you. No Homes."

Meanwhile, the couple has found out that the cost of these Homes is well beyond their means, and that Mother will be left on Medicaid after her assets are gone. Their own assets are not large, and they would be totally responsible for the costs if they hired nurses to come to the house to take care of her. They also have teen-age children, approaching college, to think about, who will need money for college and probably financial help as young adults after their education is finished.

"How did people used to handle this type of situation? Before there were Nursing Homes?" Dora had once asked her doctor. "Easy," he had replied, "People didn't used to live this long. And when they did, in this state, with no other choice in the matter, people simply left them home alone, tied down or roaming wherever they wanted to roam." That did sound "easy" to Dora, but on balance, worse than the other alternatives. They have the house to think of, too, and leaving Mother alone all day sounds like playing Russian Roulette with house and Mother both.

But their first concern is for Mother. The life projected for her does not really seem to be worth living, the more they think about it. "Putting a pillow over her face" is a dreadful thought, of course, but it is what Mother wants, and if they can't face the pillow, the Hemlock Society advertises many more humane ways to bring life to an end. Should they go the pillow route? Should they explore the "rational suicide" alternatives with Mother? Or should they insist on the Home (or put her there anyway after she is no longer organized enough to resist)? Or should they devastate their own financial resources with hired nurses? Are there other alternatives?

In our attempts to reach the good or just solution in this case, what approach do we take? Typically, if we are (personally) in the middle of cases like this, we are strongly tempted to grasp at whatever "solution" appeals to us at the moment: that is, whatever solution accords with our previous prejudices and tendencies. But one of the major objectives of the teaching of ethics is to draw us beyond that subjective stance to one where all rational persons could agree that the right course, or a right course, is being pursued. That means that we must reach a course of action that is objectively right, or at least open for public scrutiny.

What would constitute an orderly approach to such problems? First, as participants and decision makers, we should

organize our options in the situation—what alternatives are really open to us? and note the probable **outcomes** of each. What, in this situation, is it possible, and reasonable, for us to do? And what will be the likely results of each of those choices? Which of the outcomes on the list are totally unacceptable? They should be eliminated, and the rest left for further consideration at a later stage. In this step, we are reasoning *teleologically* or *consequentially,* looking to the means that will produce the most desirable ends.

The Vlasovics, in this case, have the options of

1. Leaving Mother alone—and risking her and the house. That's not acceptable, save for very short periods of time.
2. Bringing in nurses by the day. That will turn out to be very expensive.
3. One of them quitting whatever else they're doing and just taking care of Mother. That will lower the family income, for all purposes, substantially, and no one wants either the burdensome task or the loss of income.
4. Putting that pillow over her face. The thought makes everyone queasy, and they really don't want to go to jail.

5. Putting Mother in a home. She will complain, but she will be safe, and the rest of the family can continue their own lives. This may also be very expensive, until they can establish Mother's eligibility for Medicaid.

Before they act, however, they must

review the rights of the various participants, for legally protected rights, in our system, trump, or override, considerations of right outcome. We must also respect moral (usually legally enforced) **rules** that are held to be valid regardless of the consequences. That is an important point: in this step we are reasoning *deontologically* or *non-consequentially*; that which violates a rule is prohibited by that rule no matter what consequences flow from doing or omitting the act.

Two of the most powerful rights and rules confront us in this situation:

First, the *right of the individual to refuse* the well-meaning ministrations, for his or her health and safety, imposed by others without consent. Mother does not want to go into a Home, and that should settle that. Medicaid has nothing to do with it.

Second, the option preferred by Mother herself, the pillow placed over her face, violates a stringent *rule against voluntary homicide*, "thou shalt do no murder." It is not our purpose at this point to enter the emotional debate about the permissibility of assisted suicide or euthanasia, as requested by a competent patient. At the time that the pillow route would have to be followed, Mother would not be competent to request anything of the sort, and placing pillows over faces does not qualify as physician-assisted euthanasia. The act would be homicide, in fact murder in the first degree, and there are very good reasons why our society forbids it. If Pete and Dora take this option, they will have violated that rule.

Meanwhile, there are other rights to be taken into account. The minor children have an absolute right to their parents' support, for maintenance (food and shelter), affection (yes, that's a right), and provision of education. To what extent will care for Mother have an impact on them?

When we have our options clear and our rights and rules factored in, we should

determine our decision, make a **disposition** of the problem, for the moment. The situation will not wait, after all; an initial decision must be taken immediately. For the moment, Mother is rational, and peaceable, enough to be kept at home with her family; also, during her periods of lucidity, she enjoys being with them and they enjoy her. Perhaps a local daycare program can take her during school hours, and the teenagers can switch off afternoons to be with her before Pete and Dora come home from work. Nurses can be hired in occasionally to give everyone a break. The solution can't last forever; will it work at all? Note that the family *must* act, with very incomplete information. That imperative is typical of such dilemmas. Pete and Dora decide to try it. And then, in a few days, weeks, or months, they must

evaluate the effects of the decision. The decision and the action do not, as Macbeth pointed out, trammel up the consequences. The world continues. We need to follow up, to find out what results our decisions have had. The Vlasovics, in this situation, will not be able to avoid the results; Mother is still in the house with them. How much are the teenagers losing from their sacrifice of their afternoons? The answer to that will depend very much on the peculiarities of this family's situation. How is Mother responding to the new program? That depends very much on the peculiarities of Mother. The trouble with ethical dilemmas, as opposed to ethics as a discipline, is that the real solution is empirical, day to day, trial and error. Finally, we have to

review the situation, reconsider the decision, with an eye toward *revision*. Nothing, in human affairs, is ever set in stone. We make our decisions, usually, for today, knowing that the decision will probably produce a new situation, with its own new dilemmas, and we will have to take on the whole problem again. The Vlasovics' decision to keep Mother home without round-the-clock nurses, bringing her to day care as often as they can, has saved them money, but after a while it will not work anymore: most day care programs cut off when the disease renders the patient violent or incontinent, and new arrangements will have to be made. But by then, there will be a new situation, with a different set of options, and possibly, a revised set of rights. The children, for instance, will eventually leave home, and confront their parents with a completely different set of demands. The federal government, for another instance, changes its mind every month about what programs to fund for the elderly; these will have to be taken into account in future deliberations.

This decision procedure, like many others in the field of ethics, covers all necessary bases for rational decisions. I prefer it to the others only because it builds in, as others do not, the recognition that nothing is ever decided–not well, anyway–once for all. The temptation to come to resolution, to solve something forever, is enormous. Resist it. Situations change, and the more flexible our decision procedure, the better suited it is to the messy world of human conduct.

This decision procedure for ethical dilemmas can be remembered easily by its acronym, ORDER:

O: options and outcomes
R: rights and rules
D: determination, decision
E: evaluation of effects
R: review, reconsideration

While we're at getting down procedures that are easy to remember, we may take note of three preliminary steps that have to be taken before we can put things in ORDER:

First, we have to **define the dilemma** that we are facing. What conflicts make the situation difficult to deal with?

In the Vlasovics' case, the dilemma is painfully evident: how to ensure Mother's and the family's *welfare* while respecting Mother's *choices* and the most serious *rules* of our society, while allocating the family's not-abundant resources *justly* among the generations that call upon them.

Second, we have to conduct empirical **inquiries** as appropriate, discover the facts, get as much **information** as we can.

What day-care programs are available? What about support groups, for patients and caregivers alike, at the local hospital? Can we get her church involved? How fast is Mother's disease progressing? What should we know about advance directives, living wills, therapies? Our options, once the decision procedure is engaged, will depend upon what is available.

Third, we have to **sort out the stakeholders**. We mentioned above that it is important to know whose interests are to be taken into account in making any decision.

Part of the work of sorting out the stakeholders is to make sure that all whose interests are really affected are taken into account. Another part of the work is to see that *non*–stakeholders who seek to attach themselves to a decision—the nosy neighbors, for instance, who don't like the cars of the visiting nurses parked (legally) on the street—are excluded from influence on the decision. Of course, that means that we may not take *their* happiness—the satisfaction they derive from running other peoples' lives—into account. The rights of the family take precedence over the preferences of their neighbors: *rights trump likes and dislikes*. This is why mere appeals to the "greatest happiness of the greatest number" are not always sufficient to decide ethical dilemmas. On the same principle, more commonly, we do not allow neighborhoods to exclude persons of an ethnic background different from that of the current residents, even though it would make all the neighbors overwhelmingly happy to be able to do that. The right of the minority family to live where they choose trumps the preferences of the neighbors not to let minorities live there. Of course the neighbors *are* stakeholders to some extent—they certainly have a right to be protected from Mother's wandering, should it come to that. The Anglo-Saxon Common Law, of which we are the inheritors, has spent patient centuries working out the details of the rights that people have vis-a-vis the neighbors, and we must be conscious of the whole corpus of that tradition.

So our first three determinations, in any ethical decision process, are of the definition, the factual information, and the stakeholders. If it makes it any easier to remember, think of these steps as a "DIS" preface to the "ORDER" decision procedure:

D: Definition of the Dilemma
I: Inquiry to obtain all necessary Information
S: Sorting out the Stakeholders

This leaves us with a procedure whose steps are easy to remember, but leaves the field of ethics in **DISORDER**! Insofar as this DISORDERed formula helps us to remember the essential messiness and anguish of ethical dilemmas, that serves our purposes very well.

Internecine Strife

Wade L. Robison

Abstract To an outsider, those within a profession seem of a kind: historians are historians, lawyers are lawyers, doctors are doctors. But to those within a profession, the nature of what it is to be a professional of a particular kind can be contested—with interesting ethical implications. It is a continuing disagreement in law whether lawyers should be hired guns, doing whatever needs to be done, even pushing against the law, to help their clients, or whether they are guardians of the law, with a higher obligation to justice. This kind of internecine dispute is a neglected aspect of professional ethics, but worth exploring because it is unclear how we can resolve such a dispute and because different conceptions of a profession have competing ethical implications. Indeed, competing conceptions of what it is to be a lawyer are founded on competing moral principles about what a lawyer ought to do. We thus cannot simply appeal to the norms of the profession because those are precisely what is in dispute. If a professional is of such-and-such a kind, a hired gun, say, then the professional is acting properly, ethically, to follow the relevant norms, but if a professional is of another kind, then another set of norms becomes relevant.

Keywords Profession · Ethics · Professional norms

We might think we could appeal to the norms of a profession to settle any issues that may arise about the conduct of a professional, but those within any profession can disagree about what it is to be a member of a profession and disagree to the point of casting out someone who adopts a different understanding of what professionals in that profession ought to do. We shall begin with an example from historians and then provide examples of other disputes within professions about the very nature of the profession.

W. L. Robison (✉)
Rochester Institute of Technology, Rochester, USA
e-mail: wade.robison@gmail.com

E. E. Englehardt and M. S. Pritchard (eds.), *Ethics Across the Curriculum—Pedagogical Perspectives*, https://doi.org/10.1007/978-3-319-78939-2_12

Historians

Stephen Ambrose's three-volume biography of President Eisenhower became the standard in part because he claimed it rested on extensive interviews he had with the President for several hours at a time twice a week over a long period. The discovery that he never met with the President alone and only met with him and others once or twice, and not in total for more than two-and-a-half hours at most, meant that his biography instantly lost all credibility as a work of history.

The judgment implicit in that loss of credibility rests on the universal agreement that historians are to tell the truth about the past. Thucydides begins his histories by saying, 'To hear this history rehearsed, for that there be inserted in it no fables, shall be perhaps not delightful. But he that desires to look into the truth of things done, and which … may be done again, or at least their like, shall find enough herein to make him think it profitable' (Thucydides 1989, 14). By the 17th century, the understanding that historians are to tell the truth was so embedded in the profession and elsewhere that in his extensive dictionary Pierre Bayle could ask, rhetorically, after remarking that historians can upset their readers with the truth, 'Does this free a historian from the obligation he is under of relating the truth with all possible exactness?' (Bayle 1826, 214). As David Hume puts it, 'The first page of Thucydides is, in my opinion, the commencement of real history. All preceding narrations are so intermixed with fable, that philosophers ought to abandon them to the embellishments of poets and orators' (Hume 1985, 422).

However historians may agree about the necessity of an historian's telling the truth to be an historian, they are in disagreement, sometimes vehemently so, about how to tell the truth—and even what it is to tell the truth and whether that is ever possible. We would need an historian of historians for a full and accurate picture of the disagreements over what it is to be an historian, but it will be enough for our purposes to tease out just a few of the issues that can separate historians one from another.

One tension within the profession concerns the way in which historians write. As one historian puts it,

> The gradual withering of the narrative impulse in favor of the analytical urge among professional academic historians has resulted in a virtual abdication of the oldest and most honored role of the historian, that of storyteller (Foote 1975).

This lament from an historian who writes narratives must be paired with a critique of Gordon Wood, one of the most honored of contemporary historians:

> From the perception of many contemporary historians, one could argue that *Radicalism* was the moment at which Wood went from being a neo-Whig historian to an old-school Whig historian. One could also argue that the real change was from being an academic to a more-popularly-oriented historian (Hattem 2013).

This is not a minor disagreement, as it turns out, but a major dispute about how an historian ought to write. *The Glorious Cause: The American Revolution, 1763–1789*, is volume II of the Oxford History of the United States, and in his review of that work, by Robert Middlekauff, Gordon Wood is unsparing in his criticism of Middlekauff for not recognizing 'that society and culture transcend the particular aims and purposes of individuals, that people make their social and intellectual history but are at the same time bound by what they have made.' Middlekauff thus writes of the Revolutionary War, Wood observes, as though people 'are free-acting autonomous moral agents whose motives and actions have clearly defined consequences.' Personal blame and praise is thus appropriate, and what is left out are all the social and economic forces that would explain why, for instance, 'Revolutionary Americans suddenly broke from English constitutional practice and effectively barred the subsequent emergence of parliamentary cabinet government.' Wood sums up this particular criticism by saying, 'Middlekauff's account of the formation of the Constitution virtually sets back scholarship on the issue at least a century' (Wood 1982a).

This is from one famous historian about another famous historian and is an indication of the depth of disagreement about how historians are to write and, much more important, about the very nature of history. This has been referred to as the 'split in the American historical profession between traditional narrative and quantitative social science' and raises, obviously, both the hackles of its competitors and serious epistemological questions about how we are to come to knowledge of the past (Wood 1982b).

This split is not just between historians who write narratives and those who wrote 'historical monographs,' as Wood puts it, but between those who think history is created by the great individuals in history—e.g., George Washington—who make all the difference and those who think social and economic and other forces are what matter in historical change.

One obvious moral implication is that a citizenry primed with the former view is all the more likely to take seriously some self-proclaimed 'great man'—like Hitler or Mussolini—who says, 'I am the only one who can fix it.' If social and economic and other forces are what matter for societal change, then appealing to a 'great man' means ignoring the causal factors responsible for the troubles a society is having.

This split between historians is by no means the only one. Barbara W. Tuchman once wrote that she writes history 'not "to instruct but to tell a story"' (Wood 1984). Some historians think one point of writing history is to instruct and could argue that there is no way historians cannot instruct when writing history. We are all familiar with the substance of Edmund Burke's remark: 'Those who don't know history are doomed to repeat it.' The point is that we are to learn from history how to avoid the mistakes others have made.

It should be no surprise, however, to find historians criticizing books of history that are written 'not merely to impart historical information but to teach moral and political lessons.' Such books 'reflect an assumption that they are to serve as what are now called "modes of socialization," as sources of values, aspirations, and models to follow whose influence lasts far beyond childhood' (Foner 1972).

The ethical implications of such a view are obvious—with different values being hawked depending upon the differing views of individual historians. An historian who thinks Elizabeth I of England by far its best monarch is teaching moral and political lessons far different from an historian who thinks Henry VIII the best model to emulate.

We hardly need any other examples of the contested conceptions historians hold about how to write history, but here is one more. Francis Parkman wrote a seven-volume history of *France and England in North America* from 1865 to 1892, and though it was lauded at the time, it fell out of favor because, among other things, the histories were dominated by his belief in the inevitability of progress, by his acceptance of the Great Man theory of historical change, and by an anticlericalism so strong as to affect his judgment of events at crucial moments (Taylor 1983).

We see again the charge that an historian has adopted the 'Great Man theory of historical change,' but two new charges are added. One is that historians should not assume that events proceed inevitably towards progress. The other is that an anti-religious bias has affected Parkman's objectivity.

Even with this last example, we have by no means exhausted the contested conceptions historians have of their fundamental goal of telling the truth. As Gordon Wood puts it, '[T]hat we historians are telling the truth is what distinguishes us from fiction writers' (Wood 2009, 109). The seemingly universal agreement that telling the truth about the past is the role of historians—part of their role morality, what they ought to do as historians—obscures serious disputes about how historians are to tell the truth. These disputes are internal to the discipline in the sense that historians disagree with each other about what they ought to be doing as historians, but the disputes have significant consequences for the public.

Some of these are epistemic. When we discover that a famous historian like Stephen Ambrose lied about events so central to one of his major works, a lie that undercuts its historical value, we ought to be at least moderately chary of taking historians at their word. When we discover, in addition, that even they disagree about how to fulfill their overarching aim to tell the truth, we ought to wonder what we are to accept as historically accurate and what not. We outsiders have no way to judge these disputes or to determine how they affect what various historians write. So we have no way of knowing whether what we are reading has been shaped by a contested conception of history and so no way of knowing how to protect ourselves from any biases that have entered into what we are reading.

But a second set of consequences is of more immediate concern. The differing conceptions of how historians are to tell the truth about the past have moral implications. They are not just different ways of approaching the past—as though we could read a multitude of books about the Civil War, say, and simply pull their narratives together into a single coherent whole. They are different ways of understanding the past—as object lessons for what we ought to avoid, as describing individuals who model how we ought to behave, as evidence for the inevitable progress of humankind and so a directive about how we ought to behave to further that progress, and so on. These different understandings of the past have different moral implications, giving us moral guidance of different sorts.

These disputes within history are not morally neutral, that is, and in some cases at least rest on competing ethical principles about what an historian's role is. Wood is not just saying that Middlekauff fails to provide an adequate understanding of why our constitution ended up as it did. He is saying that Middlekauff's account is *harmful*, setting 'back scholarship on the issue at least a century.' That is a moral claim, based on the principle that historians are obligated to further the study of history, not undermine it through carelessness or negligence.

Such disputes within a profession are all too easy to find, and some are perhaps of more obvious concern to the public than others. We shall look at one of prime importance to public employees.

Accountants

Accountants have the seemingly impossible task of providing an objective evaluation of what a corporation, say, is worth. It seems an impossible goal because what is required for objectivity is that accountants keep books in accordance with GAAP, the Generally Accepted Accounting Principles. The advantages are obvious. There is a single standard, supposed to be used world-wide, that allows investors and governments to compare the financial health of whatever they are providing an accounting for. The standard is also objective because GAAP requires that the value of a piece of property, say, be the price at the last transaction—an objective measure of what someone was willing to pay for the property. We know that the value of a piece of property does not stay constant: its value can go up or down depending on all sorts of variables, e.g., whether the economy is growing or not, whether some undesirable neighbors have moved in, whether the property is kept up, and so on. But using the selling price as GAAP requires ensures that accountants do not have to estimate what a piece of property would sell for were it now on the market. Estimates can vary enormously, depending on what variables someone takes to be relevant and on what judgment is made about how the property's future will play out. The last selling price is a real number stating exactly what a buyer was willing to pay.

But as we know, that selling price need not reflect the true value of the property. A company may have purchased property to expand only to discover afterwards that the ground is toxic and the cost of recovery many times the original cost of the property. So if the company paid a million for it, it will be listed on the books as worth a million—as an asset. But the company cannot sell it for a profit or even for what it cost since its actual market value is a million minus whatever the cost of the cleanup—$30 million? The company could not give it away, that is, but would have to fork out the cost of the cleanup to get rid of it. The one million asset on the books would become a thirty million liability. An objective evaluation in accordance with GAAP requires accountants to use the last selling price of a property, its value on the books. But the book value need not be the market value.

Accounting firms have found this a less than trivial problem. They have been sued for failing to provide investors with the information they need to know about a company's real, i.e., market, value. Audits indicated the savings and loan institutions in the 1980s were solvent because they calculated the book value, but the books hid the losses from properties those institutions purchased at prices significantly higher, in many cases, than they were worth. The institutions just kept the properties on the books and so looked solvent when, had their real worth been audited, they would have been seen to be insolvent (White 1991). Ernst & Young had to pay $400 million in 1992 because it 'had improperly audited federally insured banks and savings institutions that later failed' (Mills 1994). KPMG had to pay $97 million to the state of Victoria, Australia after being sued by the government for a 1988 audit of 'the Tricontinental Group merchant bank [that] failed to disclose its problems'—liabilities that totaled $1.85 billion when it collapsed shortly after the audit. KPMG's defense was that it followed GAAP in its audit, but following GAAP failed to disclose the bank's market value. The state of Victoria had put money into the bank, following the auditing, and lost it when the bank collapsed shortly afterwards.

Courts effectively held that the accounting firms had failed to act in the public interest—'clients, credit grantors, governments, employers, investors, the business and financial community, and others who rely on the objectivity and integrity of certified public accountants to maintain the orderly functioning of commerce' (Code of Professional Conduct of the American Institute of Certified Public Accountants, 0.300.030). It is in the public's interest to know the true worth—the market value—of a corporation, say. Investors cannot make reasonable decisions about where to invest without that knowledge.

The tension within accounting is between providing objectivity through GAAP's requirement to use the last selling price for the value of an asset and acting in the public interest to give its market value. It appears that accountants cannot do both (Oliver and Robison 1995, 3–11).

This tension is internal to accounting and may seem unlikely to produce the kind of internecine conflict we have just examined among historians, but the same sort of problem has produced just such a conflict between members of the American Academy of Actuaries. Public pension funds keep two sets of books—the official record and the market value. The official record is the one that those with pensions and those responsible for pensions see. The book for the market value is kept private.

The two sets of books need not correspond, and problems arise when the market value is significantly less than the stated 'official' value. Carmel-by-the-Sea discovered that the market value of its pension fund was $48 million short of what it owed the city's retirees—a very unwelcome and unexpected surprise since the official value showed the debt to be a quarter of that. Having two sets of books, one 'official' and the other the true, or market, value,

> raises serious concerns that governments nationwide do not know the true condition of the pension funds they are responsible for. That exposes millions of people, including retired public workers, local taxpayers and municipal bond buyers — who are often retirees themselves — to risks they have no way of knowing about (Marsh 2016).

A small pension fund in California showed 'far more money than it needed' and so decided to convert to a 401(k) plan, only to discover that it owed 'more than half a million dollars' (Marsh 2016). The fund knew the book value, but not the market value—which the pension system kept secret. Those in the pension fund faced a risk they did not know about—and had no way of finding out until it was too late to make an informed decision.

There are two sets of books because one, the official one, calculates the value of a public pension's funds at an average of '7.6% a year' while the market value is calculated at a 'riskless rate, currently below 3%.' If a portfolio were to accrue in value by 7.6% per year, it would not take too many years for its apparent worth to be far more than it would be were it to accrue in value by less than 3%. Funds calculated at the higher rate are now one 'trillion short of the money they will need to fund pension credits that workers have already earned. But if pension systems were required to use a riskless rate, currently below 3%, the shortfall would soar to more than $3 trillion.'

Using an optimistic rate of return has consequences far beyond keeping everyone but those who manage the pensions in the dark. Jeremy Gold is one of the authors of a paper that was to be published by 'a 14-year-old task force on pension financing' of the American Academy of Actuaries and the Society of Actuaries (the AAA and SOA). As he noted, 'Consistent lowballing of pension costs over the past two decades has made it easy for elected officials and union representatives to agree on very valuable benefits, for very much smaller current pay concessions' (Malanga 2016).

The thesis of the paper Gold and others wrote was that 'many state and local retirement systems calculate their obligations using overly optimistic future rates of return. The authors want states and municipalities to adopt new valuation standards that would make projecting the cost of future benefits more predictable' (Malanga 2016).

The task force was shut down by the AAA and SOA and denied permission to publish the paper. The paper was 'a work product of the joint PFTF,' they said,'intended to be published by the academy and the SOA as a jointly owned and copyrighted paper.' But the paper will not be published or endorsed by either group, and,

> Because the paper was the work of the joint task force, we do not think it would be appropriate for members of the task force, as individuals, to take the existing paper and simply publish it somewhere else. We recognize that some of the individuals who have been on the PFTF have their own personal ideas and views on these topics, and the academy and the SOA encourage those individuals to express those ideas in other forums. But they cannot use the existing paper, with the particular expressions of ideas as developed by the task force, as the vehicle to do so (Burr 2016).

An internal tension between how to calculate the value of pensions has thus split actuaries, some arguing that overly optimistic projections of the value of public pensions leave governments and those who are to get pensions ignorant of the real value of their pensions, others arguing, it appears, that optimistic projections are justified because 'governments don't go out of business the way private companies do,... [they have] a much longer window to recover from bad investments' (Malanga 2016).

Whatever the cause of the tension, it has significant moral implications. We would rightly object on moral grounds were a pension fund to lie to those expecting pensions, telling them they would be paid in full, say, when the fund was about to go insolvent. Those expecting pensions would retire and discover that the money they expected to live on in retirement was not there, exposing them and those dependent upon them to poverty and all the attendant ills. But the public pension funds might as well be lying given the effects of its failure to disclose the true value of funds. The perception everyone would have rests upon what is made public, and what is made public is an optimistic projection of what the fund could be worth if it paid a high interest rate, rarely achieved.

We can see the effects of such optimistic projections in various municipalities across the United States. Dallas has ‘a hidden pension debt of almost $7 billion,’ the result of a decision in 1993 by state lawmakers who ‘sweetened police and firefighter pensions beyond the wildest dreams of the typical Dallas resident. They added individual savings accounts, paying 8.5 percent interest per year, when workers reached the normal retirement age, then 50. The goal was to keep seasoned veterans on the force longer.’ The actuaries involved pointed out that the assumptions the legislature was making ‘were shaky.’ So they did their duty. The legislature did not, and the mayor of Dallas now says ‘his city appeared to be “walking into the fan blades” of municipal bankruptcy’ (Walsh 2016).

The problem regarding accountants with which we began this section is far more difficult to solve than this tension regarding pensions. We cannot figure out any easy—or even complex—way to solve the accounting problem. Accountants cannot obtain objectivity by guessing the worth of a piece of property were it to be sold, but when they appeal to the price at which it was obtained, they achieve an objectivity which does not reflect the current value. The value they provide is the value the property did have, not what it does have now.

Accountants are not going to split into opposing factions because of this problem. It is internal to accounting itself: the norm for achieving objectivity, GAAP, guarantees that accountants will not obtain an accurate accounting of a company’s current worth. So appealing to the norms of accounting is not going to help accountants any more than appealing to the norms of history will help historians settle the question of how to tell the truth.

The dispute between actuaries is of the same sort as that between historians. It cannot be settled by appealing to the profession’s norms because what is at issue is what the norm for calculating the value of pensions ought to be. That the dispute runs deep is evidenced by the AAA and SOA not only refusing to endorse a report of its own task force, but also insisting that the authors are not to publish it on their own.

It is more than a little unfortunate that the split has such implications for the public’s ability to make informed decisions about the future—especially for those individuals whose life savings are tied up in pensions. Their being able to live with some comfort in their old age depends upon knowing how much has really been put aside and thus whether they need to save more.

Other Professionals and Lessons Learned

The split among actuaries, as with the split among historians, cannot be settled by appealing to the profession's norms. Those norms are just what are in question. Is it a feature of an actuary's role to provide an accounting of public pensions that reflects a realistic rate of return or one that reflects, on average, an unrealistic 7.6% rate of return? Despite that question having a rhetorical ring to it, its answer will depend upon actuaries coming to grips with what it is to be an actuary—what obligations an actuary has to the public and how those obligations are best fulfilled. The answer depends not upon a clear understanding of an actuary's role that we have at hand, but upon the results of a reasoned discussion within—and without—the profession about that role and an examination of the ethical implications of choosing one understanding of that role over another.

This sort of tension between actuaries and between historians can be found in many other professions. When DNA revolutionized biology, biologists trained in the old style found themselves marginalized. A biologist at the University of Wisconsin complained that he had been a professor of biology, but after DNA-savvy biologists came to dominate the department, he found himself relegated to a 'field biologist'—a downgrading of his status, he thought, and a source of tension about the nature of biology. We are all familiar with the dispute among lawyers as to their fundamental role in defending clients: should a lawyer be a hired gun, doing whatever can be done to defend a client, including, for instance, allowing a client to lie before a jury or judge, or should a lawyer hold the integrity of the legal process of more importance?

Such a dispute about the role of lawyers obviously has ethical implications, and the dispute is grounded on either side on moral principle. If the presumption of innocence has any importance, it is argued, then surely the burden must be on the state to prove guilt beyond a shadow of a doubt—within the existing legal framework. As long as a lawyer does not break the law, anything goes in defending a client, it is claimed, because it is a moral imperative that guilt must be proved. Otherwise innocent people may be convicted. On the other hand, it is argued, although the legal system is an instance of imperfect procedural justice, with some guilty being found innocent and some innocents found guilty, the integrity of the system depends upon the trust that no one is trying to game the system and that a defendant's lawyers are doing their best to prevent a proof of guilt—within the confines of the existing legal framework. They are not to bribe the judge or members of the jury or coerce any witnesses into recanting their testimony or making up new stories. Our trust in the integrity of the legal system depends upon our trust that members of the legal profession—the lawyers, the judges, the court stenographer—are doing what they ought to do to maintain and further that integrity.

Such a dispute about the role of lawyers comes to a head when a lawyer is told by the client of an intent to lie upon the witness stand. The lawyer ought to try to convince the client that lying is a mistake, that the prosecuting attorney is bound to catch the client out, and that the consequence will surely be a conviction when the jury becomes convinced the client is lying. But if the lawyer is unable to persuade the client, what ought a lawyer to do?

Lawyers disagree, some arguing that a lawyer has what amounts to an absolute obligation to defend a client even if the client lies under oath, others arguing that as an officer of the court a lawyer cannot let a client lie without telling the judge of the client's intent, effectively ensuring a mistrial since the attorney/client relationship has then been breached. We cannot appeal to a legal norm to determine who is right because it is the norm that is in question.

We can provide one more brief example. Physicians are trained to learn how to look at patients as mechanisms, as it were, like bicycles that have some failing or other or have been in some accident and are mangled. We want that objectivity from physicians. The last thing we would want is for our physicians to turn red with embarrassment upon seeing us undressed for an examination. But physicians must also relate to us as human beings, not as mechanisms, with a bedside manner quite at odds with the attitude of a bicycle mechanic telling us about what needs to be done to our bike to make it serviceable again.

It makes a moral difference, obviously, how our physicians treat us. If our physicians cannot see us as mechanisms subject to breakdown and breakage, they will not assess our condition properly and will not provide us with the medical care we need. But if our physicians cannot communicate to us, person to person, with the care and compassion we rightfully expect from a caregiver, we will properly feel objectified—not a patient, but an object for inspection. The necessary communication between physician and patient will suffer accordingly.

There is more to say about such internal disputes within the medical profession and legal profession and, indeed, many other professions, but we have enough to conclude that:

- Not all disputes can be settled by a profession's norms since there can be internal disagreements within each profession about its nature and thus about its norms.
- These disagreements are not ethically neutral and, indeed, often rest on competing ethical principles about what a profession's role is.
- How these disputes are settled, if they are, is a matter of ethical concern not just to those within a profession, but to the public at large since no matter how they are settled, they will have a great impact on the public, with some settlements causing great harm.

The bottom line is that the norms of a profession do not provide definitive answers to what those within a profession ought to do. As we saw with historians over their disagreements about how to tell the truth and with actuaries with what rate of return is reasonable to project, the norms can be contested. Such contests cannot be decided by appealing to what is itself being contested, the norms of the profession. Those contests must be determined on other grounds, the reasons for adopting one conception over another.

Perhaps a more dramatic way of putting this is that appealing to the codes of ethics of the various professions will not, in itself, settle any moral issue. The codes are at best a guideline to what, at some particular time, those, or some of those, in a profession think the profession and members of the profession ought to do. That a profession has reached enough of a consensus at a particular time to agree on a code of ethics is no guarantee that the code is exhaustive of all the ethical issues those in a profession will meet or even a guarantee that what it says about any one ethical issue is definitive.

The requirement that a profession reach agreement about its code pushes its provisions towards an acceptable generality that is of little help in specific cases. Even a provision that may seem obvious on its face—as we saw with the fiduciary rule that financial advisors are to act in their clients' best interest—can be contentious and contested within a profession.

Whenever we have a serious ethical issue, we cannot just appeal to the norms or the code of ethics of a profession. We must instead provide reasons for making whatever decision we make, reasons that presuppose an objective understanding of what is at issue. Getting that objectivity can be itself the most difficult of undertakings. Serious ethical issues generally involve cases where both sides in the case have what they take to be good ethical reasons for deciding in favor of their side rather than the other. Objectivity requires that we hoist ourselves out of our own position to put ourselves in the position of those on the other side who think themselves in the right, letting go of the passion that underlies our commitment to our side to see how others could be as passionate about theirs. Judges do this in considering cases where two parties care enough about an issue to spend the time, the energy, and the money to go to court and risk losing it all. A judge is obligated to understand thoroughly the reasons each side has for its view and then adjudicate between the two. Making moral judgments in serious cases requires the same of us.

References

Bayle, P. (1826). Scepticism. In *An historical and critical dictionary, selected and abridged from the great work*, (Vol. III, p. 241). London: Hunt & Clark.

Burr, B. E. (2016). Actuarial leaders disband task force, object to paper on public plan liabilities. *Pensions and Investments.*

Code of Professional Conduct of the American Institute of Certified Public Accountants (2017). ET Section 51, Article II, Section 01 (June 1)

Foner, E. (1972). Is this history necessary. *New York Review of Books.*

Foote, S. (1975). The Great American Butchery. *New York Review of Books.*

Hattem, M. D. (2013). *Where have you gone, Gordon Wood?*. The Juno: A Group Blog on Early American History.

Hume, D. (1985). Of the Populousness of Ancient Nations. In *Essays: Moral, Political, and Literary*. Indianapolis: Liberty Fund.

Malanga, S. (2016). Covering Up the Pension Crisis. *The Wall Street Journal.*

Marsh, M. W. (2016). A Sour surprise for public pensions: Two sets of books. New York Times.

Mills, J. (1994). Accounting settlement in Australia. New York Times.

Oliver, B., & Robison, W. L. (1995). Moral issues in accounting. *Professional Ethics, 4*(2).
Taylor, W. P. (1983). Repossessing America. New York Review of Books.
Thucydides. (1989). *The Peloponnesian War,* (T. Hobbes & D. Grene, Trans.). Chicago: University of Chicago Press.
Walsh, M. W. (2016) Dallas stares down a Texas-size threat of Bankruptcy. New York Times.
White, L. J. (1991). *The S & L Debacle.* New York: Oxford University Press.
Wood, G. (1982a). Star-Spangled History. New York Review of Books.
Wood, G. (1982b). Writing History, In response to Star-Spangled History. New York Review of Books.
Wood, G. (1984). The March of Folly: From Troy to Vietnam. New York Review of Books.
Wood, G. S. (2009). History as Fiction. In *The Purpose of the Past,* (p. 109). New York: Penguin Books.

Philosophy's Role in Ethics Across the Curriculum: Failures, Successes, and Suggestions for the Future

Phyllis (Peggy) Vandenberg

Abstract This paper is a report on the academic discipline of Philosophy and its engagement in practical/applied ethics—primarily including the teaching of ethics across the university. The involvement of philosophers individually and philosophy as a discipline in civic and public engagement or anywhere practical ethics is being discussed or taught will also be considered. This includes making policy in government, corporations, professions, or in any societal discussions where there is need for practical determinations of an ethical nature. In keeping with a common philosophical practice—the grounding of scholarship in the history of philosophy, tradition and the accepted canon—the chapter begins with the philosophical and public engagement of Socrates in 428 BCE. The conclusion argues that 18th century moral sentimentalist, David Hume, has insights into human nature and moral development that are helpful to effective philosophical engagement in the 21st century.

Keywords Practical ethics · Applied ethics · Public philosophy
Professional ethics · David Hume · Moral sentimentalism · Ethics and pedagogy
Philosophical ethics

Introduction

This chapter is a report on the academic discipline of Philosophy and its engagement in practical/applied ethics[1]–primarily including the teaching of ethics across the university. The involvement of philosophers individually and philosophy as a discipline in civic and public engagement or anywhere practical ethics is being discussed or taught will also be considered. This includes making policy in gov-

[1]I will use applied ethics and practical ethics interchangeably in this chapter.

P. (Peggy) Vandenberg (✉)
Grand Valley State University, Allendale, USA
e-mail: vandenbp@gvsu.edu

E. E. Englehardt and M. S. Pritchard (eds.), *Ethics Across the Curriculum—Pedagogical Perspectives*, https://doi.org/10.1007/978-3-319-78939-2_13

ernment, corporations, professions, or in any societal discussions where there is need for practical determinations of an ethical nature.

Brief History of Moral Philosophy and Higher Education

I begin with the philosophical and public engagement of Socrates in 428 BCE and end with the state of philosophy and philosophers' engagement in the 21st century. Beginning with Socrates specifically is useful because he modeled at least two approaches to the teaching of ethics. It seems the involvement of philosophers in practical ethics has an interesting history. There have been times over the years that philosophers have thought that practical ethics was a move away from philosophy. Practical ethicists were less of a philosopher than the theoretical ethicist. In some ways practical ethicists had to work at being accepted and respected in their own academic departments. As the demand for ethical education at the college level grew, philosophers in many universities were not interested in being part of it. One of the reasons for this is that philosophers tend to discuss dilemmas for centuries without a solution. I believe this to be the case when considering the teaching of practical ethics. Some believed that studying moral theory covered the practical without articulating or applying the moral theory specifically to actual professions or situations. My point is that some philosophers want to be involved in ethics but not the practical, situational type. Yet, there is an historical tradition for philosophers doing practical ethics.

To make this case, I will use Michael Pritchard's work where he claims an historical precedent for philosophers doing practical ethics. Specifically, Pritchard uses the words of Henry Sidgwick, Thomas Reid, and Socrates to put philosophers in the practical philosophy discussion. Starting with Socrates, Pritchard uses the *Euthyphro* and the *Crito* as illustrations of Socrates doing two different kinds of philosophical ethics (Pritchard, 15–19). In the *Euthyphro*, Pritchard sees Socrates as moving the discussion back to premises and more foundational issues in the determination of the good. In the *Euthyphro* Socrates meets Euthyphro on the court house steps where Euthyphro is about to take his father to court because of his father's role in the death of a slave. Who or what makes the good, good is what Socrates inquires and discusses with Euthyphro. The discussion does not really aid Euthyphro in the determination of whether he, in fact, is correct in taking his father to court. Socrates questions Euthyphro's assumption that he knows what is right; and he wonders at Euthyphro's certainty. Philosophers have taken this modeling in the *Euthyphro* seriously as proper philosophical discussion. Philosophers work to develop this ability that Socrates had to spot assumptions not clearly grounded or supported, to take moral conversation in a sense backwards, sometimes as far back as to be moving into the realm of metaethics. There is no conclusion in this discussion with Socrates that helps Euthyphro to know whether putting his father on trial is truly the right thing to do. Euthyphro may be more informed about the questions surrounding the sources of the good not necessarily being from the gods.

But that doesn't really help him decide about his father. The *Euthyphro* is philosophical dialogue at its very best, the philosophically respected meta-ethics discussion. Even though it is still the favored philosophical approach to ethics (discussed more thoroughly below) it is not really helpful to knowing the solution to an actual problem in real time facing real people that needs to be decided quickly and ethically. It is a good model for helping students understand what a decision can imply about the assumptions taken and we will address that case below. Thankfully though it is not the only way Socrates models philosophical dialogue. The *Crito* can be a much more helpful model where an actual resolution and action determined is needed. Again with Pritchard's help (18–19) we will see how that works below.

First a walk through the history of teaching practical ethics in the university explains how meta-ethics may have become the most respected form of ethical studies in the discipline of philosophy. The respect for meta-ethics can be traced to Socrates but the respect and primacy given meta or theoretical ethics was not yet hierarchically superior to practical ethics. How did this happen?

Moral Philosophy in undergraduate studies during the 1800s was the center of the curriculum taught by the university president with an objective to equip "graduating seniors with the ethical sensitivity and insight needed if they were to put their newly acquired knowledge to use in ways that would benefit not only themselves and their own personal advancement, but the larger society as well" (Sloan, 2). Along with this objective, which can also be described as character building, was that moral philosophy was used as a unifying principle across the disciplines (Sloan, 4–5) and a universal moral identity for society, developing an intellectual harmony across religion, science and morality (Sloan, 6). However the desire for social harmony, according to Sloan, led to the avoidance of conflict and in the late 18th to early 19th century moved moral philosophy away from character building and individual development to a more conservative approach in support of the status quo, avoiding critical criticism of the social issues of the time. Academic philosophers took the role of harmonizing to the extreme and stayed clear of controversy (Sloan, 8). Moral philosophy also moved from the 19th century center of the curriculum to by 1905 a course among many in the college philosophy department (Amherst illustration, Sloan 9).

Sloan explains that the emphasis went from student centered character development to academic discipline specialization and scientific research replaced reform and social criticism. Value-free inquiry became the standard (Sloan, 12–3). The social sciences such as psychology and sociology also moved further away from their own moral philosophy tradition to a purer empirical, scientific base so by the early part of the 20th century moral philosophy became primarily the domain of philosophy departments (Sloan, 18–20). Although there were exceptions, by 1925, "the pattern becoming prominent among universities and colleges alike was for ethics as such to be offered as an elective within departments of philosophy (Sloan, 22).

In spite of Dewey and Whitehead's warnings of the problems with abstract theories in philosophy, philosophers such as G. E. Moore and Ludwig Wittgenstein moved the central task of philosophical ethics to, "one of analyzing the meanings of ethical terms and judgments and their justification," in other words metaethics (Sloan 32–35). This philosophical move toward metaethics and the continual development of disciplinary boundaries and specialties caused the teaching and study of ethics to become more isolated. By the 1960s, "a sampling of 100 college and university catalogues… found that only 27 institutions 'required any philosophy at all for graduation….'" "By the mid-1960' the teaching of ethics was in deep trouble" (Sloan 41).

Hume Scholarship: An Illustration

I will use scholarship on David Hume's ethics as an illustration of this affection for meta-ethics in philosophical scholarship, but Hume scholarship is certainly not unique in this process. Hume scholar, Geoff Sayre-McCord describes metaethics in the Stanford Encyclopedia this way,

> The range of issues, puzzles and questions that fall within metaethics' purview are consistently abstract. They reflect the fact that metaethics involves an attempt to step back from particular substantive debates…. Some meta-ethicists early in the twentieth century went so far as to hold that their own work made no substantive moral assumptions at all and had no practical implications. (Sayre-McCord)

Hume scholarship may be unintentionally relegating Hume's moral theory to history with a few guardians of its purity rather than applying Hume's ideas to the issues of the day. In my own quick survey of the publications in Hume Studies since 1975, the topics noted numerous times included causality, induction, inference, relations, necessity, modes, identity, Spinoza, reason, self, imagination, time, miracles, skepticism, religion, atheism. There are, of course, articles on Hume's ethics but they tend to be discussing meta-ethical concerns rather than using Hume's process and theory to solve actual practical concerns. There are exceptions such as works by Annette Baier but the practical application of Hume's theory is not primary among Hume scholars.

Nicholas Capaldi seems angry below as he criticizes Hume scholars who continue to analyze the moral theory in Hume's Treatise, which is more complex and abstruse, and ignore his more understandable and later written enquiry on morality. As Capaldi explains, these scholars in order to make their arguments **have to ignore even Hume** who preferred the second enquiry to his earlier written Treatise.

> [I]t is necessary to disregard his [Hume's] assertions in the ***Enquiry Concerning the Principles of Morals*** that there were errors in the ***Treatise*** and that the second ***Enquiry*** represents his mature view, and hence to disregard consideration of progressive modifications of his views by addition and deletion from the ***Treatise*** through the ***Enquiries*** and beyond; (e)it is necessary to disregard any and all of his writings on practical/moral historical topics that do not fit with the epistemological orientation of the Enlightenment Project. What authorizes these exegetical practices is the analytic notion that the unity of science leads ***aufin*** to the rejection of the agent self and to assuming that there is a knowledge of the object-like substructure of subjects which permits us to overrule the

> agent's (or author's) understanding of his own action (or works). A more honest approach would be to admit that the analytic reading is deliberately ignoring Hume's architectonic for its own purposes. But even here we can protest, for by ignoring Hume's own programme analytic readers are evading one of their most powerful critics. This is not just bad scholarship; it is bad philosophy. (Capaldi, 133–4)

This is even the case when Hume scholars are teaching practical ethics. That was the case with me. The texts didn't include it and I wasn't sure exactly how to or what parts of Hume to use. Yet, the process of discussing a difficult moral situation and resolving it is exactly the way Hume and his fellow sentimentalists, Francis Hutcheson and Adam Smith, explain that moral foundations and determinations are developed. Conversation, an intersubjective activity that is the heart of morality is what we do and who we are. Hume scholars understandably continue to mine and analyze Hume texts to make sure his genius is not misinterpreted. But shouldn't an effort also be made to make Hume more accessible, especially his moral theory and his explanation of the way people are indeed the real makers of morality in the company of others through conversations and experiences?

I think this tendency in Hume scholarship keeps non-Hume scholars from understanding and using Hume's theories in ways that can help solve or perhaps understand the disagreements we suffer from in our contemporary world and add Hume's insights on human nature to practical ethics. I address this more thoroughly below.

Philosophy Colleagues

The place for ethics in university philosophy departments is a given but can be complicated. Theoretical ethical discussions and the history of ethics is welcome and typically every philosopher teaches or can teach an introductory course in ethics. Offering courses in and the teaching of practical ethics is not as straightforward. When there are courses in practical ethics, some faculty believe any philosopher can teach them also, even when the course is an upper-level offering. Recently in my philosophy department while hiring an adjunct professor to teach an upper-level professional ethics course, the faculty doing the hiring had an argument over whether any philosopher could in fact teach professional ethics. One side of the argument wanted to hire someone who had done research and written on practical ethics. The counter claim was that anyone who says they can and wants to teach practical ethics, indeed can. For these colleagues, practical ethics was not actually or need not be a specialization. This would not be the case with an upper level course in the philosophy of language, for instance. Our department would make sure that the hire was doing research in language or a related field. This is not always the case with professional ethics because in some philosophy departments a course in practical ethics is not as respected as theoretical or historical courses are. Many of our faculty, for instance, wish the professional ethics course was not a part of our offerings and there is some question of whether practical ethics belongs in a philosophy department and whether practical ethicists are doing rigorous philosophy. I do not think my department is unusual. I also think that, like other philosophy departments, we are changing and moving toward a proper respect and inclusion of practical philosophy and philosophers in academic philosophy.

And there is good reason for this emerging respect, as there is a tradition for the inclusion of practical ethics as the proper domain for philosophers. For that historical precedent, Pritchard takes another look at a Socratic dialogue, specifically the discussion in the *Crito* for a more practical type of philosophical inquiry (Pritchard, 18–9). Socrates is in prison and condemned to death. Crito visits him and tries to persuade Socrates to escape. A decision has to be made, an action taken or not taken. This is a situation that demands resolution. Should Socrates escape from prison or stay and accept his punishment, the hemlock, graciously? It is instructive for us to examine the way Socrates does philosophy to determine what he should actually do. He tells Crito that they both need to agree on the resolution. They need to talk it out and consider what is at stake in their decision. They proceed to do just that. So I agree that this is, in fact, a helpful model for philosophers or anyone in a conundrum concerning an actual decision needing resolution.

Pritchard also sees historical significance for the place of practical ethics in philosophy in the writings of Thomas Reid (18th century). Reid offers the practical philosopher advice on the importance and the place of philosophy in the determinations and discussions of practical philosophy. Pritchard notes the importance that Reid placed on practical ethics over that of the theoretical in the following Reid statement from Reid's *Practical Ethics*. "There is in Ethicks as in most Sciences a Speculative and a practical part, the first is subservient to the last" (Reid, 110). So the importance of the theoretical was in its usefulness to the practical. For Reid, Pritchard points out, the problem for deciding on proper choices and actions is not the lacking of overriding master principles (that philosophers seem so determined to provide). It is, according to Reid, the biases and prejudices that get in the way of good judgment that causes the errors in making moral determinations (Pritchard, 10–12).

A Seat at the Ethical Decision Making Table

Pritchard also argues that those with specific knowledge of the practical or, in the case of professional ethics, those who are the professionals in the field are an important part of the ethical decision making process when there are specific ethical situations that arise in their professions that demand practical determinations and resolutions (Pritchard, 2). I am in agreement with Pritchard that these non-philosophers should be seated at the practical ethics table as a part of an ethical deliberation. But are non-philosophers seated with philosophers or have philosophers been moved away from the actual decision making table and, in some cases, lost their seat completely. It may be that while some philosophers were trying to decide whether practical ethics was the proper domain for serious philosophy or what constitutes the "good," their seat was taken. Non-philosophers stopped listening to them. Or on the other hand, philosophers who wanted to include professionals and those with specific expertise such as medical professionals and technicians in medical ethics, made their case so convincingly that philosophers were no longer understood as needed in the discussion and again lost their seat at the table. Surely practical ethics is properly a conversation that should include all those who have a stake in the outcome or have some knowledge and input that helps the proper resolution of an ethical problem. And it seems that there is a place for someone who always points to the inconsistencies and poorly made assumptions, which is what philosophers do so well. But the philosophical spirit of no resolution is not helpful when a decision must be made and acted on. If the philosopher is there to prove that no resolution is the correct one, she impedes the possibility of resolution. It must be realized that usually a resolution, even an imperfect one, must at some point be found. Like in the Crito, none of the options were the best outcome. But a decision needed to be made. In my Introduction to Philosophy classes, I am very comfortable when students write in their papers that no matter how in depth they investigate an issue they can't decide on one side or the other. This happens often in free will/determinism papers and in the problems of personal identity. But in my professional ethics class, this won't do. They are in the real world now and can't let the decision hang in midair. A decision must be made, acted on and consequences dealt with. Certainly philosophers have been in these ethical discussions since Socrates and their input is valuable. But their involvement should be taken as important, but not as the central participant and certainly not the lone deliberator or deliberators conversing only with other philosophers. We have an important seat at the table when it comes to practical moral decision making as long as we are helpful to the process. If we aren't helpful we should lose our seat until we can behave and give the kind of insights that well trained philosophers are so good at and know when to be quiet and listen. It may be too late though in some cases because we have misbehaved and may not be invited to lots of tables and committees and advisory boards. Let me explain.

Awhile back, I was at a conference in Detroit and met a woman who worked for a think tank in Kalamazoo, Michigan. They work on resolutions to ethical dilemmas that are produced by advances in technology. She often puts together workshops and study groups where she hosts persons who can help them in their ethical determinations. In the past, she has invited philosophers whom she thought could provide them with valuable insights. But she gets resistance from other participants when she does. One of the reasons for the resistance is that in the past philosophers spoke in terms that only mystified and confused the other participants. Some saw the philosophers as arrogant (who'd have thought?) and thought they might consider the other participants as lacking. Experts the philosophers were, not participants. So she invited the philosophers less and less and has had great success without a philosopher at her table. The philosophers lost their seat at that table because of what I will call inappropriate but not atypical behavior of past philosophers. And this is not the only place where philosophers have lost their seat. The presence of philosophers at other tables such as hospital ethics boards and in discussions of ethical codes for particular professions and in academia across the disciplines is not as common as one would think and certainly not considered necessary in many cases. More and more professional schools and disciplines prefer to teach their own ethics courses, not leaving philosophical principles out as they include them, but certainly leaving philosophers to deliberate their abstruse theories in the sanctity of their offices and articles that few non-philosophers read, at least, not actual decision makers. The practice of philosophers having the last unproductive word that more than not can silence conversation has led to their losing their seat at many tables. Maybe the same thing is why the pejorative term "intellectual elite" caught on with the masses and the media. Perhaps there are some common experiences with academicians arrogantly speaking over their heads and pointing to their own ideas as lacking, sometimes directly and insulting them, or indirectly but every bit as demeaning and insulting.

Two years ago I was asked to be the lone philosopher on a hospital ethics committee. I was quick to accept the three-year appointment and anxious to be engaged in real time dilemmas and decisions. As I have been participating I am learning that we are not only dealing with what is going on in the present but we are also developing policies that can hopefully ease future dilemmas. The medical world is a foreign place for a philosopher (at least to me) as the discoveries and technological advances are coming at them at a speed unknown or experienced by philosophy. Check out how many faculty in Philosophy departments still do not have a cell phone or still have "dumb" ones. Not moving forward in technology is still an option for them. This is not an option for other professionals. While a philosopher deliberates, many decisions have to be acted on. So I realized from the beginning I may be in trouble, as my training may not work well in this real world context. Like a properly trained philosopher, I can stew over the proper wording in a paragraph for days. So I sat at the table quietly and wondered whether I could be helpful, but I wasn't sure how. I even kept my mouth shut when a doctor declared that he practiced paternalistic medicine and will continue to in spite of modern policies and regulations to the contrary. My mind quickly was thinking of autonomy, moral agency, personal identity, consciousness, justice, rights, etc. I didn't know where to begin so fortunately I stayed quiet, and the longer I sat there, the more I got it. The discussion harassed me for days afterward. I realized that I, too, practiced paternalistic teaching in my own profession. I had never noticed that before. I started thinking after three meetings that I may not, in fact, have anything to offer this committee and I was getting more and more convinced of that. I was grateful that I did not voice my philosophical arrogance that dominated my mind. I hadn't offended anyone as yet. Perhaps when I did figure out what I had to offer it would be received well.

Again as a good philosopher, I looked to history. I wondered whether I could find philosophical models for my participation on this committee. It seemed obvious to me to start with Socrates as Mike Pritchard did. I imagined Socrates in my seat at the table. I quickly saw the image as not at all helpful. My goodness, Socrates was quite arrogant and surely irritating; and when he complimented people it was to trap them in embarrassing self -realizations. I decided, in spite of the *Crito* dialogue, Socrates was not a good image or model for my participation. I will note it would be a good way to explain to my colleagues that I was in fact doing philosophy in this setting but lacked in modeling what I hoped my participation would actually look like. Luckily, Pritchard did have another model for me in Henry Sidgwick (Pritchard, 9–10). According to Pritchard, Sidgwick emphasizes the need for philosophers to use their common sense in the deliberation of practical ethics, to find middle axioms rather than exceptionless rules and principles. Pritchard is impressed with Sidgwick's advice for philosophers who are doing practical ethics to continue to play a major role, but not to do it alone. Pritchard passed along these considerations from Sidgwick to philosophers participating in practical ethical deliberations:

1. Philosophers cannot be expected to have access to all the facts that inform good judgment. They need to take into consideration all the complexities and variety that is a part of actually lived life.
2. Philosophical judgments need to be aided and checked by moral judgments of persons with specialized knowledge not philosophy.
3. Sometimes ordinary people have a better sense of what ought to be done than they can articulate reasons for, an unconscious reasoning process but correct judgment (Pritchard 9–10).

I certainly agree with the first two points but I do think that they could still possibly encourage a philosopher to act as or think of themselves as the most important voice at the table, the third is a helpful attitude to take when listening to medical personnel. A philosopher needs to understand that seemingly incorrect reasoning or inaccurate articulation can still produce a proper solution. I know this can be troubling, but the philosophers' ear should be more sophisticated and listen more carefully, letting go of the critical analysis that we have been trained to use against each other. As I sit on the hospital ethics board, I have to continue to remind myself to listen charitably and resist pontificating. I am one voice with one seat at the table.

Next I wondered at whether Henry Sidgwick's student, Bertrand Russell could help with the role of a singular philosopher on an ethics committee. I think that Russell is a good model for civil engagement, particularly because he was an accomplished and respected logician and philosopher and his work is still respected by the discipline overall and every astute philosopher that followed. And yet he also wrote and commented on social issues and was able to write for non-philosophers alike on issues of marriage, civil law, and what he is most known for, his writings on war and peace. He is certainly a public philosopher and a giant in everything he participated. That last part was what I thought would work against my lone participation on my committee. He was too large, as were other public philosophers, Noam Chomsky and Cornel West, for instance. I needed a quieter approach.

For the first few meetings, I said very little. I like to think I behaved myself as the members realized I was not there to judge or to be the authority on decisions. I didn't have to pretend either, as the dilemmas they face daily are quite difficult to resolve. And yet resolve them they must and sometimes quickly and decidedly. When we talk about a specific case I really don't have much to say generally. I am behind on the details, as I do not understand all of the language of medicine. For a while I didn't believe I would ever be helpful. But sometimes I do ask for clarification. Once we discussed a patient who would not let them take a port out of her arm, wanting to keep it. I was puzzled about why they wouldn't just let her keep it and why on earth would she want it. I thought perhaps she was frightened about the pain she would suffer. It was the first time I spoke up. I asked why she wanted it and found out it was for her drug use. Then I understood something I had not thought of. That was, that sometimes the medical personnel and patient/s have much different goals based on entirely disparate value systems. Patients do not always want health. If that were the case, then the only disagreements would be about how to support them in this pursuit. But there are so many more ways to be at odds in the medical realm. I am learning some of them slowly.

Often times I thought that perhaps I do not have the proper background because of my lacking the medical education and experience. Perhaps the answer is that philosophers get medical training or the opposite that medical personnel get philosophical training. That seems to be a trend actually with people combining degrees, legal and medical, biology and philosophy. Go to school longer and get experienced in many professions or at least two. Or the other trend toward interdisciplinary degrees seem to be catching on. But I think the table with all these aspects represented, if we all speak from our own expertise while respecting others, would be my preferred model. I am so happy to have come to this conclusion as I really do not want to work in the medical field or get another degree in order to be a part of the ethical conversations.

After a few monthly meetings, I was asked to give a quick explanation of the basic moral tenets of medical ethics. I have given a number of lectures, i.e. on autonomy, non-maleficence, utilitarianism, etc. The 15-minute presentations were well received and I was asked to continue with one each month. I added presentations on benevolence, justice and then virtue and care ethics. My presentations did not seem to solve particular situations succinctly and in some ways didn't seem helpful at all. But all of the presentations sparked important discussions and I learned first-hand how principles help and hinder resolutions. Staffers from various departments expressed different perspectives and approaches in response to the presentations. We all learned from each other how possible and sometimes difficult it was to function in accordance with set principles, of autonomy, for instance. They helped me see the limits of theory first hand and how helpful it can be in some ways even if only to spark and define a discussion. I have so much to learn from them and I think they learned that I listen without judgment. This model for philosophers teaching theory is becoming quite common. I have a colleague who has developed a number of short education modules in theory and practice for a hospital system and facilitates hour long discussions for hospital staff in their continuing education program. These kinds of programs along with colloquia hosted by hospitals are certainly places where philosophers have something to offer.

Also as a member of my hospital committee I went to an all-day medical ethics conference organized by another local hospital. Much of the medical staff in attendance earned credits for it. There were nurses, physicians, medical technicians, social workers, lawyers and ministers. I know that for sure because one of the speakers asked specifically for a raise of hands by occupation but never asked if there were any philosophers or academic ethicists present. I seemed to be the only one out of the 200+ people. While listening to the first speaker, I was sure that I would have to accept that academic ethicists do not have much to offer medical ethics committees. But I think that conclusion was coming from a belief that I had to have something to say about everything. I think that is a characteristic of not just philosophers but of academics over all. But as the day progressed and I calmed that voice of misplaced arrogance, I changed my mind. I did have something to offer. It was a small thing and I was encouraged by those at my table to bring it up in the question/answer time.

I realized that day that there was a key to understanding my place as a philosopher and maybe a model for other philosophers in any practical ethics discussion. The following is what a philosopher can add in addition to briefs and presentations on scholarship and research background.

1. Remember that as a philosopher you have an essential piece but it is no more essential than others at the Table
2. That essential part is to reflect to the people at the table the implications of the policy or resolution, what assumptions they are making that they may be unaware of.
3. Point out potential biases and/or prejudices
4. Make your points but as information not resolution. Practical application of your input is not necessarily part of your scope so be careful or leave it to the practitioner.

I am not indicating that this should be used when a philosopher is acting as a citizen and community member in a discussion on problems of social justice or fairness. To be heard, as a voice in the political arena and in civic discussions as Russell, Chomsky, and West model is an activity of a very different nature than serving on an ethical advisory committee where the philosopher is serving as a consultant. I also believe that this type of attitude should be used when working in academic settings. As the teaching of practical ethics is becoming the domain of the particular discipline, Communication Ethics, Business Ethics, Bio-Ethics, etc., we philosophers may be able to get back into the conversation but not by being the authority in the discussion but rather being a helpful resource for those disciplines and professions and a part of the discussion.

Public Philosophers

There are often situations and times when philosophers believe they can inform or help the public with issues and decisions. In some instances, moving a superficial conversation to consider all aspects and understanding the implications for beliefs or policies is helpful. This happened in my department. Impressively a colleague set up a series of talks for the public to help make sense of the political climate with some insights from academia. It was interdisciplinary with scheduled talks by professors from psychology, political science, history and, of course, philosophy. The series was aptly named, "Are you trying to figure out what is going on? Back to Basics: The liberal arts and sciences as common ground for meaningful engagement." I was impressed and liked that it was advertised with a welcome to all members of the community. Then I read the title of the first talk, "The Theory of Knowledge: Epistemic Pornography and Epistemic Poison." This is clearly a philosopher being a philosopher, but not being a public philosopher. It isn't the salacious "pornography" term but the title that excludes even the well-educated public. Philosophical titles can put people off. Not many people wish to sit and be confused by technical terms and language that is not readily understandable. It doesn't matter whether the talk itself was accessible and it actually was more so than the title. If philosophers want to make an impact on certain things that can only be done by conversing with the public, and they have to make an effort to communicate in ways that they can be heard and understood. We have to know who we are speaking to if we are going to be heard. Communication in every way is complicated, but being able to speak with people knowing or being aware of commonalities is necessary. The series title caught the commonality of frustration, "Are you trying to figure out what is going on?" and then the talk title promised more confusion.

Philosophers must strive to stay in conversations concerning ethics, all conversations, yes, really all conversations without intimating and confusing their audience. This is not just the case with philosophy, it is the case with mathematics, psychology, history, political science etc. Each of the liberating arts are necessary in every conversation particularly when ethics, practical ethics especially, is involved. The human community needs constant reminders of our commonalities, particularly the ability to reason. We also need to understand moral foundations and agreements and the necessity of social agreements and organizations. The list is long and it will take each of us to be in the conversation and understand our audiences. There is a process for what I am suggesting and it might come as no surprise that I think the Scots had something to say about it, particularly, David Hume, which I will explain below.

Hume's Approach: Pedagogy and Public Engagement

There are many ways philosophers can teach practical and professional ethics and still maintain their status as a philosopher in good standing. Most importantly philosophers can teach practical ethics and be doing good philosophy at the same time. Some approaches are quite common: theory to application (in some cases specific to a profession) or case study to theory. Both of these standard approaches are fine and I will not expand on them because there are many articles on how to do these and the benefits of each. Instead I would like to suggest another approach to teaching practical ethics that would appeal to philosophers and would work well for a practical ethics course. It is also an approach that would work in ethical discussions outside of the classroom, in boardrooms, public policy meetings, or wherever ethical determinations are being made. Basically I would like more philosophers to consider teaching Hume or at the very least be aware of the Human process of ethical development.

I think Hume's moral sentimental theory is worth a look by faculty across the curriculum who are teaching practical/applied ethics courses and philosophers serving on committees and advisory boards. Typical theories used in practical and professional ethics text books, whether written by philosophers or other academicians or practicing professionals, are those of Aristotle, Immanuel Kant and John Stuart Mill. This emphasis on rules, individual autonomy, agency and rationality leaves out the importance of our conversations and interactions with others. The writings of the moral sentimentalists developed the case that we are profoundly interrelated and form our knowledge of ourselves and develop our morality in the company of others. Recently while judging the College Ethics Bowl finals, I kept track of theories mentioned by the teams in their argument. There were numerous mentions of Aristotle, Kant and Mill but only one team referred to Hume and sentimentalism. This is in spite of the fact that the ethics bowl process is certainly Humean, in the sense that it is an ethical conversation. This differs from debates in the past where students had to support arguments they disagreed with. They can agree and expand and give other arguments for the same position. This is a Humean process of interaction that expands our moral understandings and evaluations.

Hume sees individual agency as isolating, and describes an essential interrelating that involves people necessarily with each other so that we can understand ourselves and the world around us. Hume did not claim that it should be this way; rather he observed his experiences and the experiences of others and then described how we come to know ourselves and make moral evaluations. Others, Hume observed, are not the enemy or the competition, something to overcome. Hume was clear that relating to others with the help of our instinctual sympathy is necessary and integral to our developing humanity and the cultivation of our moral tastes. Hume argued that the solitary life is isolating and wretched; that we learn about ourselves in the community of others and that our moral sentiments change and develop in response to our relationships with others. This moral psychology aspect to Hume's theory is especially appealing, as its empirical base rings true for students. For Hume, our moral evaluations involve three natural concerns, two we are born with—caring for ourselves and caring for others—and a third concern that we develop soon after birth—for our society, again a psychological approach. These are characteristics easy for students to see in their own lives. Combining an understanding of these characteristics with the understanding that we make moral evaluations gives students a sense of individual responsibility and confidence to act and respond morally in their lives and professions. Also worth noting to students and readers is Hume's emphasis on our conversations, observations, and experiences with others that are influential in the development of our moral sensitivities and moral sense faculty. That our moral making ability develops as we learn about others and their concerns encourages the kind of interactions with others in our professions and work places. This psychological understanding of human nature is helpful to effective functioning in many ways, in addition to resolving the moral decisions and dilemmas often faced.

I made a case in "A Humean Look at Feminist Ethics," (Vandenberg) that the feminist movement was indeed a Humean process towards acceptance and developing morality in light of feminist thought. I also believe that the moral acceptance of gay marriage and environmentalism are also Humean processes of moral development through experiences and conversations that caused the change of sentiments, particularly moral sentiments. The process of discussing a difficult moral situation and resolving it, is exactly the way Hume and his fellow sentimentalists, Francis Hutcheson and Adam Smith, explain that moral foundations and determinations are developed. Conversation, an intersubjective activity, that is the heart of morality and is what we do and who we are. Hume scholars understandably continue to mine and analyze Hume texts to make sure his genius is not misinterpreted. But shouldn't an effort also be made to make Hume more accessible, especially his moral theory and his explanation of the way people are indeed the real makers of morality in the company of others through conversations and experiences.

Using Hume's sentimentalist theory in a practical ethics course can help students to identify the process by which they determine moral approbation and disapprobation. This is not only empowering their own moral attitudes and sentiments, it can help them to explore how and where moral disagreement develops. Understanding that their moral sentiments play a role and that it is their own responsibility to inform their sentiments with facts and experiences puts the responsibility on them making them less dependent on others' moral judgments. They become more deliberate about moral understanding. The only dependency is an intersubjective one, as they learn to use information and experiences as input and not doctrine. The doctrinal acceptance of moral judgments is the way people, who do not understand the importance of their own experiences and critical thinking, passively inherit their moral stances. It is the role of the educator to empower students by making them aware of their own natural moral making faculties. Hume's sentimentalist claims described to students has this kind of influence. It isn't necessary to try to convince the students either. Just the description of what Hume explains as the moral sense faculty alerts students to their own moral feelings and realization of their own moral approvals and disapprovals.

As work and community relationships expand and include more cultures and religions our morality can expand also to include elements from each of them. Awareness of this process improves our relationships and our understandings of others and we see ourselves more clearly and understand what it means to be human. As our understanding of humanity and its potentiality and differences increase, our moral foundations change and adapt to this understanding. So the more we interact with each other across the differences, the more our morality will reflect that. It is only in the isolation of doctrinal and narrow certainty where we stay morally rigid and are unable to interrelate or understand differing perspectives. So the message from the eighteenth century is to talk to each other, listen and interact and a common morality can or will come from that interaction. Philosophers are great at facilitating just this type of discussion process in the classroom and in the community. Modeling the process for our students when using any of the approaches for teaching practical ethics can be the best form of pedagogy.

So consider that this is what is happening. While philosophical ethicists are analyzing principles and looking for indisputable standards and in turn teaching ethical principles and standards, the rest of the ordinary people of the world are involved in making moral values. The philosophers are making values, too, not only in their offices, classrooms, and their diligent writing but on their way to work, in their family and collegial relationships just like everybody else. We operate and continue to see and understand more and more what it is we are and can be in relationships with others. We inform and are informed in relationships. This is what humankind does to learn about themselves—what they are and can be—and how they formulate morality. Morality is not ever statically defined or stuck in past moral judgments or misunderstandings. We are not victims of unchangeable social constructs but part of the process of individuals relating to others, learning about what we are and what we can be as continually developing moral beings. Philosophical conversations have and should continue to play an important role in this dynamic process of moral development and the practical determinations of morality.

References

Capaldi, N. (1992, November). The dogmatic slumber of Hume scholarship, Hume studies, (Vol. XVIII, no 2, pp. 117–135).

Pritchard, M. S. (2006) *Professional ethics: Thinking ethically*, (pp. 15–19). University Press of Kansas.

Sayre-McCord, G. (2014) Metaphysics. In N. Z. Edward (Ed.), *The stanford encyclopedia of philosophy*, (Summer 2014 edition). URL = https://plato.stanford.edu/archives/sum2014/entries/metaethics/.

Sloan, D (1980). The teaching of ethics in the American Undergraduate Curriculum. In C. Daniel & S. Callahan (Eds.), *Chapter one in ethics in teaching in higher education*. Plenum Press. (Sloan's chapter documents the history of practical ethics in the college curriculum. For a more thorough list of references and historical documentation, I recommend this chapter and its list of resources).

Vandenberg, P. (2013, August) "A Humean Look at Feminist Ethics," *The European Legacy*, (Vol. 18, No. 5). Routledge Press.

Part III
Topics Across the Curriculum

Research Ethics Education Changing the Culture of Science and Engineering: Past is Prologue

Brian Schrag

Abstract This chapter addresses two issues: (1) how to do effective research ethics education in the sciences and engineering, and (2) identifying the barriers and opportunities in the university setting regarding the provision of effective and widespread research ethics education for faculty and their students. The first four sections address the first issue, the last section addresses the second issue. This chapter incorporates revised versions of previous papers: Schrag, Brian "Teaching Research Ethics: Changing the Culture of Science" *Teaching Ethics,* Volume 8, Number 2, Fall 2008 and Schrag, Brian "Introduction: The Challenge of Research Ethics Education in the University setting: A response to NIH and NSF Regulations" *Teaching Ethics,* Volume 12, Number 2 Spring 2012.

Keywords Practical moral reasoning in research · Moral recognition in research Ethical issues in research · Teaching research ethics · Graduate Research Ethics Education Project · Research ethics education in the university

Introduction

From 1995 to 2006, I was Project Director of a collaborative effort of the Association for Practical and Professional Ethics for a project, Graduate Research Ethics Education (GREE) funded by the National Science Foundation (Grants # SES-9817880 and SBR 9241897) to teach research ethics to graduate students and postdoctoral fellows in the physical and natural sciences, social sciences and engineering. I want to share some reflections on the nature of research ethics education (REE) drawn, in part, on that experience as well as from an important

B. Schrag, Founding Executive Director (retired).

B. Schrag (✉)
Association for Practical and Professional Ethics, Indiana University, Bloomington, USA
e-mail: bschrag@indiana.edu

E. E. Englehardt and M. S. Pritchard (eds.), *Ethics Across the Curriculum—Pedagogical Perspectives*, https://doi.org/10.1007/978-3-319-78939-2_14

seminar held in 2011 on the current barriers and opportunities in the university setting for doing REE. The first section is a brief sketch of the historical context for teaching research ethics. The second section includes my own account of the ethical tasks of researchers. The third section discusses the implications of those ethical tasks for the pedagogical objectives of REE. The fourth section summarizes what we did and what we learned from the NSF project. The fifth section discusses the current barriers and opportunities for effective REE in the university setting.

Historical Context

By way of introduction, I provide a very brief overview of the development of research ethics and the context for the development of teaching research ethics. The growth and development of interest in research ethics extends over the past 50 years and has often been spurred by public concern about various events and issues in scientific research. I briefly catalogue those concerns.

Human Subjects

In the 20th century, the concern about research ethics first gained worldwide attention in the Nuremberg Trials with revelations of Nazi doctors' experiments on Holocaust victims. That incident resulted in the development of the Nuremberg Code for Research on Human Subjects, which called for a prohibition on experimentation on human subjects without their knowledge or consent.

That awareness and concern for research on human subjects was heightened in the United States by revelations of a series of human experiments that were conducted in this country without the subjects' knowledge or consent. The first of these was the Tuskegee Syphilis Experiment, conducted on a population of African-American men in Macon County, Georgia. That experiment began in 1932, well before the Nazi experiments, and continued until 1971, long after the Nuremberg Code was established. The 1950s and 1960s saw disclosures of other experiments involving research on vulnerable human populations. For example, nontherapeutic studies of hepatitis were conducted using institutionalized mentally retarded children at Willow Brook State School in New York. More recently, we have learned of human radiation experiments conducted in the 1950s on U.S. soldiers and civilians.

Attention to these early cases led to the creation of the National Commission for the Protection of Human Subjects in 1974 and its issuance in 1979 of the influential Belmont Report, which is a statement of ethical principles and guidelines for research on human subjects. That document, in turn, shaped subsequent federal guidelines on government-funded research on human subjects not only in the natural and biological sciences but also in the social sciences.

Research on Animals

Public awareness of ethical concerns dealing with scientific research has expanded beyond research on human subjects to include research on non-humans. Considerable public attention has been given to alleged abuse of animals in research, the proper use of animals in research or any use of animals in research. That has led to the development and refinement of federal guidelines and institutional policies on the use of animals in research. Not all animals are covered by such guidelines, however, and many species remain completely unprotected.

Scientists now generally agree that wherever possible, animals in research should be replaced with nonanimal alternatives such as computer models or cell lines; the number of animals used in research should be reduced as much as practicable; and experiments should be refined to reduce the pain and suffering of animal subjects wherever possible. Discussion continues about appropriate guidelines on experimentation with animals.

Research Ethics and the Practice of Science

Some public concerns have to do with the impact of ethical/unethical behavior of scientists and engineers on the very practice of science and engineering. One area involves conducting and reporting research. Over the past thirty years, we have witnessed a steady drumbeat of cases of fabrication of data and fraud in research reporting. In some instances, work of other scientists based on fabricated data has been wasted. In others, use of results based on fabricated data has resulted in injury or threat of injury to the general public. Some scientists, trying to call a halt to fraud and fabrication, have stepped forward, blown the whistle, and paid with significant damage to or loss of their careers. Not surprisingly, both the awareness of fraud and fabrication and the treatment of whistle blowers raised questions in the public's mind about the credibility of work in science and the level of public support deserved by science.

I have mentioned a few areas of concern in research ethics, but the field involves a wide range of topics including the web of relationships in laboratories and research communities, relationships of faculty with each other, relationships of faculty and mentors with graduate students, and relationships among graduate students and postdoctoral fellows. It involves conflicts of interest, conflicts of commitments of faculty, and issues of institutional responsibility for education in ethical research. Some of the causal factors that may have accelerated ethical breaches in research have to do with increasing pressures on researchers to obtain funding and to publish or perish. More recently businesses have exerted pressure to control the publication of research they have funded.

Ethics and the Culture of Science

Scientists themselves are increasingly concerned about issues in research ethics. That trend can, I think, be partly explained by a shift in the intellectual framework over the past 50–75 years. In the 1920s, an intellectual position known as logical positivism developed. It claimed to justify a sharp distinction between facts and values and encouraged the notion that scientists' and engineers' work was value free and therefore scientists did not have to worry about ethical issues in their disciplines.

In my view, that intellectual framework also helped to create and support a scientific culture over the past seventy-five years that led many scientists to be indifferent, skeptical and even antagonistic toward the role of ethics in research and, consequently, toward the need for ethics education for scientists.

Logical positivism, in various forms, was shown to be indefensible 50 years ago (at least to the satisfaction of many philosophers). Consequently, logical positivism's influence in philosophy has waned considerably, but even today it seems to have residual effects in other disciplines, including both the natural and social sciences (in my view, especially in the social sciences). Nevertheless, an awareness of the deficiency of its extreme forms has filtered down to other disciplines, and its support has gradually crumbled. Over the past 20–30 years, the scientific community's growing awareness of the deficiencies of logical positivism, has, I believe, led to more openness and added legitimacy to the discussion of ethical issues in science and engineering. The combination of concern about ethical lapses in science and the collapse of logical positivism as an intellectual bulwark against taking seriously the ethical issues in science have now opened the area for discussion. The federal government is now pressing scientists either to conduct research in an ethical manner or risk sanctions, and new federal regulations now require ethics education in federally funded science research.

Ethical Tasks of Researchers

In addition to the general context of research ethics, it is important to be clear about the ethical tasks faced by researchers and the kind of ethical deliberation in which they must engage. That is essential in determining the kind of pedagogical objectives one ought to pursue in ethics education for researchers.

Scientists and engineers are practitioners engaged in a kind of practical activity. During the course of that activity they routinely have to solve problems involved in carrying out research. Thus they may have to address such general problems as "How should I design this research study?" or more specific problems such as, "How can I discover the DNA profile of this Native American tribe?" or "How can I find out if the nature of syphilis differs by race?"

There is often a moral component to that problem solving activity whether the researcher recognizes it or not. In pursuing the research activity, for example, a researcher may consider whether or not to deceive a subject in order to pry out his/her most deeply held thoughts and feelings on a very sensitive subject or whether to lurk on the Internet and observe a subject's behavior without the subject being aware of the researcher's presence.

The researcher's decision making process, both its moral and non-moral components, involves making a decision within the context of the objectives, standards, and the current empirical knowledge of that particular science practice. Because it is a practical activity, that decision making also involves making decisions within the constraints of limited time and imperfect knowledge. This focus on decision making for action, incidentally, distinguishes it from the philosopher's sometimes activity of deliberation on an ethical issue in no particular context and with no pressing need to solve a particular problem.

The nature of the researcher's practical activity has pedagogical implications for teaching research ethics. Ethical theories, as such, do not provide determinate decision making procedures. That is why ethics education for researchers is not well served in a course in research ethics by simply canvassing the standard ethical theories. (*Note: In this essay, I will use the terms "ethical" and "moral" as interchangeable and the terms do not indicate a substantive difference.*)

Ethical Tasks of Research Practitioners

As researchers engage in research activity, they will inevitably encounter practical ethical issues related to their research that they will need to address in order to be able to perform their research. Practical ethics for researchers involves at least four distinct kinds of ethical tasks: (1) recognition of moral problems; (2) solution of moral problems; (3) judging moral actions; and (4) engaging in preventive ethics.

Recognition of Moral Problems

Experience does not come stamped with the ethical components clearly identified. Not surprisingly, the history of scientific and engineering research includes many examples of researchers' failure to recognize the ethical dimensions of scientific research at all or to recognize more specific ethical issues in particular research projects.

The researcher's capacity for moral recognition must function at several levels. Researchers must realize that there is a moral component to scientific research. They must also be able to recognize that a particular situation involves ethical issues and when the situation is ethically problematic. At minimum, the researcher should have a "gut instinct" that something is wrong in a particular case, even if he or she cannot immediately articulate precisely what the ethical issue is. Finally, researchers need to be able, perhaps after reflection and consultation, to identify in specific detail the precise nature of the ethical issue at hand in a particular research project. Thus the researcher needs to have his or her moral radar turned on, have an early warning system and a more precisely tuned "local" moral radar.

The sources of moral blindness in researchers may be varied. To begin with, researchers must be able to recognize that there are moral dimensions to the scientific research enterprise. A particular epistemological paradigm, such as logical positivism, can contribute to moral blindness at this level. Secondly, a researcher may be in the grip of inadequate moral theory. Thus, one could argue that a close reading of the correspondence of the researchers in the Tuskegee Syphilis experiment reveals that the researchers were so in the grip of utilitarian theory that they could not recognize the moral weight of the respect for persons due their subjects. Sometimes the moral blindness is due to a lack of awareness of an appropriate moral concept. Thus, the ethical treatment of animals in research was long delayed by a lack of awareness of the notion of moral respect for sentient creatures. A fourth source of blindness can result from researchers employing moral concepts that are inadequate or in need of being modified or reformed. The current notion of informed consent may work well in medical research but less well in observational research or in research involving indigenous groups. The concept of privacy may need rethinking when it involves research on subjects in the "public" sphere. Finally, compliance education can lead researchers to think that as long as they are in compliance with all the standard rules for ethical research, then they have exhausted the moral considerations regarding their research project. Even worse, sometimes the rules are themselves misguided. Sometime that is a result of the research community's blindness to certain moral issues. Capacity for moral recognition must go beyond compliance education to enable researchers to recognize, when appropriate, the moral inadequacy of current regulations.

Solving Practical Moral Problems: The Moral Agent's Perspective

Once a researcher recognizes that there is an ethical issue involved in the design or execution of his or her research project, he/she faces the question, "What should I do about it?" The researcher is now faced with the task of solving a moral problem from the perspective of a moral agent.

Hampshire (1949) distinguishes the perspective of the "moral agent" from that of the "moral judge" or "critic" or "spectator." The ethical task of the moral agent differs from that of the moral judge. The objective is different; the mode of deliberation is different. The moral agent must fashion a solution in order to solve a moral problem with the research and get on with the research. The activity focuses on developing a judgment to guide future action as opposed to merely making a retrospective judgment of actions already taken. By contrast, the task of the moral judge, Hampshire argues, does not involving devising a solution to a practical problem but is an act of classifying moral acts or conduct for the purpose of blame or praise.

The task of the researcher, qua moral agent, is to find a solution that allows him/her to conduct the research in an ethical manner. "Reasonable Moral Pluralism" (Cohen 1993) is a central feature of practical activity which the researcher must be prepared to deal with in determining what to do. As a matter of practical ethics, there may be conflicts between competing moral considerations, such as "respect for persons" and the value of the research, with no clear priority of the values or moral principles involved. It may be that not all the values can be satisfied simultaneously. The researcher may have to decide what to do in the face of legitimate competing moral considerations. To make matters worse, there may be competing legitimate legal, prudential, social and technical considerations as well. (Illustrations of researchers who have engaged in such a practical reasoning process include archaeologists who work on Native American archaeological sites in the context of the Native American Graves Protection and Repatriation Act; researchers seeking to do biomedical or health research on Native American groups who have their own Institutional Review Boards and legal authority to control all research done on their lands; researchers who engage in collaborative research with Native American communities in a relatively new approach called action research.)

Because of this plurality of legitimate competing moral and nonmoral considerations, a justified solution may involve compromise, arbitration, negotiation, or reconciliation processes as well as reasoning about the moral considerations. The researcher must use moral imagination to identify a range of alternative solutions to the problem. Some proposed solutions, upon reflection, will be unacceptable. Some may be acceptable, and some might be quite good in meeting as many of the legitimate considerations and constraints as possible. It is important to realize that there may be no uniquely best or determinate solution. It is also important for researchers to realize that there may be some research that, all things considered, cannot be morally justified and should not be done.

Although there is a kind of moral reasoning involved in arriving at a solution regarding what to do, it is not a simple deductive process from principles; it may involve analogical and inductive reasoning or casuistic reasoning. Neither is it a linear process. What the researcher assumes to be the relevant facts, the relevant moral concepts or principles may change as the researcher thinks about possible action steps. It is also the case that the agent's reasoning process involves a temporal dimension. In the course of dealing with the problem and even after the agent acts on a proposed course of action, new facts may emerge or the action may have unanticipated consequences that pose new challenges for the researcher.

Judging Moral Actions

A third kind of ethical task for the researcher is reasoning to morally evaluate past research behavior, rather than reasoning to find a solution to an unresolved ethical problem. Such a task lacks both the unique moral agent perspective and the temporal dimension of the problem solving focus. It is the task of what Hampshire calls the "moral judge" or "spectator" or "critic." It is the kind of task required when the researcher is called to evaluate charges of research misconduct against a fellow researcher or a program. It involves an attempt to determine if a particular action or practice falls under a specified standard ("Did the researcher commit fraud?") or in some cases, attempting to determine which standard ought to apply to a case ("Should this research protocol have required informed consent or did it qualify for a waiver?"). In either case, it is a kind of moral classification problem rather than a problem solving task. Awareness of research rules and guidelines is a necessary condition for this sort of task. Compliance education which focuses on teaching codes and regulations addresses this sort of task. I would argue that the reasoning task that researchers more frequently engage in is that of moral problem solving rather than moral judging and, if they are to themselves avoid being the subject of such judging, the education for the ethical problem solving task may be the more important challenge in REE.

Preventive Ethics

As researchers learn, sometimes painfully, it is often easier to take steps to avoid landing in an ethical difficulty that it is to deal with an ethical problem after it develops. This approach is often referred to as preventive ethics. Thus, it may be better for an ethnographer to find a way to take subjects of observation into his or her confidence at the beginning of the research project than to deal with the ethical issues created by a failure to do so. Here the reasoning is neither an effort at ethical classification of a past action, nor an effort to resolve an actual current problem, but is aimed at crafting a policy or practice that avoids such problems in the future. This task may not have the immediacy or specificity of an actual moral problem but may require devising a strategy to be employed by a moral agent or research organization. Thus, one way to avoid some ethical issues with graduate students regarding a question of who owns the data in a lab is for the lab to make it a practice to discuss lab practices and expectations regarding data management and ownership at the beginning of a graduate student's career at that lab rather than address that issue as the student is ready to leave and take data with them. Discussion of ethical cases, especially ones with stages, is a good way to raise awareness of the need for thinking about preventive ethics.

Relation of Ethical Tasks to Pedagogical Objectives in Research Ethics

These four ethical tasks have implications for pedagogical objectives in research ethics.

It is illuminating to consider pedagogical objectives for research ethics in the context of a more general discussion of pedagogical objectives for ethics education. Daniel Callahan provided a seminal discussion of such objectives in an essay, "Goals in The Teaching of Ethics" (Callahan and Bok 1980). The essay is part of a collection that grew out of a systematic study of the teaching of ethics in higher education conducted by leading scholars and teachers of ethics education across disciplines and sponsored by the Hastings Center.

Callahan argues that ethics education ought not to be focused on acquiring factual information or the mastery of a body of literature. (Note the contrast with much of compliance education.) Rather, it should focus on the development of several distinct capacities/dispositions. These include: (1) stimulating the moral imagination; (2) recognizing ethical issues; (3) eliciting a sense of moral obligation; (4) developing analytical skills; and (5) tolerating or reducing disagreement and ambiguity. In what follows I will suggest ways in which these objectives can connect to the ethical tasks of researchers. Although I will use Callahan's categories, I will adapt them to my own purposes and make no claim to a careful or representative exposition of his own account of these objectives. I will comment on the first four of these objectives.

Pedagogical Objective: Stimulating Moral Imagination

Adam Smith, in the *Theory of Moral Sentiments*, argued that we are not only rational creatures but also empathetic creatures. We have a capacity to imagine ourselves in another's place. That capacity, coupled with our emotional capacities for sympathy, fellow feeling, and moral outrage allow us to identify with the other and allow us to imagine what it is like to be on the receiving end of unethical treatment and come to recognize our equality with others as moral beings, to take the moral point of view. This is the motive force, as Callahan puts it, of a "drive to get straight on ethics." Hence, he argues, ethics education needs to stimulate the moral imagination. This capacity is important to nurture in researchers in particular, since it undercuts a culture of emotional detachment that tends to exist in research. It allows the researcher to more vividly imagine what it would be like to be on the receiving end of the researcher's activity. That allows the researcher to weigh more carefully and accurately the moral consequences of alternative actions the researcher is considering in a research design or as solutions to a moral problem one has encountered in research activity.

Taking the moral point of view also helps researchers to understand that as researchers they are still moral agents. As I noted above, earlier generations of scientists, influenced by logical positivism, would have said that science is value free; the scientist's only responsibility is to do good science. Today, researchers are still sometimes socialized into the view that there is a disconnect between their activity and the web of broader moral relations and that, qua researchers, scientists are not moral agents; they need not concern themselves with the moral implications of their work.

A related view, not strictly the view that research science is amoral, is the view that the value of research activity is privileged and automatically trumps other more general moral considerations (e.g., the value of knowledge is the highest good and overrides considerations of the subject's welfare, whether human or other animals). Thus, a researcher may argue that if, in the course of their work with early teens, they observe subjects engage in life-threatening behavior, then, unlike the person on the street, they have no obligation to act on that information. Doing so would interfere with the research program and impair the validity of the study. Their first obligation is to the research.

The recognition that researchers also have a wider range of moral responsibilities beyond those of the research allows them to recognize that there can be a conflict between moral responsibilities in research and values of the research on the one hand and wider moral responsibilities on the other. The wider moral responsibilities are not automatically overridden by responsibilities of the research.

There is a different sense of moral imagination that is worth mentioning. Sometimes when faced with an ethical issue, what is required is to think outside the box and come up with imaginative solutions regarding what action to take. This is not so much moral imagination but imagination in the service of moral action. Whistle-blowers, for example, sometimes find themselves in a difficult situation because they can conceive of only limited alternative actions, blowing the whistle on the corporation or going along as an accomplice. It sometimes takes moral imagination in this sense to conceive of other morally defensible alternative actions. A cultivated habit of asking oneself if there are other reasonable alternative actions can lead to more imaginative and positive action steps. The Seven Step exercise in analysis mentioned below is intended to cultivate that habit.

Pedagogical Objective: Recognizing Ethical Issues

One task of practitioners is to recognize the moral dimensions of their practice and consequently, one task of ethics education is to prepare researchers for moral recognition at both the general level and at a more detailed level. Capacity for moral recognition also helps with the researcher's task of making decisions for action as well as judging moral actions.

Callahan argues that there is a fine line distinguishing moral imagination and recognition of ethical issues. The difference is cognitive. As we argued earlier, a researcher might have an uneasy feeling that there is something not quite right in their research project, or that of a colleague, but not be able to articulate it. There are really two tasks for the researcher here that require pedagogical attention. The first is the researcher's need to be able to do a kind of ethical assay of all the ethical issues raised by a situation (e.g., the design of this ethnographic project involves deceiving subjects, invading their privacy, inflicting insight on them, and poisoning the well in terms of the possibility of future research in this community or geographical/cultural area.) See for example, "Piercing the Veil: Ethical Issues in Ethnographic Research" (Schrag 2009). The second is clarification of particular ethical issues, e.g., is obtaining informed consent from individuals adequate when the subject of research is an ethnic group with distinctive decision making procedures?

The process of moral recognition in such cases is a cognitive one in the sense that it is in part a classification process, placing the action or issue within the moral landscape. For example, the failure to reveal to subjects that one is actually a researcher studying them and their behavior is a form of deception. The pedagogical challenge is to provide the researcher with a sense of that moral landscape. Unless one has an awareness of the moral landscape, that recognition will not occur, at least not in an especially useful manner.

Capacity for moral recognition can be promoted by exposure to general moral theories and middle level principles for research ethics such as those contained in the Belmont Report, as well as historical accounts of research ethics (e.g., James Jones, *Bad Blood: The Tuskegee Syphilis Experiment* or Robert N. Proctor, *Racial Hygiene: Medicine under the Nazis*.) Viewing video accounts by scientists who have struggled with ethical issues in their research or dealt with research misconduct may also be helpful. Writing case studies and writing commentaries on the cases can also aid in the development of a capacity for recognition of ethical issues. A common phenomenon observed by teachers of ethics is that students, when first exposed to a systematic discussion of ethics, including ethical theories and major ethical concepts and distinctions, suddenly see ethical issues everywhere. Compliance education, which typically makes the researcher aware of basic rules governing research ethics, can also help. However, as indicated above, compliance education can cut both ways. The outcome of the development of moral recognition should be a shift in the investigator's dispositional thinking from, "How can I investigate this phenomenon?" to "How can I investigate this phenomenon in an ethical manner?"

Pedagogical Objective: Reasoning About Ethical Issues

Callahan conceives the development of reasoning capacity as the development of analytic skills, including clarification of concepts, analysis of the meaning, consequences and coherence and consistency of moral rules and moral principles. He takes this objective to be essentially the development of logical skills reasoning. The analytic skills Callahan identifies have a role to play, but are not sufficient.

I take a broader view of moral reasoning in this case and the pedagogical challenge in teaching it. As indicated above, two tasks of the researcher that involve ethical reasoning are solving practical moral problems and judging ethical actions. I have already articulated some of the capacities to be developed in each of those categories.

The kind of reasoning process involved in solving a practical problem can perhaps be illustrated by the "Seven Step Moral Reasoning Model" originally developed by Patricia H. Werhane for the Arthur Anderson Project on Integrating Ethics into the Business School Curriculum. For an exposition of this approach as applied to research ethics see Swazey and Bird (1997).

In this model, given a situation that requires the ethical solution to a problem, the moral agent works through the following seven steps:

1. What are the relevant facts?
2. What are the ethical issues?
3. Who are the primary stakeholders?
4. What are the possible alternative actions of the moral agent?
5. What are the ethical issues and implications raised by each alternative?
6. What are the practical constraints on each alternative?
7. All things considered, what action/actions should be taken (would be morally justified by the agent)?

As indicated above, this is not a linear reasoning process. Deliberation at any one stage may force the agent to revisit previous stages. Having researchers actually work through this process with cases can be an effective way of teaching this reasoning approach.

For the GREE project, we believed that the activities of having participants write and discuss cases as well as write commentaries about cases were effective ways of nurturing participant's capacities to reason and make practical moral judgments in research activity. In working on the resolution of such problems we thought it more effective to let ethical theory lurk in the background and bring it to bear only as necessary. For case analysis to be effective in teaching reasoning to solve practical problems, it was crucial and intentional that the cases be cast in terms of the need for a specific agent to make a specific decision regarding a particular action step in a research setting. One often sees cases without that feature and for some purposes that case format may be appropriate, but not for our purposes. We also, for the reasons articulated above, encouraged participants, where appropriate, to cast the case in stages as the case developed over time. In our experience, scientists and engineers were initially surprised and then pleased to discover, as they grappled with cases, that the discipline of ethics could bring to bear such a rich literature in concepts, skill in conceptual analysis, as well as the level of rigor in reasoning to justify conclusions. They recognized it as the equivalence of good evidence in science.

Pedagogical Objective: Disposition to Responsible Ethical Behavior

Can ethics education increase the disposition of researchers to ethical behavior in their research? There are no guarantees in ethics education. However, a few considerations may be appropriate. Many researchers, like others, have a desire to do the right thing in their professional life as well as their personal life. But a desire does not yet rise to the level of a disposition.

The historic culture of science with its view of science as an amoral activity creates a barrier to such an ethical disposition by researchers. A larger barrier may be the enormous pressure on scientists to get grants to carry on their research. Taking time to explore the ethical dimensions of their research can be seen, at a minimum, as an unwarranted cost of "time away from the bench." (Indeed, a number of our GREE participants indicated that was the reaction of their advisor or lab supervisor when the participant raised the idea of applying to attend the GREE workshop.)

Can REE reinforce that desire to do the right thing and raise it to the level of a disposition? If one is successful in meeting the pedagogical objectives outlined above, the researchers will be clearer that they do indeed function as moral agents in the research setting and that research activity does indeed have a moral component embedded in it. He or she will be clearer that, qua researcher, he or she has moral responsibilities to others in research and moral obligations above and beyond those within their research activity which may at times override considerations of research activity. This awareness by itself does not rise to the level of a disposition to act ethically.

However, if education succeeds in helping researchers to think through the ethical issues they now see in their research activity and to develop some skill in ethical decision making, then researchers will have more confidence in doing that thinking and less resistance to doing it. It will also give the researchers more confidence in their ethical analysis and judgment and more capacity to resist those who urge unethical behavior.

Having once developed the capacity to recognize ethical issues and think through the ethical issues raised, it will be difficult to ignore those capacities. (One can't go home again, so to speak.) The more one engages in ethical reflection, the more habitual it becomes. That reflection reinforces a habit of looking for the ethical elements in a research situation and trying to determine the right thing to do. Habituation, by actually trying to do the right thing repeatedly, also reinforces the desire to do the right thing. All of this increases the likelihood of developing a disposition to ethical behavior.

Finally, there may be instances in which, as a researcher, acting ethically actually results in doing better science at the individual level. There is even more likelihood that acting ethically is good for the enterprise of science as a whole. If, as a scientist, one identifies with the practice of science, then that will reinforce even more the disposition to act ethically.

Graduate Research Ethics Education Project (GREE)

Project Goal

It is in this context that our work began with graduate students and post-doctoral fellows in science and engineering. From the beginning our intention was to make some contribution to changing the culture of science, particularly in how research ethics is perceived, practiced and taught in the science and engineering community. As ethicists and teachers of research ethics, it was our conviction that this was best done by teaching scientists and engineers not merely a code of ethics or a set of compliances rules to be followed, but by teaching them an understanding of the nature of ethics and ethical reasoning, set in the historical context of research ethics and with the understanding of the centrality of ethics to doing science and engineering.

We recognized the power of the mentoring tradition in the sciences. Graduate students have the strongest identification with and motivation to learn from their advisers or mentors. The socialization forces on graduate students to fit into their professions are enormous, and students understandably look to their faculty advisers and mentors for guidance. Most current faculty in science have learned research ethics through mentoring relations, but those mentors have generally not been trained in ethics or ethics education. Not all faculty recognize the legitimacy of ethics education in science. Those who do are likely to be novices in ethics education and tend to provide a provincial view of research ethics—the view from my lab/bench/ discipline. When approached from that perspective, it is sometimes difficult to be self-critical about research practices in one's own bailiwick. Sometimes the ethical perspective is better gained from a broader interdisciplinary view, where one can learn from the experience of other practices or disciplines. It is also the case that, unlike earlier generations of researchers, the practice of the physical and natural science now often involves large laboratories with much less contact between lab supervisor or advisers or mentors and graduate students; hence, mentoring may be less effective.

Our strategy was to work with young researchers in graduate school or postdoctoral students at the beginning of their careers, before they were permanently socialized into a perspective which discounted or disregarded the significance of research ethics and its role in science. We believed that by developing in these participants a habit of taking the ethical questions as a normal component of their research, we could not only influence their own behavior but also make them much more conscious teachers of research ethics to the generations of students they would have in their own laboratories over the course of their careers, producing a multiplier effect.

We hoped to build a small community of young scholars who had an understanding of research ethics, would be committed to practicing research ethics themselves throughout their own career, and would pass that commitment on to their generations of students and colleagues as well as play a role in the dissemination of the practice of research ethics throughout their profession. Thus we hoped to have a leavening influence on the participants, their students and colleagues, on REE in various departments and universities or in industry and ultimately on the culture of science. Consequently we wanted to try an intervention to provide ethics education in research where it often did not exist.

Faculty

We consciously selected a team of GREE faculty, including scientists and engineers as well as ethicists with expertise in research ethics and ethics education. It was our belief that it was desirable to have a combination of ethicists and scientists; that ethicists may not have sufficient understanding of the culture of science in different disciplines and that scientists may not have adequate expertise in ethics and ethics education. It was also our belief that, given the emphasis on mentoring in science, the presence of scientists would give more credibility to the enterprise. It was also our belief that it was important that more than one area of science ought to be represented in the faculty, since the practice and the issues in research ethics raised have a somewhat different appearance from the perspective of different disciplines. We followed this same principle in the selection of faculty for the program for the social sciences.

The faculty for the first six cohorts included Vivian Weil, Michael Pritchard, Brian Schrag, Deborah Johnson, (all in Philosophy and Research Ethics but additional background in the sciences and engineering), Karen Muskavitch (Biology), and Aarne Vesilind (Engineering). The faculty for the seventh cohort included Vivian Weil, Michael Pritchard, Brian Schrag, Barry Bull, (Education), Chip Colwell-Chanthanaphonh (Center for Desert Archaeology), Peter Finn (Psychology), Frederika Kaestle (Anthropology), Ulica Segerstrale, (Sociology), and Stuart Offenbach (Psychology), who conducted program evaluation surveys and longitudinal survey.

These faculty members were supplemented by guest lecturers from various areas of science. In all seven cohorts, the faculty developed the curriculum, selected readings, selected participants, made presentations, assisted with development of participant cases and participated in the collaboration project.

Selection of Participants

We attempted to select young scientists and engineers from a variety of fields who showed some promise of leadership in their own fields as well as an interest in research ethics. We required each participant to write an essay saying why she or he wished to participate and each was required to be nominated by a faculty advisor or mentor or other department faculty member. We worried from the beginning about participants returning to unreceptive departments with no support for their new-found enthusiasm for research ethics.

That worry turned out to be well founded. We later learned that several participants came to the program over the strenuous objections of their faculty advisors, who argued that the program would be a waste of time and would do nothing to advance their careers. Several others returned home to quite a hostile reception when they attempted to share what they had learned.

For the seven cohorts of participants selected for this endeavor (1996, 1997, 1998, 1999, 2000, 2001, 2005), we advertised for applicants from all over the United States to apply for participation in the program. We wanted students who had actually had experience at the bench or in field research and hence had begun to experience first-hand some of the ethical issues that may arise in research. Participants were required to have completed at least two years of graduate work or be postdoctoral fellows. In the first six cohorts, the applicants were almost entirely limited to the natural and physical sciences and engineering. Each year we selected a cohort of 15–18 participants from universities all over the country from an annual applicant pool of 40–70. Over the seven cohorts of participants, we worked with 114 participants from 59 different universities. In the first six cohorts virtually every branch of science was represented as well as 16 engineers from various areas of engineering. The final cohort, a pilot project to do the same for graduate students and postdoctoral fellows in the social sciences, included participants from 9 different disciplines or sub-disciplines in the social sciences.

It was our conviction that the challenges of teaching research ethics in the social sciences were sufficiently different that, in the first six cohorts, participants from the social sciences were, with a few exceptions, not included. In the seventh cohort, we tried a pilot program in which we did focus on participants in the social sciences and our experience convinced us that there are significant differences and challenges in teaching research ethics to researchers in those fields, as I will indicate below. In the seventh cohort, we had 18 participants from 11 different universities; the disciplines included sociology, psychology, archaeology, anthropology, education, and information science.

Specific Teaching Objectives

Our specific teaching objectives for these workshops were conceived in the context of the goals articulated in the Hastings Center Report (1977) on objectives for ethics in higher education and included efforts to: (a) develop in participants a capacity to recognize moral issues in research, a capacity for moral reasoning about those ethical issues, a capacity for moral imagination, and a disposition to act in morally responsible ways in research; (b) develop a multidisciplinary perspective on issues of research ethics; (c) introduce participants to an understanding of ethics and provide a historical perspective on misconduct in science; (d) enhance their effectiveness as teachers of research ethics; and (e) create a network of young scientists and engineers who will, over their careers, act as a catalyst to help create a scientific and engineering culture in which ethical considerations are simply considered a part of doing good research.

These are objectives not likely to be achieved in many of the minimalist programs developed by universities to meet NIH/PHS training requirements, for example attending a lecture on compliance or completing a short set of questions on a web site.

Pre-workshop Activities

One provision of the project was to provide participants with a working library in research ethics. Each participant received a library of 15–20 books and a number of articles related to research ethics. After being selected in early spring, the participants were asked to do a substantial amount of reading in preparation for a workshop in early summer. Readings focused on the history of research ethics, ethical theory and particular issues in research ethics.

Participants were also asked to write an initial draft of a case involving some incident in research that raised ethical issues they found especially troubling, interesting or puzzling. Participants were given materials on how to write up the case. Many drew from their own first hand experience in the lab and described an incident they had observed or experienced themselves. Those cases were sent to us in advance for review and suggested changes before participants arrived at the workshop. The idea was to use the case writing to get participants to begin to reflect on their own experience and articulate that in light of their readings in research ethics.

Workshop Activities

Participants attended a very intensive four and one half day workshop that ran each day from 8 a.m. to 9 p.m. We deliberated about the minimum length for the workshop and concluded that if we lengthened it, we ran the risk of not getting any bench scientists to apply since their advisers or lab supervisors probably would not approve a longer stay away from the bench. Our poll of participants confirmed that concern. Pedagogically, a longer and less hurried format might have been preferable, but we had to be practical.

The workshop included lectures and discussions of issues relevant to research ethics; discussions of cases and the teaching of cases and research ethics teaching. We consciously chose to not spend time systematically lecturing on ethical theory but rather introduced ethical theory or concepts as they arose naturally in the discussions. Different styles of teaching ethics were consciously modeled during the workshop. An evening discussion series of videos, focused on various aspects of research ethics, provided participants with an opportunity to discuss and synthesize what they were learning as well as making them aware of pedagogical tools they might be able to use in their own teaching.

Participants met in small groups early in the workshop to discuss their own cases and see how they might be modified or improved. This was another way participants came to realize how differently individual disciplines viewed the same research ethics issues and how distinct were some of the issues facing specific disciplines. Participants were then encouraged to refine their cases during the workshop. On the last day, all the participants and faculty assembled to review all the cases. Each case was discussed to improve it as a pedagogical tool. The substantive issues raised by the case were also discussed. This exercise was another opportunity for participants to learn from the perspectives of those in other disciplines.

Developing Case Studies

After each workshop, participants worked on-line for several months to further refine their cases, based on the feedback from the workshop. The cases were then posted on a list serve, and participants and faculty were asked to provide further feedback to each participant. The e-mail conversations among participants usually generated several hundred exchanges between participants during the summer and fall. After submitting a final case, participants were then asked to write an ethical commentary on the issues their case raised. Each faculty member in the workshop was also asked to write commentaries so that each published case was accompanied by two commentaries. The faculty commentaries frequently allowed the introduction of appropriate ethical considerations or theories or pedagogy in more detail.

We had several pedagogical objectives in having the students develop cases in research ethics and write commentaries on the cases: (1) To help them recognize ethical issues in their research experience; (2) To help them to reason about those issues and bring to bear appropriate ethical principles and considerations in that reasoning, along with historical perspective which they had gained during the workshop; (3) To encourage the use of moral imagination in finding ethical solutions to the issues they encountered; (4) To think about the role of preventive ethics in research; (5) To help participants begin to think of the pedagogical objectives they might pursue in teaching such cases; (6) To encourage them to think about cases in which someone did the right thing.

Those cases and commentaries, together with faculty commentaries, were printed in Cases and Commentaries in Research Ethics: Volumes 1–7. All seven volumes are described and made available at the Association for Practical and Professional Ethics. The first six volumes have been made available for posting on the Online Ethics Center.

Taking GREE Home

Participants were also asked to engage in some project at their home institution to share what they had learned. The format varied with each individual. Some led a brown bag discussion with their laboratory colleagues. Some offered a seminar for their departments. Some taught a session in a faculty member's course. Some developed and taught a course of their own. Some played a role in planning a REE program for their department or a larger, campus wide effort. Some participants assisted faculty in revising practices in their lab or department. We asked faculty to send us a letter indicating the participants' impact on the local campus. It was clear that they had much more impact than we initially anticipated.

Planning and Teaching Research Ethics for the Social Sciences

A request for an extension of the project was granted which allowed for a planning conference in 2004 to consider how to extend to the social sciences the approach we used to teach research ethics to graduate students in the physical and biological sciences. A group of social scientists were invited to the Indiana University, Bloomington campus to discuss with three of the ethicists from the project (Vivian Weil, Michael Pritchard and Brian Schrag) some of the research ethics issues they faced in the social sciences and how the teaching needs for research ethics of graduate students and postdoctoral fellows in the social sciences may differ from those in the physical and natural sciences. It became clear that in the social sciences, there would be more concern with the ethical concepts of ethical relativism, ethnic group consent, privacy, and issues dealing with surveillance research, use of data bases and dealing with IRB's.

Seventh and Final GREE Workshop

This workshop was carried out June 1–5, 2005. This time the workshop focused on ethical issues in research more relevant to the concerns of social scientists as identified in our 2004 planning meeting with social scientists and included faculty from the social sciences. Eighteen participants were selected from a national pool of applicants. Disciplines represented in the selected group included Sociology, Psychology, Social Psychology, Education, Anthropology, Child and Family Development, and Information Science. Participants included eleven females and seven males. Two of the participants were Native American, one was African American. This was the first cohort in which some participants were from the same institutional departments as some of the faculty on the project staff.

Findings

We believe that, in the course of this project, we have learned a number of things about teaching research ethics to graduate students and postdoctoral fellows in the physical and natural sciences and to a lesser degree in the social sciences.

We made every effort to track and maintain contact with the participants from all seven cohorts, a challenge given the highly mobile nature of participants at this stage of their careers. Of the 114 participants, in 2008, 23 remained in graduate school or medical schools or residencies, 13 were in postdoctoral positions, 28 were in faculty teaching positions, 11 were in research positions, six were in research in the private sector, six were in government positions related to science and seven had gone into other professions including law and medicine. At that point we had lost contact with 21 participants.

These findings draw on our observations during and at the end of this project as well as evaluations and feedback from participants in a variety of forms. We collected evaluations from participants immediately after each workshop and asked for a statement from participants as well as a letter from advisors commenting on their advisee's home presentation. One participant provided me with a nine-page page reflection on his experience and two other shared letters addressed to others reflecting on their GREE experience. There have been hundreds of informal e-mail messages from participants over the years regarding the GREE experience. We did an evaluation of the project at the end of the first grant and completed a longitudinal survey of all seven cohorts in the form of a telephone interview regarding their GREE experience. This included 35 of the 86 participants in the seven cohorts with whom we were in contact and who remained active in science. Fifteen were in faculty teaching positions, five were in faculty research positions, four were in the private sector, one was in government, and three were in postdoctoral programs and the remaining seven in graduate school.

1. ***The Value of an Interdisciplinary Faculty Team***

We found the collaboration between those trained in science and those trained in ethics and ethics education to be essential. By and large, science faculty do not have training or experience in teaching ethics and ethics faculty do not have the detailed knowledge of the culture of the sciences. Over the seven cohorts, the faculty taught each other a great deal. For example, the ethicists and sometimes the scientists were surprised to learn about the variability of research practices and culture between labs, between disciplines, between universities. The scientists were sometimes surprised to learn about competing views on the nature of ethics and the teaching of ethics and the difference between compliance education and ethics education. The presence of both ethicists and scientists in the teaching forum served as a check and balance on the perspectives offered. There is a danger of provincialism if research ethics programs are done by either ethicists or scientists alone. A team approach provides some protection from that provincialism of perspective. Since it is not uncommon for the teaching of a research ethics course to be assigned to a single individual who may have expertise in science or ethics but not both, we believe our experience suggests it is worth the extra expense and effort to use an interdisciplinary teaching team in any program to teach research ethics to graduate students and postdoctoral fellows.

2. ***The Value of the Interdisciplinary Nature of the Participant Group***

Our longitudinal survey of the participants in the seven cohorts revealed an almost unanimous agreement of the value of the interdisciplinary experience. Years after they participated, almost all participants identified this as the experience that stood out most for them. Having such an interdisciplinary group of participants produced extremely rich discussion on many issues in research ethics, which cut across disciplines. This discussion included questions such as "What counts as raw data?" "Who owns the graduate student's data?" and many questions of authorship practices; supervisors' practices in dealing with graduate students and postdoctoral fellows; and institutional practices for educating graduate students about ethical conduct of research.

Although responses to such questions may vary with individual disciplines, the discussion was very effective in broadening participants' perspective on research ethics. As participants learned about research practices from other laboratories and other disciplines, they came to realize the strengths and weaknesses in the approaches of their own laboratories and disciplines. That enhanced their capacity to be self-aware and self-critical of their own lab or practice, much as traveling in a foreign country enhances understanding of one's own culture. For example, students trained in the medical sciences who had received some bioethics education in their medical training were surprised at the narrowness and limitations of their own ethics education as they listened to the ethical challenges of other researchers. The value of this interdisciplinary experience is not one that students are likely to receive at their own universities if all their REE is done within the department or school.

We would raise a cautionary note regarding interdisciplinary courses, given our experience. At the beginning of the project, we had debated including equal participation from the social sciences and as well as the natural and biological sciences and elected not to do so. There were however, relatively few participants from engineering and the social sciences in the first six years of cohorts. In our longitudinal survey, we found some acknowledgment from those particular participants that they had experienced some difficulty relating to the experience of other participants. As one engineering participant noted in the longitudinal interview: "I was the only engineer in my [cohort]…I did not have that experience of bonding with my crew and saw that had to do with the disciplines. They all had a lot in common with their research.

After observing the seventh cohort, which included only social science participants, we realized that the gap between the research experience, methods and focus of problems of those in the social sciences and those in the natural and physical sciences is sufficiently different that the pedagogical value of an interdisciplinary mix may be diminished if these two groups are taught together. For example, ethical issues of relations in the lab, especially student-advisor relationships and entanglements with industry, were very important in the cohorts of physical and biological sciences. Those issues were much less interesting to the social scientists. Having a cohort composed entirely of social scientists allowed us to devote much more time to issues such as ethical relativism, the problem of working with IRBs perceived as dealing with social science research from a bioethics perspective, or the theoretical and practical issues involved in group consent, or surveillance research.

Thus, it is possible to attempt to be too interdisciplinary and attempt to create a one-size-fits-all research ethics course for graduate students in all the disciplines. The net result can be a course that does not address the issues in a context relevant to some in the audience or fails to appeal to the research experiences that are recognized as familiar to some of the participants. Either one ends up appealing to the experiences of those in select fields, or one simply addresses issues such as IRB review or authorship common to all fields but leaves out the depth of detail that engages the participants for their own field. A failure to address research concerns of participants in their discipline can "inoculate" them against the value of research ethics. As one of our participants noted, in one institution he attended after the GREE project, he found graduate students in ecology dismissing research ethics as irrelevant after they had participated in a required course in research ethics at their institution that focused on biomedical research and placed all other issues in the biomedical context.

We believe the strength of our interdisciplinary approach worked in part because we limited the fields to those which had at least some common ground. Universities developing REE programs on their own campus may wish to consider separate programs for the natural and biological sciences on the one hand and the social sciences on the other.

3. ***The Mixed Value of an Off-Campus Course***

One of the "empowering" elements of the workshop as reported by participants was the freedom for participants to discuss ethical issues they experienced in their own research group or department or discipline or university. Being outside their own domain freed them to say things and raise issues many would have been reluctant to do on their own campus. Partly this was a result of being able to discuss concerns they had without fear of reprisal from a faculty member or department, and to learn that others had similar concerns and how they had handled them. Partly it was because of the opportunity to compare practices and policies at their own universities with those at other universities. In some cases, they discovered with pride that their own university was something of a leader in enlightened practice; in other cases they found ideas for policies they could take back to their own university.

Our participants did have to return to their home campuses and our project, by itself, could not do much to enhance their local campus support for research ethics or where needed, long term institutional change in the research education process. Some participants noted that coming back to a campus which had no ethics program was isolating and alienating. Several of the participants (especially in the early years) participated in GREE over the explicit and sometimes hostile objections of a mentor or advisor who objected on the grounds that the project was either irrelevant or a waste of time or both. Some GREE participants received a hostile reception by their department or faculty when they returned home to share something of what they had learned. However, in general, faculty were supportive and appreciative of the training. Some actually collaborated with their GREE participant to institute changes in their departmental program.

An alternative to our off campus model was an NSF project (Grant # SES-0115480) conceived by Michael Pritchard at Western Michigan, which drew on our experience, and is, in a sense, a second-generation ethics education initiative that extends the work we were able to do in our project. It was designed to be located on a single campus but combined interdisciplinary components that reach beyond individual departments and allowed interdisciplinary dialogue within the institution. The team pairing of graduate students and faculty within the departments creates the possibility of a forum for safe and nonthreatening dialogue and the opportunity to develop collegial student-faculty relations within departments. Both features provide an opportunity for departments in the same institution to identify, modify and improve efforts in REE and research practices. The structure of the project also institutionalizes feedback and dialogue with the administration. All of these features enhance the possibility of long-term institutional change and renewal of REE and the conduct of responsible research. As such, this project could turn out to be a model for institutional change and renewal in a way that our project could not.

4. ***The Value of the Graduate Student Perspective***

Working with graduate students and post doctoral fellows has helped us to see how different their perspectives are, compared to faculty, on many of these issues; it has given us a glimpse of issues often not talked about or admitted by faulty in discussions of research ethics. Graduate students are much more attuned to problems of relationships in the laboratory between students, advisers and faculty; problems among faculty; and problems among graduate students. Faculties need to be listening more to their graduate students' concerns in these areas. Our work has underscored the need for safe forums for graduate students to have regular discussion of such problems with faculty and has revealed the relative lack of forums in many settings. It has also enabled us to see a wide variety of department or laboratory approaches for ethics education, relations with mentors, and faculty conflicts of interest, and to assess the strengths and weaknesses of those approaches. Some of our participants were very happy with their departments' practices, and some were very angry. What we saw has underscored the value of preventive ethics in these settings. Some problems appear over and over because institutional arrangements have not been created to address these issues. Because of this difference in perspective, institutions should not assume that a research ethics course geared to faculty will be adequate to meet the need of graduate students.

5. ***Preventive Ethics***

We were particularly struck by the impact on participants of the concept of preventive ethics. As they struggled with many difficult cases, particularly those involving mentor relations, other laboratory relations and relations in the research community, they grasped the value of avoiding an ethical difficulty when possible, rather than trying to resolve it after the fact. As one of the early participants noted, reflecting on the GREE experience: “I began to understand not only the ethical dilemmas that being a practicing scientist presents but also steps that could be taken to minimize these situations. I became a fan of preventive ethics.”

As we listened to participants share experiences in their graduate programs and discuss their own cases, we noted that many of the ethical difficulties that arise in research departments or laboratories could be avoided if departments self-consciously set out clear expectations for graduate students at the beginning of their programs and provided vehicles for regular and open discussion of those expectations.

One participant helped us appreciate one preventive mechanism for helping maintain a healthy balance in the power relations between faculty and students, thereby avoiding many of the ethical issues that can arise in that relationship. That is the mechanism of having portable national fellowships for students. If some outside agency such as NSF controls the funding for a graduate student’s career, that student is much less vulnerable to a faculty member’s unethical behavior. If the department environment is unethical or if the student is treated unethically, the student can take their funding and move to another institution.

6. ***Value of Case Studies***

Cases were initially prepared before the workshop, reworked during the workshop with two feedback sessions, and polished over the summer and fall with feedback via e-mail from peers and faculty. The extensive revision and collaboration process was intended, in part, to develop an extended conversation among participants and faculty on research ethics issues and teaching research ethics and to create a community of colleagues who trust each other enough to raise these issues and discuss them frankly.

Developing and discussing case studies drawn from their own experience, and writing commentaries on their own cases proved to be a very effective pedagogical tool for helping participants learn to recognize ethical issues. The intellectual difference between writing one’s own case and reading and discussing one written by someone else, is significant. Experience does not come with “ethics problem” stamped on it. To move from raw experience with its amorphous and confusing mixture of impressions, emotional reactions and relevant and irrelevant facts to the recognition of an ethical issue is itself a significant intellectual activity. Considerable reflection is often required to determine that a problem is indeed an ethical issue and to identify and enumerate the ethical components of the situation. Many participants had the experience of beginning with a gut instinct that something was wrong and eventually being able to articulate the ethical issues. Others found they began with one notion of the ethical issue in a situation and came to see either that it was not an ethical issue or not the one they thought it was, or that there were other ethical issues they had not recognized or considered.

Writing commentaries on their own cases particularly challenged participants to reason about the cases, to articulate concepts and relevant differences between their cases and standard cases they had studied or discussed, to discern where relevant concepts and principles applied. The work with case studies and their discussion is a mode of indirectly teaching reasoning about practical ethics. The focus on crafting ethically justifiable solutions to problems challenged them to develop their capacity to put themselves in the place of the recipients of their proposed actions and thereby developed their moral imagination.

Discussion also spurred reflection on teaching objectives and techniques as they thought about how to teach the cases. Cases proved to be a pedagogical tool with which participants were comfortable and one they could use in a wide variety of situations when they tried their hand at teaching research ethics on their own campuses. Participants and faculty can use cases to initiate discussions of ethics without having to assume the intimidating position of ethics expert or having to deliver a scholarly lecture on ethical theory.

Feedback from students indicated that the use of cases for discussion in their labs and departments provided a nonthreatening means for faculty and students to raise and discuss issues in a case that were in fact impeding work in their own lab. It allowed students and faculty together to discuss a case that indirectly addressed issues in their own labs, yet because it came from an independent source, faculty were not put in the position of defending the behavior, and students were not seen as criticizing the local practice directly. It also provided a vehicle for students to raise questions about local policy they may not have felt free to raise otherwise.

We were surprised that participants came up with a new range of cases year after year. Topics included perhaps the predictable issues of authorship, data ownership, collegial relations, experimentation on humans and animals, and deception in research, but also the less familiar issues such as compliance with laws in archeological research, use of credit ratings to track down experimental subjects, use and abuse of research data in developing environmental policy, the impact of industry funding on integrity in research, ethical guidelines for research in other countries, the responsible use of engineering modeling in forensic engineering, and expert testimony.

In the sixth cohort, we experimented with encouraging participants to write up cases in which someone did the exemplary thing. Case writing tends to focus on problematic behavior. The motivation for writing such cases may often be the author's moral outrage at someone's behavior. Morally outrageous behavior can also be pedagogically effective in gaining and holding the reader's attention. In that sense, the cases are easy to teach. The downside of such cases is that focusing only on "bad behavior" may give students a skewed perception of scientific practice and may lead to cynicism about the behavior of others as well as their own. Regular use of cases of "bad behavior" may also condition students to assume that the behavior of agents in cases is always unjustified. We gave participants the option of writing up cases that might display exemplary behavior. Many had difficulty doing that. They found it easier to write about bad behavior. Others wrote up cases that reflected standard practice in their area, which they assumed was therefore exemplary. Further discussion and analysis led them to conclude they had in fact a case of morally unjustified behavior.

A good format for exemplary cases turned out to be one in which they described the character's behavior without identifying it as exemplary and asked the readers to discuss whether the agent's behavior or some alternative was ethically more defensible. This exercise requires readers to go through the same ethical analysis and justification of alternatives that they would in other cases and hence recognize the moral justifiability of the agent's behavior. We expect that this format will help with the teaching difficulty of the lack of drama in cases of exemplary behavior compared to those involving unethical behavior.

We initially thought that we should not list the authors of the published cases since many of them drew on their own experience or situations in their home institutions. The wisdom of that insight was quickly reinforced by a couple of unfortunate incidents involving discussion of participants' cases on their home campuses. The development and discussion of these cases in our own workshop often allowed participants to engage in a full airing of ethical concerns that they were not really free to discuss on their own campuses. As noted, that is one advantage of ethical training that takes place outside the home department or institution.

7. ***Cohort Seven: Research Ethics in the Social Sciences***

This last cohort was taught with an attempt to see if the approach and pattern of teaching we used in the physical and biological sciences could be effectively extended to the social sciences to determine what issues were of most concern to social sciences and if the issues that needed to be addressed were sufficiently different that there would be any benefit in segregating the participants in the social sciences in their own workshop.

Based on our planning conference with social scientists in 2004, and the collaborating faculty in 2005, it was clear that the readings needed to be substantially different. The readings related to fundamental ethical issues remained the same but were supplemented with readings on ethical relativism, and substantial readings in archaeological and anthropological research ethics as well as readings on informed consent, in research on groups, participatory research and research on children, and the use of statistics and data bases.

The participants were clearly struggling in their own research in dealing with the IRB review process in a way we did not see with former groups. One participant had been caught in the cross-fire between those in the discipline of history who denied the authority of IRBs over their research and oral historians who accepted it. Others felt their IRBs did not appreciate the difference between the kinds of human subjects research done by social scientists and its special needs in informed consent compared to that done in the biological sciences. Others were struggling with the issue of informed consent, particularly with ethnic groups or those from another culture and the idea of participatory research. Still others were dealing with the ethics of surveillance research, the ethics of deception research and the use of large social science data bases (in education). These issues are at the core of research in these disciplines and are quite unlike those of concern to previous cohorts. It was clear that had we tried to include in a research ethics workshop an equal number of social scientists and physical and natural scientists, we would not have been able to provide the depth of discussion for either group that we were able to do by segmenting them. While addressing the substantive issues for one group, the other group would be disengaged (as the experience cited earlier of ecology students forced sit through a human subjects training course illustrates).

That said, the general approach to the workshop of advanced readings, varied presentations, small and large group discussions of cases and a requirement that the participants take GREE home worked equally well with the social scientists. We eliminated the evening video series of scientists talking about ethics in research, in order to give participants more time to interact formally. That was, on balance beneficial, but something was lost in not hearing scientists talk about their professional responsibilities.

It was our hope for the GREE project that it would provide a solid grounding in research ethics for the GREE participants. Beyond that, we hoped to enable the GREE alums over the course of their careers, to provide leadership in their disciplines, in research ethics and REE for their students, their faculty colleagues and their universities across the wide range of sciences and engineering represented by the GREE alums.

We hoped to provide for the universities a model of research ethics education that is distinguished from and goes beyond mere compliance education which has now settled in as the norm in many universities. We hoped to encourage universities to think more carefully and deeply regarding their university and administrative obligations to provide REE that goes beyond compliance education for their students and faculty and the most effective means for doing that. We hoped to provide through the GREE participants a modeling of the collaborative nature of REE in the university.

The Challenges of Research Ethics Education in the University Setting

In 2011, the Association for Practical and Professional Ethics organized an important one day seminar, "The Challenges of Research Ethics Education in the University Setting: a Response to NIH and NSF Regulations" held in conjunction with the Association for Practical and Professional Ethics Annual Meeting. The seminar was important because it brought together in the same room, diverse groups involved in research ethics education/compliance education who have much to learn from each other regarding REE but historically have not had an occasion to interact (each group spending their careers in their own professional silos). The following comments are drawn from my published introduction of that seminar and the published papers of the seminar (Schrag 2012).

Over one hundred participants attended and included research and teaching faculty across the sciences and engineering; faculty outside the sciences with expertise in ethics and ethics education; university administrators charged with oversight of REE and compliance education for students and faculty in the sciences and engineering; as well as representatives of ethics education organizations within the university including ethics centers and Ethics Across the Curriculum programs. The intention of this proceeding was to give voice to those working in the trenches in REE and to speak to the larger community that is required for effective collaboration to provide sound research ethics education, and includes best practices and continuing challenges.

The central focus of the day was a series of four panels, which included: (Panel 1) "The Challenge of Research Ethics Education in the University: The View from University Offices of Research;" (Panel 2) "Collaborative Possibilities Between Ethics Centers, Ethics Across the Curriculum Programs and University Offices of Research;" (Panel 3) "Research Ethics Education in the University: The View from Early Career Faculty;" (Panel 4) "Special Focus: Research Ethics Education in the Social Sciences". A complete list of all the seminar participants can be found in my introduction along with the papers from the four panels.

The Unique Perspective of the GREE Participants

Of special note at the seminar was the presence of fourteen GREE alums, including eight who participated in the four panels and six who participated in other ways on the program. All are in positions of leadership in REE on their own campuses. The GREE alums had a unique perspective, compared to most other participants in the seminar. The GREE participants, though not formally trained ethicists, were provided substantial REE by seasoned ethicists. By the date of this seminar, some of the GREE participants had already become tenured faculty or were working their way through the tenure process. Others were completing their graduate program. Collectively, their graduate work and faculty experience cover a broad range of disciplines at a variety of universities. Since the GREE experience, they have all had the opportunity to view the remainder of their graduate career (and for many of them, their early faculty careers) through the lens of having already had a solid grounding in explicit research ethics and REE and a very strong desire to contribute to REE in their respective universities. They offer a rare resource for their respective universities. Consequently, their observations on the challenges and opportunities that they have encountered regarding research ethics and ethics education in their universities provide an especially valuable voice for research faculty and University Offices of Research to hear. (I will abbreviate University Offices of Research as UOR.)

Panel III: Research Ethics Education in the University: The View from Early Career Faculty

Here we have young faculty with more training in explicit research ethics and ethics education (particularly face-to-face education) than most of their colleagues or research administrators, and they are highly motivated to share that training. Yet they found all the institutional incentives and pressures against them for doing so. It is especially instructive to hear how that expertise and motivation has been thwarted in their institutional setting.

One GREE alum, Dudycha (2012) noted that at his institution the criteria for tenure are very explicit. (1) The criteria do not reward tenure track faculty for engaging in explicit REE. (2) In general there are no resources or incentives for faculty to provide explicit REE. Petruska (2012) observed that there is an institutional barrier, what he calls a “Pickle of Credit,” at the heart of the problem of tapping faculty expertise in REE. He argues that there is a disconnect between the UOR/compliance programs which do have a responsibility for REE and that of the academic departments or research programs which do not have a direct charge to engage in REE. The “Pickle of Credit” will not be solved unless universities, the UOR, in collaboration with senior faculty and departments, do two things: (1) Make individual faculty efforts at teaching research ethics count toward teaching/service; and (2) Provide some form of compensation to the departments whose faculty are involved in REE. Only the University and the faculty can solve this “pickle of credit,” but it is not clear how many are working on it. Both Dudycha and Petruska note that young faculty are more likely to be sensitive to the nature and need for REE but also are most vulnerable (in terms of the standards for tenure) if they engage in that activity. Senior faculty are the least vulnerable in terms of pressure to perform but are also least likely to see the need for explicit REE and least likely to feel comfortable teaching REE. That implies the need for much more faculty development in REE.

Even if there were no institution- wide impediments in terms of credit or lack of funding, there would remain the issue of faculty culture. The eagerness of young faculty can be thwarted by the lack of understanding by senior faculty of the significance and importance of face-to-face REE. Freeberg (2012) commented that he found that some faculty at his university continue to think that such an approach is at best “unnecessary,” (perhaps assuming that mentoring is sufficient) and at worst that it takes away from research or other course work. That view may also be a result of the lack of understanding of the difference between compliance training and explicit ethics education and the failure of some senior faculty to “own” research ethics. That points again to the need for further faculty education on research ethics. Wilson (2012) indicated that engineering faculties are also inclined to believe that engineering graduate programs do not have time for REE and that mentoring is enough.

GREE alums mentioned new challenges of REE in specific disciplines. Sara Wilson mentioned that engineering is quite new to REE and provides research evidence for how really different the research approach and experience is for engineering students and the implication of that for modifications in what is taught in REE. One size does not fit all. Ryan (2012) emphasized some of the unique research ethics issues that are raised in informatics research and the very serious issue that IRB members may not be keeping up with the innovations in informatics that are relevant to assessing the ethical appropriateness of some research designs in that field. She argued for the urgent need and value for collaboration between faculty and IRBs to identify ethical concerns and guidelines in the use of online technology in research.

Panel I: The Challenge of Research Ethics Education in the University: The View from University Offices of Research

The institutional origin of these challenges was addressed by participants in panel I. All are from UOR. All agreed on the need to move beyond compliance training. They discussed, from their various perspectives, the challenges of identifying precisely the aims of Responsible Conduct of Research (RCR) programs and the cost of such programs. A common theme was resource constraints, both time and money. (Note: "RCR has sometimes been used to refer to REE, sometimes to compliance education and sometimes to both. Below I will substitute REE for RCR where appropriate.)

Research administrators have had responsibility for RCR at least since the inadequacy of ethics education by mentoring has been recognized. That education has often focused on compliance training since administrators have responsibility for ensuring university compliance with government regulations in research. Historically most such administrators have not had formal degrees in ethics or experience in ethics education. A more significant number have had a legal background which has perhaps made it natural for them to focus on compliance and a compliance approach. Consequently, many in UOR have not been in a good position to recognize the difference between compliance training and REE, nor the difference in goals, outcomes or resources needed. Panicker (2012) in her essay for this collection, notes how frequently the terms "research regulation" and "research ethics" are used interchangeably in many education programs including NIH and CITI, and the damaging impact that has had.

As UOR administrators, they also have a responsibility for the cost of REE programs and are sensitive to faculty complaints about time taken away from research to do RCR education. That has led increasingly to the reliance on the use of online RCR education, which is essentially a form of compliance training. It costs less and takes less time for faculty than more effective face to face RCR ethics education. Panelists noted that university officials at the highest levels perceive that compliance education is imposed on the university by the government hence not central to the mission of the university. They are motivated to install compliance training simply to provide the university with legal liability protection. Hence the university administrators seek to provide it for the least possible cost. Many university officials fail to recognize the difference between compliance training and REE in terms of objective, content, and mode of delivery. Consequently there is a failure by many administrators who set university budgets to recognize that REE, unlike compliance training, is a core part of the university educational mission and fail to fund it appropriately.

Ratterman (2012) notes that the modes of compliance training to satisfy many of the regulations and the means for doing so, including quizzes and online offerings, do not achieve the aims of REE, nor are the measures of compliance training adequate measures of REE. There is a tension between the UOR need to fulfill the "annual metrics of office performance to justify continual support" and the need to foster "true change in the institutional culture" in terms of REE. Ratterman urges that UOR must resist the tendency to conflate the tasks of compliance training and REE. The focus must not be merely on documenting compliance training, but on significant REE. Ratterman makes clear that, unlike compliance education, in REE, "one size does not fit all." REE ethics education must be tailored: to address the levels of sophistication of researchers from the level of trainees to senior faculty; to the sorts of research being done; to the cultural variations in fields as different as engineering and action research. All of these concerns imply the need for greater collaboration between UOR and faculty and ethicists and a significant shift in approach from many current programs in UOR in many universities.

Panel II: Collaborative Possibilities Between Ethics Centers, Ethics Across the Curriculum Programs and University Offices of Research

Participants in the second panel make more explicit what was implicit in Panel 1. Many high level administrators and members of Offices of University Research lack the background in ethics to recognize the distinction between compliance education and REE in terms of educational goals, teaching methods, modes of delivery, educational outcomes and resources required.

This panel included those with a background in academic administration, faculty research, research ethics, ethics centers, and ethics across the curriculum. Some have formal training in ethics and ethics education. It is significant that one rarely finds a group with this diverse background working together on an actual university campus to provide leadership in REE ethics education. They argue for the need (not often recognized) for collaboration on campus between administrators, research faculty, and ethicists to provide for REE education.

Wueste (2012) sets out a crucial distinction between the notion of compliance and that of ethical responsibility. He notes that compliance has to do with and focuses only on the compliance of a researcher with federal or legal guidelines, whereas the notion of responsible conduct of research *preceded* any legislative guidelines. For example, the ethical issues of falsification, fabrication, and plagiarism were recognized in research long before the existence of any governmental regulations regarding such concerns and hence their "wrongness did not then and does not now depend on the existence of governmental regulations." These are the norms implicit in scientific research. He observes that compliance may be the province of those with a legal or regulatory background but the ethical norms of a practice are the ethicist's expertise.

Wueste notes the implication of this distinction for ethics education. Compliance training focuses on making researchers aware that something must be done to comply with a regulation. Ethics education focuses on why these things should be done. This involves making explicit the underlying norms of science. Making the norms explicit is necessary for researchers to "own" the norms governing research. As ethicists know, that is best done by the pedagogical approach of dialogue in a face-to face setting, not in an online course. He observes that this is the sort of teaching activity which practical ethicists do for a living in many areas of professional practice and which ethics centers with a focus on practical ethics facilitate on many university campuses. Therefore, ethicists and ethics centers on campus could make a significant contribution to REE in the university; both with the ethics education of science trainees as well as with helping research faculty integrate research ethics into their courses.

Drawing on his experience in teaching business ethics, Wueste notes the impact of a compliance mentality in teaching a target audience. The focus on compliance training leads the target audience to treat policies and training as hoops to be jumped through and creates an authoritarian, adversarial relation between those in charge of the education and those receiving the education. (That explains the resistance of research faculty to courses in compliance training as a substitute for REE.)

A focus on REE, on the other hand, results in an understanding of conduct that is owned and freely chosen. The outcome of REE should result in respect for and adherence to the educational norms inherent in research. This echoes the observation in the essay by May (2012) regarding an educational focus on group decisions based on shared norms of the research enterprise. Wueste adds that recognizing the links between REE education, Ethics across the Curriculum, and Academic Integrity movements on campus would provide some opportunities for tapping the ethics education expertise of practical ethicists elsewhere on campus. Wueste concludes, "Those opportunities, for the most part, have been missed by RCR programs."

Bird (2012), both a scientist and ethics educator, concurs with Wueste that trainees and faculty must "own" the ethical norms that underlie research and for that to happen, both trainees and faculty must become explicitly aware of those norms. Compliance training does not do this; mentoring does not do this adequately. Mentoring may address the ethical issues implicitly but not explicitly. The problem with the mentoring approach is that the rationale behind exemplary behavior is not always obvious and thus the norms are not made explicit.

Bird notes that a topic that is missing in much of the RCR literature involves the goals of ethics education. Early in this paper we discussed the views of Callahan and Hampshire regarding such goals which we incorporated in the design of the GREE project. Bird observes that a proper understanding of these goals raises serious concerns about any of them being met by compliance training or online ethics training. Neither compliance training or online ethics training is effective in providing ownership in the norms or experience in making and defending ethical decisions in the practice of scientific research. Bird, Pritchard (2012) Essigmann (2012) and GREE Alum McCafferty (2012) all provide excellent accounts of collaborative efforts to pursue these goals in a concrete workshop and how positive is the faculty response to that sort of ethics education. Face-to-face workshops are more expensive than online training, but far more effective for meeting the goals of REE than is compliance training. Most campuses do not have science faculty who are also well-versed in practical ethics. That is why there is a need for collaboration with practical ethicists, ethics centers and ethics across the curriculum personnel who can provide that expertise.

Giffels (2012) addresses a crucial but rarely discussed issue if there is to be such collaboration. There are cultural and cognitive gaps between administrators, ethicists, and researchers that must be acknowledged as real and which pose real challenges to collaboration.

Ethicists have not always paid their dues in terms of understanding the details of the practice of scientific research in different disciplines or the guidelines that have developed in research ethics. Neither are they aware of the responsibilities and constraints under which research administrators must operate. Ethicists cannot be expected to be invited to make a contribution to REE armed only with grand ethical theory but ignorant of practical ethics in general and the practice of science and engineering in particular (not just medical ethics).

There are, fortunately, a significant number of practical ethicists and ethics centers who are members of the Association for Practical and Professional Ethics who have done their homework and are involved in collaboration with UOR on their campuses. Many were participants in this seminar. Others remain untapped on their local campus because their expertise is not recognized by research administrators or research scientists.

Giffels clearly makes the point that what can be said of ethicists can also be said of the research administrators and the research scientists. It is one reason the confusion between compliance training and REE has prevailed for so long. He rightly argues that the cultural and conceptual gaps between these three groups must be narrowed before meaningful collaboration will be possible. All these groups coexist on the same campus but in fact the silos in the university create difficulties and resistance for collaboration. The university culture rewards and reinforces faculty to focus on ever increasing specialization in their discipline. The press of administrative duties discourages administrators from taking the time and energy to acquaint themselves with the details of research practice or the nature of practical ethics. There is no arena in which these interactions are regular or easily available.

Giffels suggest the creative and practical idea of establishing a shadowing program on campus to allow members of each of these groups to develop some sense of the cultural and conceptual differences between them. Such a program would be coupled with group meetings to encourage the processing of the experience and building a sense of community. There are some cost implications in terms of time but it is a doable proposal, if the will exists. Leadership for such a program will likely have to come at the level of the University Office of Research.

Our own GREE program and the collaborative programs mentioned above have demonstrated the effectiveness of such programs in doing REE with faculty and students. Faculties, when exposed to such programs, embrace them and become advocates for them in their disciplines. In order to establish adequate support for REE in departments and disciplines, a much larger percentage of faculties need to be educated. More UOR need to be educated on the crucial distinction between compliance education and REE.

GREE alum Frugoli (2012) emphasized what happens when, although the university formally assumes responsibility for REE, it does not, for whatever reason, allocate adequate financial resources to do that. Frugoli makes an absolutely foundational observation, key to the entire discussion of this seminar. Budgetary allocations are within the purview of the university and represent the value choices of the upper administration, ultimately the president of the university. For example, the University of Notre Dame allocated the entire $14.5 million received from its appearance in the 2006 Fiesta Bowl to support academic purposes, not the athletic program (Yost 2010). The budget choices that administrators make are always challenging and almost always between good things. However there are resources that could be tapped for REE. Nebeker (2012) mentions that in the past 10 years her institution has taken in more than one billion dollars in research grants. Imagine what an additional 1% of indirect costs from that amount, set aside for REE could have provided over the past 10 years.

REE must be a core focus of research education. Failure to adequately fund REE at each university is, at best, a result of ignorance of the importance of a part of the core mission of the university. At worst it represents a deliberate choice at the highest levels of the university to place budget priorities on programs less central to the core of the university mission.

References

Bird, S. J. (2012). Involving faculty in teaching the responsible conduct of research. *Teaching Ethics, 12*(2), 65–76.

Callahan, D., & Bok, S. (eds.) (1980). *Ethics teaching in higher education* (pp. 61–80). New York: Plenum Press (Chapter 2).

Cohen, J. (1993). Moral pluralism and moral consensus. In D. Copp, J. Hampton & J. E. Roemer (Eds.), *The idea of democracy* (pp. 270–291). Cambridge, UK, New York: Cambridge University Press.

Dudycha, J. L. (2012). Research ethics: Education from the view of a junior faculty. *Teaching Ethics, 12*(2), 87–92.

Essigmann, J. M. (2012). Addressing the need for templates for teaching responsible conduct of research at research universities. *Teaching Ethics, 12*(2), 83–87.

Frugoli, J. (2012). Be careful what you wish for. *Teaching Ethics, 12*(2), 103–105.

Freeberg, T. M., & Moore, T. M. (2012). Research ethics education challenges in a psychology department. *Teaching Ethics, 12*(2), 107–111.

Giffels, J. (2012). First steps: Minding and mending the gaps. *Teaching Ethics, 12*(2), 77–81.

Hampshire, S. (1949). Fallacies in moral philosophy. *Mind, 58,* 466–482.

May, T. A. (2012). Developing diverse and robust research ethics education. *Teaching Ethics,* 12(2), 35–42.

Nebeker, C. (2012). One size does not fit all. *Teaching Ethics,* 12(2), 53–57.

McCafferty, J., et al. (2012). New NSF and NIH RCR guidelines: A three phase plan.

Panicker, S. (2012). Research ethics education in the behavioral and psychological sciences: A mistaken case of one size fits all. *Teaching Ethics, 12*(2), 137–140.

Petruska, J. C. (2012). Research ethics education in the university: The pickle of credit. *Teaching Ethics, 12*(2), 95–102.

Pritchard, M. S. (2012). An institutional change model. *Teaching Ethics, 12*(2), 81–82.

Ratterman, A. G. (2012). Developing diverse and robust research ethics education: One offices approach. *Teaching Ethics,* 12(2), 43–47.

Ryan, M. (2012). Federal ethics regulations governing internet research: New educational challenges for universities. *Teaching Ethics, 12*(2), 127–136.

Schrag, B. (2008). Teaching research ethics: Changing the culture of science. *Teaching Ethics, 8*(2), 79–110.

Schrag, B. (2009). Piercing the Veil: Ethical issues in ethnographic research. *Science and Engineering Ethics*, 15(2).

Schrag, B. (2012). Introduction: The challenge of research ethics education in the university setting: A response to NIH and NSF Regulations. *Teaching Ethics,* 12(2), 1–22.

Swazey, J. E, & Bird, S. J. (1997). Teaching and learning research ethics. In D. Elliott & J. E. Stern (Eds.), *Research ethics: A reader*. Hanover: University Press of New England (Chapter 1).

Wilson, S. (2012). Research ethics education in engineering: The view from early career faculty. *Teaching Ethics,* 12(2), 119–126.

Wueste, D. E. (2012). RCR: Some splendid opportunities. *Teaching Ethics, 12*(2), 57–64.

Yost, M. (2010). *Varsity green: A behind the scenes look at culture and corruption in college athletics* (p. 83). Stanford: Stanford University Press.

Ethics Across Early Childhood Education

Michael D. Burroughs

Abstract Early childhood ethics education has been of longstanding interest for philosophers, psychologists, and those with interests in child development and education more generally. The significance of early childhood education remains vital today, with an expanding focus on ethical, social, and emotional education in pre- and primary classrooms. Taken together, and given the confluence of several areas of development in early childhood—cognitive, moral, social, and emotional—this period of life presents robust opportunities for ethics education. I conceptualize ethics education in early childhood in two broad ways: first, as an educational process embedded in the child's experience of the school as a *sociomoral* environment that can provide important, if indirect, opportunities for ethical learning and development. Second, I take up ethics education in the child's participation in specific ethics and social-emotional learning programs for classroom, home, and broader community use. While presenting a range of approaches to early childhood ethics education, I argue for the benefits of methodological pluralism, identifying the many continuities and opportunities for collaboration across the theoretical and practical divisions set up in the field of ethics education, and call for collaborative teacher-researcher partnerships in order to develop effective ethics education programming.

Keywords Early childhood · Ethics education · Constructivist education
Character education · Social-emotional learning

Early childhood ethics education has been of longstanding interest for philosophers, psychologists, and those with interests in child development and education more generally (professional educators, parents and siblings, and community members). This concern stems, perhaps most fundamentally, from what Arendt (1954/1977) characterizes as the *natality* inherent to childhood and, in turn, the obligation that this natality, this new-ness of each generation, entails for human society. As Arendt

M. D. Burroughs (✉)
California State University, Bakersfield, USA
e-mail: mburroughs1@csub.edu

E. E. Englehardt and M. S. Pritchard (eds.), *Ethics Across the Curriculum—Pedagogical Perspectives*, https://doi.org/10.1007/978-3-319-78939-2_15

writes, "education belongs among the most elementary and necessary activities," an activity that "never remains as it is but continuously renews itself through birth, through the arrival of new human beings" (Arendt 1954/1977, 182). In response to this "necessary activity," some of our greatest minds—from Plato, Aristotle, and Rousseau, to Dewey, Montessori, and Piaget, to name just a few—have grappled with the best methods for introducing children into the world via educational systems.

The significance of early childhood education remains vital today, with an expanding focus on ethical, social, and emotional education in pre- and primary classrooms. Many skills and competencies essential to ethical life begin to emerge during this time of life (years 2–8 of the human life span), including (but not limited to) emerging awareness of ethical concepts such as fairness, reciprocity, harm, and the welfare of others; the experience of moral emotions such as compassion, empathy, and guilt; and a burgeoning understanding of the perspectives of others with distinct feelings, intentions, and beliefs from one's own. In addition, in leaving the home and entering the social world of the school, young children begin to encounter ethical issues in interactions with peers that must be navigated (e.g., sharing, conflict resolution, and peer interactions more generally) that, in turn, can foster the development of social, emotional, and ethical skill sets. Young children's encounters with ethical concerns can also include confrontation with significant social and political issues in our world beyond the classroom, including prejudice and political unrest (Coles 1986; Burroughs 2016). Taken together, and given the confluence of several areas of development in early childhood—cognitive, moral, social, and emotional—this period of life presents robust opportunities for ethics education.[1]

In discussing aims and practices of ethics education in early childhood, and given the emphasis of this volume on ethics across curricula, I will focus primarily on the context of schooling and related programing.[2] I conceptualize ethics education in early childhood in two broad ways: first, as an educational process embedded in the child's experience of the school as a *sociomoral* environment (including its rules, norms, conflicts, peer and child-teacher interactions, etc.) that can provide important, if indirect, opportunities for ethical learning and development. Second, I take up ethics education in the child's participation in specific ethics and social-emotional learning programs and curricula adopted for classroom, home, and broader community use. While presenting a range of approaches to early childhood ethics education, I argue for the benefits of teacher-researcher

[1]Despite the unique opportunities for ethics education in early childhood a recent survey article on forty years of scholarship (1971–2011) in the *Journal of Moral Education* (arguably the leading academic journal in the area of moral education and related research), notes a "near absence of papers on moral education in childhood and early learning" (Lee and Taylor 2013, 423). Given this absence in the *Journal of Moral Education* and the comparative importance of education in early childhood, this chapter will contribute to filling an important gap in ethics education literature.

[2]For discussion of moral development in other significant contexts, including within the family and home, see Berkowitz (1992), Dunn (2014) and Kuczynski and Knafo (2014).

collaborations and methodological pluralism. As Marvin W. Berkowitz, a leading character education researcher, argues, our primary aim as ethics educators and researchers should be "to help kids become good people" and, to this end, there are benefits to identifying the many continuities and opportunities for collaboration across the theoretical and practical divisions set up in the field of ethics education (Berkowitz 1992, 44). Thus, I will discuss some of the beneficial connections to be found between current approaches to early childhood ethics education and, further, suggest that greater attention be devoted to teacher-researcher collaborations in order to develop effective ethics education programming.

Ethics in Early Childhood: Foundations in Social Experience

As numerous researchers and practitioners have shown, ethics and ethical concerns are not the exclusive, privileged territory of adult life. Rather, from a young age, ethical concerns and questions are present in the lives of young children (Damon 1988; Gussin Paley 1997; Piaget 1932/1997; Nucci 2001). Providing examples of ethical concerns in the lives of preschoolers, DeVries and Zan (1994) write:

> While the content of moral issues in the lives of children differs from that of adults, the basic issues are the same. Children worry about how people (first of all, themselves) are treated long before they can understand the Golden Rule. They worry about aggression, fair use (for example, of dress-up clothes), and equal participation (for example, in clean-up). These are issues of rights and responsibilities just as are adult concerns with crime and violence, equal employment opportunity, and the need for everyone to protect the environment (28).

The *locus classicus* for recognition of ethics in early childhood experience comes in Jean Piaget's foundational study of moral development, *The Moral Judgment of the Child* (1932/1997). Based on extensive fieldwork and interviews with young children Piaget argues that children's experience and mastery of the juridical complexity of games and reciprocal play with peers can be a primary form of ethics education and, in turn, moral development. Like the system of morality (as Piaget defines it), children's games are constituted by a formal system of rules in relation to which children—apart from and, in many cases, *despite* adult interventions—take on evolving modes of perspective-taking, judgment, autonomy, and relations of respect. Piaget writes:

> The rules of the game of marbles are handed down, just like so-called moral realities, from one generation to another, and are preserved solely by the respect that is felt for them by [children]...The little boys who are beginning to play are gradually trained by the older ones in respect for the law; and, in any case, they aspire from their hearts to the virtue, supremely characteristic of human dignity, which consists in making correct use of the customary practices of a game. As to the older ones, it is in their power to alter the rules. If this is not 'morality,' then where does morality begin? (1932/1997, 14).

Any of us who have spent time with children—or who can recall our own childhoods—will probably have in mind debates on the playground about the rules of a game, sometimes taking more time and possessing more nuance than the game itself. But in this play, or in this ethical work that is well prior to the formal lessons of the school and well beyond the rules handed down by the adult, lie the very skills, the affective and mental orientation, that is needed to live an ethical life with others: including empathy, perspective taking, dialogue, and negotiation; judgment and reason; and the increasing sense of autonomy that is required to accept and endorse ethical principles.

Piaget's articulation of ethics education and corresponding moral development via cooperation and negotiation with peers has close affinities with John Dewey's conception of "indirect and vital moral education." In his *Moral Principles in Education* (1909/1975) and *Democracy and Education* (1916/2011), Dewey eschews a narrow focus on ethics education in the form of isolated lessons or coercive, external strategies of indoctrination, instead focusing on the "development of character through all the agencies, instrumentalities, and materials of school life" (1909/1975, 4). Dewey challenges the notion that ethics education, or concerns with ethics more generally, must (or should) be something brought in from outside the experience of the child and her/his social experience. To do so, Dewey notes, risks making ethical knowledge detached and inanimate, lacking the affective connection to social life that is needed to motivate conduct and a desire to *be* ethical in one's interactions with others:

> If a pupil learns things from books simply in connection with school lessons and for the sake of reciting what he has learned when called upon, then knowledge will have effect upon some conduct – namely upon that of reproducing statements at the demand of others. There is nothing surprising that such 'knowledge' should not have much influence in the life out of school (1916/2011, 194).

Instead, ethics education should be sourced in engagement with children's social experience (of which, the school serves as a central part) and the problems and questions that are already present there. In this context, the ethics education practitioner need not engineer ethical problems or challenges in order to make them seem vital to the child, but rather, spend time listening, observing, and becoming aware of the ethical dimensions and ideas already at play in a classroom, game, conflict, or relationship, and harness these experiences for their educational potential.[3] The key for Dewey, as for Piaget and constructivist educators more generally, is to focus on the social conditions that give rise, organically, to ethical

[3]This is not to say that Dewey argues against curricula or structured education in the classroom, or neglects the role of the adult educator as such. Rather, Dewey is interested in a particular kind of educational structure, combining the adult's experience and pedagogical skill sets with the native interests and experiences of children to form productive and engaging educational experience that will be a motivating force for the child, even when a guiding adult is not present. In *Experience and Education* (1938), Dewey discusses this process in terms of the interaction of *objective* (the learning environment, past knowledge, pedagogical resources, etc.) and *internal* (students' attitudes, interests, desires, etc.) conditions of educational experience (42, 45).

questions, problems, and concerns (in short, ethical experience) and to use these experiences as means for education that is capable of animating autonomous problem solving and a personal disposition toward pursuit of an ethical life (Hildebrandt and Zan 2014). In this sense, education should not be a "mere means" to ethical life, but should *be* such a life (Dewey 1916/2011, 196).

The educative potential of children's social experience has been immensely influential for contemporary approaches to ethics education in early childhood and is highlighted across the traditions of *character education*, *moral education*, and *constructivist education* more generally (see Section "Ethics in Early Childhood: School-Based Approaches" below). With its focus on social development, this influence also extends to one of the leading areas of contemporary early childhood education and research, *social and emotional learning* (SEL).[4] SEL prioritizes the behavioral, social, and emotional building blocks of human flourishing and is defined, broadly, as "the process through which we learn to recognize and manage emotions, care about others, make good decisions, behave ethically and responsibly, develop positive relationships, and avoid negative behaviors" (Zins et al. 2004, 192). Though diverse in aims, and, at this point, many in number, SEL programs focus on the cultivation of behaviors and skills that inform the development of social and emotional competencies in early childhood (Denham and Weissberg 2004; McCabe and Altamura 2011; Elias et al. 2014). As defined by the Collaborative for Academic, Social, and Emotional Learning (www.casel.org), the leading research and advocacy organization supporting SEL initiatives in the United States, these competencies range across cognitive, affective, and behavioral areas and include *self-awareness*, *self-management*, *social awareness*, *relationship skills*, and *responsible decision making*. In essence, then, SEL programs—whether introduced across existing curricula by teachers or as external programs introduced by researchers—aim to help children develop the ability to understand and manage their emotions, form productive relationships and prosocial behaviors, and to succeed in their social and academic lives.

The SEL focus on emotional competency for early childhood populations is particularly relevant as, for young children, affective development and expression precedes higher order cognition and serves as the primary motivating force of behavior (Denham and Weissberg 2004). During this time period, and with proper guidance, several social and emotional milestones can be achieved, including increased emotion regulation, self-awareness, and perspective taking (Denham and Weissberg 2004; McCabe and Altamura 2011). In turn, these skills and competencies contribute to children's ability to form and maintain successful relationships and, further, are linked to evidence-based impacts on children's academic performance and increased prosocial behavior (Durlak et al. 2011). As a result, since the

---

[4]For discussions of and research on the effectiveness of prominent early childhood SEL programs—such as *Promoting Alternative Thinking Strategies* (PATHS), *Positive Behavior Intervention Support* (PBIS), *Second Step*, and *FunFRIENDS*, among others—see Denham and Weissberg (2004) and McCabe and Altamura (2011).

1990s, SEL programming has increasingly gained support in U.S. educational policy and practice at the federal, state, and local levels.

Alongside the tremendous success and growth of SEL programming over the past thirty years, however, some SEL approaches have received criticism for their lack of "moral depth," and, what is more, for the conflation of social-emotional and ethical competencies in SEL research and programs (Carr 2002; Kristjánsson 2006, 55; Burroughs and Barkauskas 2017). That is, a child could become socially and emotionally competent as commonly defined (e.g. coming to possess skills in self-awareness, self-management, emotion regulation, and so on) and, in turn, deploy these skills for unethical ends (e.g. coercive manipulation of others or self-control while committing a crime). For this reason, and as leading SEL researchers now acknowledge, there are benefits to combining ethics education with SEL programming (Elias et al. 2014). SEL programming provides essential training in social skills, the development of positive relationships, and emotion management that are essential for the interpersonal dimensions of ethical life. Indeed, without learning to communicate effectively, to manage conflict, and to build caring relationships with others, children will be hindered in their ethical development. At the same time, ethical guidance in understanding normative values (e.g. justice, fairness, respect, care, etc.), in moral reasoning and judgment, and, perhaps most importantly, in developing an autonomous ethical orientation, can provide valuable direction for otherwise ethically neutral SEL skill sets. Thus, Elias et al. (2014) argue for the unity and collaboration of SEL, character, and moral education programs in early childhood (and in later life) in order to educate for both moral direction and the skills to carry out ethics in practice.

Ethics in Early Childhood: School-Based Approaches

The school serves as a primary context for early childhood ethics education in at least two (and often overlapping) forms: first, the school serves as a sociomoral environment that provides important, if indirect, opportunities for ethical learning and development. The sociomoral environment is conceptualized as the "entire network of interpersonal relationships in the classroom," including child-teacher relationships, child-child relationships, adult-adult relationships, and the classroom structure, norms, and rules that serve as their foundation (Hildebrandt and Zan 2014, 180). As discussed above, a focus on children's social and moral environment and its educative potential has a long and rich history in ethics education research and practice. This tradition continues most prominently in *constructivist* (DeVries and Zan 1994; Hildebrandt and Zan 2014) and *moral education* approaches (Power et al. 1989; Nucci 2001) that stress ethical and socio-cognitive development through discussion of ethical concepts and dilemmas, perspective taking, authentic and autonomous problem solving, and interpersonal relationships in the classroom.[5]

Second, and alongside its sociomoral environment, the school—as an educational institution meant to impart specific skills, values, and traditions to children—provides opportunities for structured ethics education lessons and curricula. *Character education* programs—with a traditional focus on the transmission of societal and ethical values and norms to children (Ryan 1989; Wynn 1989; Bennett and Hague 1995)—are the most well represented form of ethics education in U.S. schools and commonly utilize curricula supported by school-based, regional, or national character education organizations. However, there is great variation between character education programs in early childhood settings and an increasing tendency toward integrated programming that combines education for selected ethical values (a common trait of character education programming) with constructivist approaches that emphasize children's ability to "construct" ethical knowledge from their experience and peer interactions, along with appropriate adult guidance.

Sociomoral Education in the School

In *Nice is Not Enough: Facilitating Moral Development* (2009), Larry Nucci describes the sociomoral curriculum as an essential element of moral education:

> Children's moral and social knowledge originates in their attempts to make meaning out of social experience. Thus, we must begin our discussion of moral education by examining how schools function as social and moral environments. Social life is not experienced as an abstraction, but confronts children in their everyday efforts to negotiate their desires and needs in relation to those of others, in the social rules and norms that structure social interactions, and the feelings that come along with those experiences...How we structure educational environments and respond to student behavior forms...the moral life of schools and classrooms (53).

Given its breadth and its foundation in the lives and interests of children, the sociomoral curriculum provides a wealth of opportunities for ethics education in early childhood. As has been noted by SEL, character, moral, and constructivist educators and researchers alike, by fostering trusting relationships in an emotionally supportive school environment young children are provided with an important

[5]I use the terms (here and below) *constructivist*, *moral*, and *character* education to mark prominent, if broad, distinctions in the field of ethics education. However, in using these distinctions I do not wish to further reify theoretical and practical separations between these approaches, nor are these terms used in uniform ways in ethics education literature. As I will argue, in both practice and theory, there are many commonalities and areas of overlap between constructivist, moral, and character education (see Section "Toward an Integrated Approach to Early Childhood Ethics Education" below).

ethical foundation: a sense of goodwill and belonging that informs their own treatment of others, their understanding of ethical concepts such as justice, fairness and care, and that increases their personal commitment to living ethically (Berkowitz 1992; Noddings 2002; Denham and Weissberg 2004; Nucci 2009). A positive sociomoral environment also supports children's basic needs in *autonomy* and *competence*, giving them authentic opportunities to make decisions, complete tasks, and manage conflicts such that they develop the requisite skills and identity to regard themselves as capable persons with internal motivation to act with fairness and concern for the welfare of others (DeVries and Zan 1994; Nucci 2009). This, in turn, is central for the development of moral autonomy, or, "doing what is right for one's own reasons" (Nucci 2009, 69) The guiding idea here is that children given opportunities to regulate their own behavior, alongside the guidance of an adult, are more likely to develop the self-regulation skills and autonomous ethical orientation that is needed when no guiding influence beyond the self is present. It is central to this process that, from a young age, children are treated with respect, as persons with ethical concerns, and provided appropriate opportunities to voice and negotiate these concerns. As DeVries and Zan (1994) write:

> The child who is given opportunities for regulating his or her behavior has the possibility for constructing a confident self that values self and others positively. By respecting the child's will, the cooperative adult can help the child develop self-regulation based on respect for others as well as for self. The child able to exercise his or her will constructs gradually a stable system of moral, social, and intellectual feelings, interests, and values (50).

Several pedagogical strategies support the related processes of ethics education and moral development through the sociomoral environment of the school. For example, Watson (2014) articulates an account of *developmental discipline* as central to the formation of an autonomous ethical orientation in children. Developmental discipline moves away from traditional (and pervasive) approaches to classroom discipline based on an extrinsic system of rewards and punishments to motivate and control student behavior. Instead, the aim of developmental discipline is, first, to build trusting relationships that support ethical behavior and a child's motivation to be caring and cooperative. If children have trusting, warm relationships with their teacher in a school environment that is emotionally supportive, they will be more likely to choose to cooperate and treat others with care (none of which is as likely in an environment of coercive or antagonistic relationships). As part of class learning activities, or when misbehavior and ethical conflicts arise in the classroom, the developmental discipliner provides opportunities for discussion between children to explore and promote perspective taking and greater understanding of moral principles (e.g. reciprocity, fairness, and care).[6] A related

[6]A vast literature on conducting moral discussions with young children, including specific lesson plans and activities for this purpose, exists in the field of *philosophy for children*. See Lone (2012), Wartenberg (2014), and Lone and Burroughs (2016). For additional resources, also see the Philosophy Learning and Teaching Organization (PLATO), a national non-profit organization devoted to supporting philosophical discussion in Pre-K—12 classrooms: www.plato-philosophy.org.

strategy, discussed by Hoffman (2000), is known as *induction* and involves a disciplinary process in which the adult (e.g. after a child has caused physical or emotional harm to a peer) speaks with a child about a peer's perspective and distress, and makes clear that the child's actions resulted in the harm in question. The aim here is to help young children—who are just in the process of developing the ability to regulate their own behaviors and desires in relation to the desires and needs of others—to understand the perspective of others and produce an empathic response that, eventually, can motivate autonomously chosen ethical behavior (e.g. "You hit him and now he is crying. When you hit others, it hurts them."). In early childhood classrooms, teachers often refer to this strategy as *scaffolding* in which children are assisted in identifying a conflict and its sources, and helped to see beyond their own perspectives in order to consider and select a mutually beneficial solution (e.g. "Given that you both want to play with the truck, maybe we can use a timer so you both get a turn. Would that work for us?"). As compared to the common punishment-reward orientation to classroom discipline (which reinforces motives centered on avoiding punishment and pleasing a superior), a motivating principle for these disciplinary approaches is that mature ethical life requires understanding and internalizing ethical reasons and motivations (e.g. as based in moral values such as justice, fairness, and care). Thus, ethics education—whether in the form of discipline, conflict resolution, or the norms and activities of classroom life—must support a personal and authentic commitment to ethical life.

Many constructivist and moral educators also provide children with opportunities to develop classroom rules and guidelines for behavior as a means for fostering children's autonomy, competence, and sense of responsibility for their community (DeVries and Zan 1994; Hildebrandt and Zan 2014). As Watson (2014) notes, "when students are involved in creating structures that facilitate the smooth functioning of the classroom their autonomy is honored and they are helped to understand why the rules and structures are necessary" (166). This can be exercised in the classroom in several, concrete ways: by having students develop guidelines for fair use of toys; decide on responsibilities during 'clean up'; develop rules for 'circle time' (a common period in U.S. early childhood classrooms for children's literature reading, discussions, and other shared activities) when both speaking and listening are important; and including students in decisions about how class 'free time' will be spent. In these ways, as opposed to simply handing down required behaviors and prohibitions to children, children can collaborate with each other and the teacher to select class guidelines. In doing so, children take on greater understanding of and motivation for following these guidelines going forward and, in turn, gain a sense of personal responsibility as members of a classroom community.

Educating for Character

Character education programs are the most well represented form of ethics education in Pre-K—12 schools, benefitting from broad policy and funding support as well as the advocacy efforts of several national organizations.[7] Since the 1990s there has also been a resurgence of support for educating for traditional values in the midst of what is interpreted by some commentators as a deteriorating moral fabric in society and increasingly high rates of youth at risk (Lickona 1989; Kilpatrick 1992; Bebeau et al. 1999; Ryan and Bohlin 1999). As compared with constructivist and moral education approaches as described above, character education programs have traditionally been aligned with more conservative and behavioral educational methods, focusing on the reduction of problematic behaviors and the transmission of societal and ethical values and traditions to children (Berkowitz 1992; Howard et al. 2004). Despite this traditional focus, many contemporary character education programs are diverse in orientation and programming, utilizing pedagogical strategies influenced by constructivist, SEL, and socio-cognitive approaches. In school settings, character education programs are commonly offered in hybrid form with other values-based education programs, including citizenship education, religious education, service-learning, anti-drug, and social-emotional learning programs (Howard et al. 2004; Althof and Berkowitz 2006; Elias et al. 2014).

Character education is defined, broadly, as "an attempt to prepare individuals to make ethical judgments and to act on them" (Howard et al. 2004, 189), with *character* understood as a set of psychological, affective, and behavioral characteristics that influences a person's ability to make ethical judgments and function morally. Core ethical values—such as respect, responsibility, honesty, fairness, and caring—are central to character education programs, serving as standards and guides for educational programming. In order to instill character in children many school-based programs emphasize selected values and character traits in lesson plans, classroom activities, reward programs, and competitions. For example, the *Character Education Partnership* encourages schools to define central ethical values as the basis for classroom character lessons, behavioral standards, teacher modeling, and school-wide socialization efforts (Character Education Partnership 2010). The *Character Counts!* program utilizes character lessons, classroom decorations, and reward programs (e.g. posting lists of children who demonstrate 'good character' and the use of 'pledge certificates') with the aim of instilling preselected virtues in participating children. Known as the "Six Pillars of Character," these virtues include *trustworthiness* (e.g. be honest; don't deceive, cheat, or steal), *respect* (e.g. follow the Golden Rule; use good manners, not bad language), *responsibility* (e.g. be diligent; do your best), *fairness* (e.g. play by the rules; take

[7]Reporting on data from the U.S. Department of Education in 2001, Howard et al. (2004) report that "forty-five states and the District of Columbia received and implemented character education pilot grants through the U.S. Department of Education…in years 1995 through 2001" (209). Character education initiatives have also been central to the educational policies of several U.S. Presidents, including Bill Clinton and George W. Bush. As a result, and due to additional efforts of national character education and advocacy programs such as *Character Counts!* (www.charactercounts.org) and the *Character Education Partnership* (www.character.org), character education programs are mandated or strongly encouraged in most states.

turns and share), *caring* (e.g. be kind; be charitable and altruistic), and *citizenship* (e.g. obey laws and rules; respect authority). Like other traditional approaches to character education, a clear emphasis here is placed on the use of stories, distinct lessons, and rewards to instill virtuous behavior and character formation in young children.

As mentioned above, however, character education programs are increasingly diverse in orientation and combine influences and pedagogical strategies from constructivist, SEL, and socio-cognitive approaches. A leading character education program, the *Child Development Project* (a project of the Developmental Studies Center, www.devstu.org), focuses on elementary schools and has had the most well documented outcomes—in social-emotional development, prosocial behavior, intrinsic motivation, and academic achievement—of any character education program in the U.S. (Solomon et al. 2000; Berkowitz 1992; Battistich et al. 2004). As opposed to isolated character lessons, the Child Development Project (CDP) (and its current manifestation in the *Caring School Community* program) deploys a whole school approach to promote student growth across social, ethical, and intellectual realms and the creation of a community that supports young children's need for belonging, competence, autonomy, and a sense of shared purpose (Sanger and Osguthorpe 2009). The program emphasizes academic skills in writing and reading comprehension alongside constructivist strategies, such as the inclusion of children in school decision-making (e.g., norm- and rule-setting in classrooms) and opportunities for authentic discussion of ethical issues and conflict resolution (Battistich 2010; Brunn 2014). The CDP, and the work of the Developmental Studies Center more generally, evidence a growing trend in character education to focus on schoolwide interventions that bring together administrators, teachers, and students (and often the family) in support of character development.

Toward an Integrated Approach to Early Childhood Ethics Education

In the introduction to this chapter I characterized education (following Arendt) as among the most "elementary and necessary activities" of human society. As Arendt (1954/1977) notes, adults must "assume responsibility for the world into which they have brought the children," (187) and education—at its core, the activity of introducing children to our world—serves as one of the primary exercises of this responsibility. In turn, the form this education should take has divided the ethics education community throughout its history. Characterizing this historical division, Hildebrandt and Zan (2014) write:

> There has been a dynamic tension between *traditional education*, where instruction is primarily teacher-centered and morality is defined by the rules and dictates of authority, and *progressive education*, where the classroom is primarily child-centered and moral development is seen as the gradual construction and application of principles of justice, equity, and compassion (181).

Broadly, the banner of *character education* has been aligned with *traditional* methods of pedagogy, deploying conservative and behavioral approaches to transmit societal and ethical values to children, whereas *constructivist* and *moral education* has been aligned with *progressive* methods centering on children's acquisition of autonomous ethical agency through engagement with a sociomoral environment (i.e. peer interaction, facilitated discussion of ethical values, and constructive reasoning and reflection).

As will be clear from preceding discussion in this chapter, however, the contemporary ethics education landscape (including in early childhood contexts) is more nuanced and integrated than broad distinctions allow. Distinctions—between *progressive* and *traditional*, or *character* and *moral* education—serve some purpose in mapping the history of ethics education and in marking defining ends of a continuum of educational approaches, but, beyond this, they risk reinforcing "theoretical chasms" that are difficult for the researcher and practitioner to traverse (Berkowitz 1992, 45). Though still commonly used at professional conferences and in ethics education literature, these distinctions do not adequately reflect the extensive overlap that, especially in pedagogical practice, exists across these approaches (Berkowitz and Bier 2014). Leading figures in SEL, character, and moral education now call for collaboration and acknowledge the potential areas of synergy between their respective approaches. For example, Nucci (2009, 2), a leading moral education researcher, argues for an "integrative approach to moral development" that takes into account emotional, cognitive, and behavioral dimensions of human life and contends that early childhood ethics education should start with a focus, in part, on emotion recognition (a skill set that is traditionally emphasized in SEL curricula). Likewise, Elias and other leading SEL researchers (2014, 274) note the "long-overdue convergence" of character education, moral reasoning, and SEL in order to educate for all dimensions of ethical life. And, as noted above, the Child Development Project utilizes a range of pedagogical approaches to support moral development in early childhood, including those inspired by character education (e.g. school commitment to clearly defined ethical values and behaviors), SEL (e.g. development of interpersonal skills and positive relationships) and constructivist strategies (e.g. developmental discipline and maximizing students' sense of autonomy through collaborative rule-setting, discussion, and cooperative learning). Taken together, then, we can see that, in both research and practice, a spirit of methodological pluralism – utilizing multiple strategies to support behavioral, affective, and cognitive dimensions of moral development—animates an integrated approach to contemporary early childhood ethics education.

The move to integrated ethics education has been paralleled by increasing calls for *multi-dimensional* programs that extend beyond the traditional 'stand-alone' classroom lesson. Informed by a socio-ecological perspective of education, many researchers and practitioners have realized that, in order to be most effective, ethics education must take into account a wide range of developmental influences in order to reach the "whole child" (Marshall et al. 2011; Brunn 2014, 264; Elias et al. 2014). Children develop and participate in ethically-relevant educational experiences throughout their lives, across academic, social, and personal realms. For example, the influence of the family is perhaps the single greatest influence on a child's ethical development, and the broader community (e.g. peers, media, and cultural values) must also be considered to fully understand the reach and possibilities of ethics education efforts (Hoffman 2000; Berkowitz 1992). As a result, several programs now include comprehensive activities and curricula for teachers alongside partnerships with parents and communities (Denham and Weissberg 2004; McCabe and Altamura 2011). For example, programs such as *CHARACTERplus* and *Research Based, Developmentally Informed* (REDI) use school-home activities (e.g. discussion activities to reinforce ethical lessons and social-emotional competencies at home) and parent involvement in program evaluation and development (e.g. determination of a school's core ethical values and participation in program training sessions) in order to develop a more comprehensive approach to ethics education (Marshall et al. 2011; McCabe and Altamura 2011).

As an ethics education researcher and practitioner, I support the increasing calls for collaboration, both across character, moral, and constructivist education approaches and between the school, home, and broader community. No single approach has a comprehensive vision of moral development or an exclusive hold on best strategies for ethics education. Child development is multifaceted and, as such, is best served by integrated approaches to research and practice that can better take into account the diverse influences that inform children's ethical growth and, in turn, that are shaped and influenced by children. This integrated approach to early childhood ethics education has animated my own work in *Philosophical Ethics in Early Childhood* (PEECh), an ethics education program for preschool students informed by research and pedagogical strategies from SEL, philosophy for children, constructivism, and moral education (Burroughs and Tuncdemir 2017). On the basis of this project, and two decades of work in Pre-K—12 schools (both as a researcher and classroom teacher), I contend that another area of integration is essential for effective early childhood ethics education in schools: dialogue and collaboration between teachers and researchers. Although a full discussion of possibilities for teacher-researcher collaborations is beyond the scope of this chapter, it is important to acknowledge early childhood classroom teachers as specialists in child care, as education professionals who often possess unique insight into the social-emotional needs of their children, understand the practical challenges and/or benefits of proposed activities and curricula, and can provide guidance on best strategies for working with young children in a classroom environment. This acknowledgment is all the more important given that, in some cases, schools, teachers, and children can be reduced to mere "data or informants" in research practices with researchers failing to be responsive to the long term needs of the schools and communities in which they work (Clark et al. 1996, 196; Duncan and Conner 2013). Alternatively, when teacher-researcher collaboration and dialogue are placed at the center of education and research projects a mutual understanding can be gained that allows for information and resource sharing and, ultimately, an enriching experience for all involved (Clark et al. 1996). Professional development for both teachers and researchers, and moreover, the effectiveness of ethics education interventions, can be enhanced by a commitment to collaboration in the form of classroom and teaching observations, collective meetings, open dialogue on the aims of a project and the culture of the school in question, and co-construction of project elements, such as classroom activities, schedules for data collection, and so on.

Taken together, the increasing moves toward integrated ethics education in early childhood classrooms is a positive one that parallels the contemporary growth of interdisciplinary approaches to research across the humanities and sciences. By moving beyond our disciplinary silos and isolated views of child development and educational practice, we move closer to achieving a fuller, integrated conception of effective ethics education. A central element of this conception should also be contributed by administrators, teachers, and parents, as well as by children themselves, as we form collaborations to contribute to the ethical development of our children.

Acknowledgements Much of this chapter was composed during my time as a visiting scholar in the Department of Philosophy, Education, and Psychology, at the University of Verona, Italy. I thank my colleagues at the University of Verona for their support which aided in my completion of this project.

References

Althof, W., & Berkowitz, M. (2006). Moral education and character education: Their relationships and roles in citizenship education. *Journal of Moral Education, 35*(4), 495–518.

Arendt, H. (1954/1977). The crisis in education. In *Between past and future: Eight exercises in political thought* (pp. 170–193). Penguin Books: New York, NY.

Battistich, V. (2010). School contexts that promote students' positive development. In J. Meece & J. Eccles (Eds.), *Handbook of research on schools, schooling, and human development*. New York, NY: Routledge.

Battistich, V., Schaps, E., & Wilson, N. (2004). Effects of an elementary school intervention on students' 'connectedness' to school and social adjustment during middle school. *Journal of Primary Prevention, 24*(3), 243–262.

Bebeau, M. J., Rest, J. R., & Narvaez, D. (1999). Beyond the promise: A perspective on research in moral education. *Educational Researcher, 28*(4), 18–26.

Bennett, W. J., & Hague, M. (1995). *Children's book of virtues*. New York, NY: Simon and Schuster.

Berkowitz, M. (1992). The science of character education. In W. Damon (Ed.) *Bringing in a new era in character education* (pp. 43–63). Hoover Institution Press.

Berkowitz, M., & Bier, M. (2014). Research-based fundamentals of the effective promotion of character development in schools. In L. Nucci, D. Narvaez, & T. Krettenauer (Eds.), *Handbook of moral and character education* (2nd ed., pp. 248–260). New York, NY: Routledge.

Brunn, P. (2014). Pedagogy for the whole child: The Developmental Studies Center's approach to academic, moral, and character education. In L. Nucci, D. Narvaez, & T. Krettenauer (Eds.), *Handbook of moral and character education* (2nd ed., pp. 263–271). New York, NY: Routledge.

Burroughs, M. (2016). Developing leaders: Children and ethical leadership. In R. Alavi (Ed.), *Ethics and leadership* (pp. 65–76). Dubuque, IA: Kendall Hunt.
Burroughs, M., & Barkauskas, N. (2017). Educating the whole child: Social-emotional learning and ethics education. *Ethics and Education, 12*(2), 218–232.
Burroughs, M. D., & Tuncdemir, T. B. A. (2017). Philosophical ethics in early childhood: A pilot study. *Journal of Philosophy in Schools, 4*(1), 74–101.
Carr, D. (2002). Feelings in moral conflict and the hazards of emotional intelligence. *Ethical Theory and Moral Practice, 5*(1), 3–21.
Character Education Partnership. (2010). *11 principles of effective character education: A framework for success*. Washington, DC: Character Education Partnership.
Clark, C., Moss, P. A., Goering, S., Herter, R. J., Lamar, B., Leonard, D., et al. (1996). Collaboration as dialogue: Teachers and researchers engaged in conversation and professional development. *American Educational Research Journal, 33*(1), 193–231.
Coles, R. (1986). *The moral life of children*. New York, NY: Atlantic Press.
Damon, W. (1988). *The moral child: Nurturing children's natural moral growth*. New York, NY: The Free Press.
Denham, S. A., & Weissberg, R. P. (2004). Social-emotional learning in early childhood: What we know and where to go from here. In E. Chesebrough (Ed.), *A blueprint for the promotion of prosocial behavior in early childhood* (pp. 13–50). New York, NY: Kluwer Academic/Plenum Publishers.
DeVries, R., & Zan, B. (1994). *Moral classrooms, moral children: Creating a constructivist atmosphere in early education*. New York, NY: Teachers College Press.
Dewey, J. (1909/1975). *Moral principles in education*. Arcturus Books.
Dewey, J. (1916/2011). *Democracy and education*. Simon and Brown.
Dewey, J. (1938). *Experience and education*. New York, NY: Simon and Schuster.
Duncan, J., & Conner, L. (Eds.). (2013). *Research partnerships in early childhood education: Teachers and researchers in collaboration*. Springer.
Dunn, J. (2014). Moral development in early childhood and social interaction in the family. In M. Killen & J. G. Smetana (Eds.), *Handbook of moral development* (2nd ed., pp. 135–159). New York, NY: Psychology Press.
Durlak, J. A., Weissberg, R. P., Dymnicki, A. B., Taylor, R. D., & Schellinger, K. B. (2011). The impact of enhancing students' social and emotional learning: A meta-analysis of school-based universal interventions. *Child Development, 82*(1), 405–432.
Elias, M. J., Kranzler, A., Parker, S. J., Kash, V. M., & Weissberg, R. P. (2014). The complementary perspectives of social and emotional learning, moral education, and character education. In L. Nucci, D. Narvaez, & T. Krettenauer (Eds.), *Handbook of moral and character education* (2nd ed., pp. 272–289). New York, NY: Routledge.
Gussin Paley, V. (1997). *The girl with the brown crayon: How children use stories to shape their lives*. Cambridge, MA: Harvard University Press.
Hildebrandt, C., & Zan, B. (2014). Constructivist approaches to moral education in early childhood. In L. Nucci, D. Narvaez, & T. Krettenauer (Eds.), *Handbook of moral and character education* (2nd ed., pp. 180–197). New York, NY: Routledge.
Hoffman, M. (2000). *Empathy and moral development: Implications for caring and justice*. Cambridge, UK: Cambridge University Press.
Howard, R. W., Berkowitz, M. W., & Schaeffer, E. F. (2004). Politics of character education. *Educational policy, 18*(1), 188–215.
Kilpatrick, W. (1992). *Why Johnny can't tell right from wrong: Moral illiteracy and the case for character education*. New York, NY: Simon and Schuster.
Kristjánsson, K. (2006). 'Emotional intelligence' in the classroom? An Aristotelian critique. *Educational Theory, 56*(1), 39–56.
Kuczynski, L., & Knafo, A. (2014). Innovation and continuity in socialization, internalization, and acculturation. In M. Killen & J. G. Smetana (Eds.), *Handbook of moral development* (2nd ed., pp. 93–112). New York, NY: Psychology Press.

Lee, C. M., & Taylor, M. J. (2013). Moral education trends over 40 years: A content analysis of the Journal of Moral Education (1971–2011). *Journal of Moral Education, 42*(4), 399–429.

Lickona, T. (1989). *Educating for character: How our schools can teacher respect and responsibility*. New York, NY: Bantam.

Lone, J. M. (2012). *The philosophical child*. Lanham, MD: Rowman and Littlefield Publishers.

Marshall, J. C., Caldwell, S. D., & Foster, J. (2011). Moral education the CHARACTER plus Way®. *Journal of Moral Education, 40*(1), 51–72.

McCabe, P. C., & Altamura, M. (2011). Empirically valid strategies to improve social and emotional competence of preschool children. *Psychology in the Schools, 48*(5), 513–540.

Mohr Lone, J. M., & Burroughs, M. (2016). *Philosophy in education: Questioning and dialogue in schools*. Lanham, MD: Rowman and Littlefield Publishers.

Noddings, N. (2002). *Educating moral people: A caring alternative to character*. New York, NY: Teachers College Press.

Nucci, L. P. (2001). *Education in the moral domain*. Cambridge, UK: Cambridge University Press.

Nucci, L. P. (2009). *Nice is not enough: Facilitating moral development*. Upper Saddle River, New Jersey: Merrill/Prentice Hall.

Piaget, J. (1997). *The moral judgement of the child*. London, UK: Free Press.

Power, C., Higgins, A., & Kohlberg, L. (1989). *Lawrence Kohlberg's approach to moral education*. New York, NY: Columbia University Press.

Ryan, K. (1989). "In defense of character education." In L. Nucci (Ed.), *Moral development and character education: A dialogue* (pp. 3–17). Taylor and Francis.

Ryan, K., & Bohlin, K. (1999). *Building character in schools: Practical ways to bring moral instruction to life*. San Francisco, CA: Jossey-Bass.

Sanger, M., & Osguthorpe, R. (2009). Analysing the child development project using the moral work of teaching framework. *Journal of Moral Education, 38*(1), 17–34.

Solomon, D., Battistich, V., Watson, M., Schaps, E., & Lewis, C. (2000). A six-district study of educational change: Direct and mediated effects of the child development project. *Social Psychology of Education, 4*(1), 3–51.

Wartenberg, T. E. (2014). *Big ideas for little kids: Teaching philosophy through children's literature*. Lanham, MD: Rowman & Littlefield.

Watson, M. (2014). Developmental discipline and moral education. In L. Nucci, D. Narvaez, & T. Krettenauer (Eds.), *Handbook of moral and character education* (2nd ed., pp. 159–179). New York, NY: Routledge.

Wynn, E. (1989). Transmitting traditional values in schools. In L. Nucci (Ed.), *Moral development and character education: A dialogue* (pp. 19–36). Taylor and Francis.

Zins, J. E., Bloodworth, M. R., Weissberg, R. P., & Walberg, H. J. (2004). The scientific base linking social and emotional learning to school success. In J. Zins, R. Weissberg, M. Wang, & H. Walberg (Eds.), *Building academic success on social and emotional learning: What does the research say* (pp. 3–22). New York, NY: Teachers College Press.

Promoting Reasonableness: Science Teachers as Moral Educators

Michael S. Pritchard

Abstract Integrating ethics into science classes requires a kind of critical thinking about values in science that can play a significant role in fostering the reasonableness of students. This chapter will offer several reasons for concluding that this is an appropriate objective of science education in the schools, even at elementary levels. Thus, an effort will be made to make a case for science teachers seeing themselves as moral educators while teaching science. This chapter will feature the work of [the late] environmental chemist Theodore Goldfarb in his efforts to help pre-college science teachers bring ethics into their classes.

Keywords Reasonableness · Philosophical inquiry · Pre-college science teachers Critical thinking

Preface

In his early efforts in the 1990s to include discussions of ethical issues in his college chemistry classes, environmental chemist Theodore Goldfarb (SUNY Stony Brook) discovered that few of his students had any pre-college experience discussing ethical issues in any of their science studies. At the same time, federal funding agencies such as the National Science Foundation (NSF) and the National Institutes of Health (NIH) were beginning to mandate that those engaged in research projects funded by these agencies provide evidence that they were well-versed in research ethics. But when should education in research ethics begin, and how should it begin? Goldfarb's disappointment with the lack of readiness of his students convinced him that serious discussion of issues in research ethics can, and should, begin prior to entry into college. So, he sought funding from the NSF to develop and implement workshops on ethics in science for middle and high school science teachers.

M. S. Pritchard (✉)
Western Michigan University, Kalamazoo, MI, USA
e-mail: michael.pritchard@wmich.edu

E. E. Englehardt and M. S. Pritchard (eds.), *Ethics Across the Curriculum—Pedagogical Perspectives*, https://doi.org/10.1007/978-3-319-78939-2_16

It was my good fortune that Rachelle Hollander, then director of the NSF Ethics and Values in Science program, was familiar with my work in philosophy for children[1] (Pritchard 1985, 1996). It was her suggestion that Goldfarb contact me to see if I might assist him in his project. After a lengthy phone conversation, we agreed to become a team. For the next several years we worked together on a series of NSF supported ethics and science projects.

Our work together ended prematurely because of Goldfarb's death in 2002, just as our project was reaching its stride. He was a scientist, and I was not. I depended on him to lead me through the scientific tangles that called for ethical reflection. The combination of his scientific and ethical sensitivity was essential to our efforts. In our last year working together we assembled a manual for teachers that contains a rationale for encouraging pre-college science teachers to include ethics in their science classes. This manual also features a substantial number of case studies and lesson plans for teachers to use (Goldfarb and Pritchard 1999).[2]

Introduction

In the early 1980s, following the lead of Matthew Lipman and his colleagues at the Institute for the Advancement of Philosophy for Children (IAPC), I supported the notion that K-12 classrooms that encourage philosophical discussions can form a community of inquiry that fosters reasonableness. In writing about this, I drew on my own experiences with children and teachers in exploring philosophical ideas. A major aim in these writings was to support the notion that a place should be made for philosophical inquiry at pre-college levels. Of course, finding room in an already crowded curriculum for yet another subject in the schools poses serious problems. However, some of these problems can be alleviated by turning to a specific area of the curriculum that already exists, the sciences. My contention is that appropriately integrating ethics into science classes requires a kind of critical thinking about values that can play a significant role in fostering the reasonableness of students. I will offer several reasons for concluding that this is an appropriate objective of science education in the schools, even at the middle and elementary

[1]This work was highlighted by my *Philosophical Adventures With Children* (University Press of America, 1985) and *Reasonable Children* (Lawrence, KS: University Press of Kansas, 1996).

[2]This manual, *Ethics in the Science Classroom: An Instructional Guide for Secondary School Science Teachers*, is fully accessible at Online Ethics in Science and Engineering under the heading, "Pre-College Materials." It was supported by "Workshops For High School Science Teachers: Ethics in the Classroom, NSF Grant No. SBR-932 0255). Included is my, "Reasonable Children: Science Teachers as Moral Educators," which forms the early basis of much of what follows in this chapter. See: www.onlineethics.org/Resources/precollege/childrenreason.aspx. I further develop the notion that science teachers can, and should, see themselves as moral educators in "Ethics in the Science Classroom: Science Teachers as Moral Educators," in *Moral Sensibilities and Education III: The Adolescent*, Wouter van Haaften, Thomas Wren, and Agnes Tellings, eds. (London: Concorde Publishing House, 2005), pp. 113–132.

levels. Thus, I will be trying to make a case for encouraging science teachers to see themselves, in part, as moral educators.

Science: Body of Knowledge or Human Activity?

If we think of teaching science as basically conveying bodies of firmly established, highly organized factual knowledge to students, it may be difficult to see how science classes can contribute to the moral education of students. Science, it may be thought, is confined to *facts* and theories constructed from them; morality has to do with *values*. Facts are discovered through empirical observation and logical inference, making them *objective*; values are matters of opinion, this thinking goes, and are inherently *subjective*. Whether or not teachers intend to convey this message, many students very early on pick up the idea that there is a fundamental *fact/value* distinction, with science cast as factual but value neutral.[3]

This sort of view of the sciences can, and should, be challenged. In brief, science can be viewed, not merely as a body of organized factual knowledge (for students to commit to memory), but as an *human activity*. Science does not create itself; it requires the careful work of *scientists*. In fact, as Catherine Elgin argues, it requires collaborative work among scientists. This, she says, gives science an "irreducibly ethical dimension" (Elgin 2011, 251).

Science, Elgin notes, has the overarching epistemological goal of developing "a comprehensive, systematic, empirically grounded understanding of nature" (Elgin 2011, 252). But she identifies two obstacles that stand in the way of achieving this aim. First, nature is so complicated that acquiring a comprehensive, systematic, empirically grounded understanding of nature necessarily requires the cooperation and mutual trust of many scientists—researchers and theorists who may never actually meet one another, let alone work together in the same laboratories and research centers. Second, even at our best, we are fallible; and so is science.

When we look closely at how scientific understanding is acquired, shared, and used, it becomes clear that this activity is anything but value neutral. In sum, Elgin's argument for including ethics in science education is that ethics is an inherent feature of science.

Science requires collaboration. That collaboration requires trust. Since trust is unreasonable in the absence of trustworthiness, scientists need to be, and to consider one another, trustworthy. Since trustworthiness is a moral characteristic, morality figures in science. That being so, science education should inculcate trustworthiness and an appreciation of the ways it contributes to and figures in science (Elgin 2011, 252).

---

[3]Readers may notice a frequent shifting back and forth between 'ethics' and 'morality' without a substantive distinction being made between them. This is intentional.

Given this, it can be seen that the advancement of science depends on the integrity and cooperation of scientists as they share their results with each other and build on one another's work. Departures from this not only can cause serious problems within the scientific community, they can also cause serious public harms–harms to health and safety, economic harms (e.g., if studies have to be repeated, or if large sums of money have been unwisely invested in certain lines of research), and public erosion of trust and confidence in the work of scientists.

Although this is not a topic in Elgin's paper, various possible uses of science can raise ethical questions (e.g., whether cloning humans is acceptable, whether nuclear waste is being properly handled, whether certain environmental risks are reasonable, whether we should be taking global warming seriously now). Further, the very selection of areas in which research is done (supported by either public or private funds) reflects value choices. Scientific and technological developments can raise ethical issues (e.g., regarding their effects on the quality of life, both for present and future generations, or regarding questions about the just distribution of scarce resources). Finally, concepts found in science such as health, disease, pollution, waste, and sustainability are value-laden.

In short, when we look at science as a human activity, especially in regard to the impact it has on society, thoughtful persons are bound to raise ethical questions. This has been most obviously the case in biomedicine in recent years. But, a great deal of attention has also been focused on various uses of chemistry (e.g., industrial waste, workplace safety), biochemistry (e.g., pesticides, biochemical weapons), and physics (e.g., nuclear energy). Television has been quick to pick up on all of this, both in its news presentations and in fictionalized drama. Newpapers, magazines, movies, and the internet have done likewise. This is part of the out-of-classroom world of students. In that world science and values are inseparably intertwined. What would happen if this intertwining were to show up in the pre-college classroom, too?

Reports from the 75 teachers who participated in our Long Island "Workshops for High School Science Teachers: Ethics in the Classroom" indicate that the results can be striking. Students seemed genuinely interested in the ethical issues and were eager to discuss them. At the same time, they also realized that in order to address these questions thoughtfully, they needed to learn something about the relevant sciences. So, a pleasant bonus for those teachers who want to bring ethics into their science classes is that it can actually increase the motivation students have to study science.

Actually, this should not be surprising, for it is the expected consequence of humanizing science, a subject that for many students seems distant, dry, and designed more for memorization than active engagement. However, the point of bringing ethics into science classes is not simply to liven them up. It is that the ethical issues associated with science are a part of science itself, and they are left out of science studies at the expense of a fuller understanding of science as an activity, not just a body of knowledge.

What I have stressed so far is that there are good reasons for including some ethics in science classes. The basic argument is that this is necessary if we are to do full justice to the role that science plays both in the lives of scientists and in society in general. Insofar as ethics and science are joined in the classroom, this promises to enhance the moral education of students to at least some extent. However, at the outset of this paper I identified a more specific benefit that could come from this joining–viz., promoting the reasonableness of students. I now to turn specifically to that topic.

What Is Reasonableness?

Reasonableness has two kinds of features that are especially applicable to science: one is that science has inherently *social* dimensions, and the other is that it embraces *uncertainty* (Pritchard 1996). Both are important in making clear that the kind of rationality supported by scientific inquiry is quite distinct from those kinds of rationality that are highly individualistic and self-centered. An individual might attempt to impose his or her way of thinking on others while ignoring what they might wish to say in response. In some circumstances this might be regarded as irrational behavior, but in need not be. It may be egocentric, or self-centered; it may often be selfish or domineering. Still, as a strategy for getting what one wants, it may often work. So, in this respect, at least in the short run, it may qualify as a kind of rational behavior–or, in any case, not irrational behavior. However, it is *unreasonable* behavior. Its demands are excessive, unfair, and one-sided in ways that others find unacceptable.

In contrast, reasonableness is a positive, *social* disposition. As Lawrence Splitter and Ann Margaret Sharp put it: "[T]he reasonable person respects others and is prepared to take into account their views and their feelings, to the extent of changing her own mind about issues of significance, and consciously allowing her own perspective to be changed by others. She is, in other words, willing to be reasoned with" (Splitter and Sharp 1996, 6). This does not mean that a reasonable person simply gives into the views of others. It does mean that one accepts rational constraints that require that one be prepared to give reasons for one's beliefs that can be subjected to public scrutiny, not simply private confirmation.

A second feature of reasonableness is its acceptance of some degree of *uncertainty*. This may be uncertainty about whether one's own views are necessarily right, or even uncertainty that anyone's views are. This is not to be confused with skepticism. As Max Black says, although we may be able to show that some actions or judgments are more reasonable than others, we are very seldom able to choose a single action as uniquely reasonable (Black 1972, 202). What this means is that it is quite possible for reasonable people to come to different judgments on some matters, and further discussion might not bring them closer to agreement. But it might.

Scientific Inquiry and Reasonableness

For scientific inquiry to exhibit reasonableness, it must remain open to new evidence, and even new ways of proceeding. It must also satisfy standards of good argument, look for supporting reasons, avoid unwarranted assumptions, reject drawing conclusions too hastily, and avoid the suppression of relevant data. Ideally, scientific claims are subjected to public examination through a peer review process that carefully examines the arguments put forward. Finally, as noted by Elgin, scientific advancement depends on scientists working together, sharing results, building on one another's work, and observing standards of honesty with each other in the process.

Although good science depends on the sort of reasonableness just described, it does not follow that scientific practice always meets these standards. In fact, concerns about scientific misconduct have resulted in funding agencies such as the National Science Foundation and the National Institutes of Health requiring colleges and universities that seek research funding from them to establish and enforce policies on scientific misconduct for both faculty and students who will be engaging in this research. Of course, this is but a small minority of students in colleges and universities. But even for these students, discussions of research ethics may be their first serious academic encounter with ethics.

I say *academic* encounter because, obviously, it is not their first encounter with ethical questions. These questions begin early in life–in the home, in religious institutions, on the playground, in the streets, and in viewing movies, television, and the like. But ethical questions also arise from the moment children enter school, with its rules, requirements, and evaluative standards.

Natural Curiosity: Scientific and Philosophical

Science studies can take advantage of the natural curiosity of children. My work with the Institute for the Advancement of Philosophy for Children (IAPC) programs has convinced me that the philosophical wonder of children and their general curiosity about the world around them go hand in hand. It may be noteworthy that prior to the emergence of highly differentiated scientific disciplines in the 20th century (including the social sciences), many texts made a distinction between *natural* and *speculative* philosophy. Natural philosophy focused on the most general features (e.g., explanatory laws) of what today we call the natural sciences (e.g., biology, chemistry, physics, and the earth sciences). Speculative philosophy focused on the sorts of general issues that make up what today we call the discipline of philosophy.

Both natural and speculative philosophy seem to have counterparts in early childhood–viz., the curiosity and wonder that, sadly, often diminishes rather than flourishes as children progress through school. The big 'why' questions young children ask may have scientific answers, philosophical answers, or no answers. We may try to parse them into different kinds. But they seem similarly motivated in the child, regardless of how we (and later, they) might sort them out. It is a curiosity and wonder that is philosophical in spirit. Furthermore, although we adults may feel frustrated because of our inability to answer the large questions children sometimes ask, it seems that children often delight in their open-endedness.

At the same time, young children may be ready to begin structuring their pursuit of answers in ways that good scientists (and philosophers) do. It is the combination of their natural curiosity and this readiness that makes sense of bringing science into the elementary school curriculum. Unfortunately, by the time they reach high school, if not much earlier, many students find themselves agreeing with the sentiments of David Benjamin and Jeremy Scott, then 10th graders. "In high school there is a common system of 'learning' that goes something like this: listen, take notes, memorize, and regurgitate facts. Each high school subject seems to show the world through a distinct window unconnected to the windows presented by other classes" (Benjamin and Scott 1989, 29).

These thoughts were inspired by reading philosopher Thomas Nagel's *What Does It All Mean?*[4] What especially impressed them, they say, is that Nagel's book "made us see that as you obtain more knowledge, you find that there is more knowledge to be obtained. Answering questions brings about more unanswered questions, and thus a point of complete and final knowledge cannot be reached" (Benjamin and Scott 1989, 29).

David Benjamin and Jeremy Scott's concern was that the schools may encourage a serious misunderstanding of how we should depict the quest for knowledge. The schools, they say, invoke the image of a wise guru sitting on a mountain top in possession of *all knowledge and a complete understanding of the world.* This image may encourage the thought that, "from the guru's mountain-top, with complete knowledge, the world can be simplified and viewed clearly and accurately". Sadly, they conclude, "We have found that high school reinforces this fantasy (Benjamin and Scott 1989, 29). They make a plea for introducing philosophy at least by the time students are in high school:" Philosophy would help high school students to link and understand their knowledge. The guru may understand his knowledge, and he may in fact be a wise man, but in believing that he knows all, he lacks the open-mindedness and critical questioning we discovered through philosophy" (Benjamin and Scott 1989, 29).

---

[4]This book was recommended to them by Martin Benjamin, David's father, then a professor of philosophy at Michigan State University. He shared with them Matthew Lipman's children's novel *Harry Stottlemeier's Discovery*. Delighted with the philosophical explorations of Harry and his friends as 5th graders, David and Jeremy waited several years for philosophical thinking to be encouraged in their regular school classes. Frustrated, as 10th graders they approached David's father for a reading list. That list included Nagel's book.

Science Teachers as Moral Educators

Just as philosophy in the schools would encourage open-mindedness and critical questioning, so would the inclusion of ethics in science education. This inclusion is not the only way to portray science as receptive to open-mindedness and critical questioning. But it is an effective way, and it places science squarely in the context in which it actually operates in society. In addition, the very methods of inquiry and standards of public reasoning that good science requires can make a valuable contribution to the moral education of students, beginning whenever the study of science begins.

It seems clear that a reasonable approach to an ethical question requires carefully attending to, and seeking out, the relevant facts. Deliberately screening out information because it may make it more difficult to support one's favored position is contrary to reasonable ethical reflection; and it is contrary to good scientific reasoning. The scientific caution against generalizing from an unrepresentative sampling can help explain the shortcomings of stereotyping (common in racist and sexist thinking, e.g.). The scientific importance of looking at things from as many relevant perspectives as possible can help students understand and resist egocentric thinking, one of the most formidable barriers to reasonableness in social relationships. And the power of analogical reasoning in science can enhance ethical reasoning as well.

This last point is worth focusing on for a moment. Some years ago I was invited to visit a number of elementary and middle school classes near my university to discuss some philosophical ideas with the students. Typically a series of three 30 min sessions was arranged for each class. For my first meeting with one 4th grade class, I decided to invite the students to think about what we are likely to assume, or take for granted, often without even realizing it. I gave the students a few puzzles that can be solved only if we become *critically* aware of our assumptions. For example, 6 toothpicks can be put end to end to form 4 equilateral triangles–but only if we realize that the configurations do not need to be on a flat, two-dimensional surface. (Answer: form a pyramid with three toothpicks forming the base.) Although none of the puzzles had any social content, one 4th grader approached me after class with a puzzle of her own. She told me a fictional story she had heard about a father and his son being involved in an accident and taken to separate rooms in a hospital. The doctor entered the room of the son and said, "I cannot perform surgery on this child; he is my son." The question is: How is this possible? Having seen this story presented on the Archie Bunker "All in the Family" TV program, I knew the answer: The doctor was the son's mother. Writers of the program were counting on most viewers initially having trouble with the question because they would be assuming that the doctor is male.

It was immediately apparent that this student had understood very well the philosophical point I had intended my puzzles to illustrate. Her follow-up is striking in two ways. First, she applied this point to a social context–something none of my examples had done. This shows the reach of the sort of thinking I was encouraging the students to engage in, as well as the ability of this 9-year-old to demonstrate this with an example of obvious social and moral importance—even without this being on the agenda. Second, she did not bring up this example until *after* the class was finished, suggesting that possibly she did not think that the classroom was the place to bring it up. But the example occurred to her anyway, illustrating that perceptive students do not necessarily restrict their thinking to the specific topic at hand. So, we might ask, what if such thinking were *openly encouraged* in the classroom–especially the science classroom?

Admittedly, this example is only anecdotal. However, there is ample evidence that even elementary age students are capable of quite sophisticated reasoning (much more so than Piaget's stage theory of cognitive development would suggest).[5] It should not be assumed that cognitive readiness implies moral readiness. Nevertheless, there is considerable evidence that elementary school age children, and certainly middle school level children, are quite sophisticated within their range of experience.[6]

Of course, school science programs need to take carefully into account the extent to which students may be ready to have some ethics included in their science studies. Arguably, insofar as ethics is an integral part of scientific activity itself, it would seem that some attention could be paid to this when science makes it first appearance in K-12 education. However, certainly by high school age students should be able to appreciate the inclusion of some ethics in their science classes. In fact, in the mid-1990s the National Academy of Sciences/National Research Council's *National Science Education Standards* (Washington, DC: National Academy Press, 1996) made a special point of urging that high school science classes include some consideration of the importance of ethics in the sciences. At present, there is still relatively little material explicitly designed to assist K-middle school teachers who wish to integrate ethics into their science teaching. Although there is a growing literature on moral education at these levels, there is a need for much more that explicitly addresses the science curriculum. At the high school level there is a growing body of material available to assist teachers in bringing ethics into their science classes.[7] Some of this, for example, concerns the ways in which scientists are expected to conduct research–sorting out evidence, testing hypotheses, making reliable inferences, accurately reporting data, working cooperatively with others, and the like. The report also includes suggestions for units on personal and social perspectives on science and on the nature of science and science inquiry. Just how any of these objectives might best be met in various levels of the school curriculum needs to be worked out with considerable care; but statements of science education goals and objectives at both the state and national level urge that this task should be undertaken.

---

[5]On Piaget's underestimation of children's cognitive abilities, see, e.g., Donaldson (1979), Astington (1993), Matthews (1980, 1995).

[6]I attempt to support this in my *Reasonable Children*. Damon (1988), provides a useful summary of relevant research on this topic.

[7]A particularly useful resource for high school science teachers is *Ethics Primer*, published online by the Northwest Association for Biomedical Research (NWABR), available at www.NWABR.org.

Why Promote Reasonableness?

This chapter has emphasized ways in which including ethics in science education might promote reasonableness. It is important to note two possible aims of promoting the reasonableness of children:

- To assist children in becoming more reasonable now, while they are still children
- To assist children in growing up to be reasonable adults in a democratic society.

Both should be supported. We should be as interested in respect for children's capacity to be reasonable as children as in their capacity to become reasonable adults. Respect for children requires both.

It is important to credit elementary age children with the capability of understanding, and engaging in, some scientific practice–and in grasping moral dimensions of science (ranging from some appreciation of the importance of honesty in scientific work to understanding why recycling rather than simply discarding waste materials is important). Furthermore, when combined with thoughtful attention to the moral dimensions of science, the kind of reflective inquiry science requires will have a positive spillover effect on their other studies as well as their daily lives. That is, science education will contribute to students' moral education well beyond science.

Michael Martin has argued that good science education and moral education are mutually supportive[8] (Martin 1986, 99–108). On the one hand, good science education contributes to moral reflection and decision-making by promoting inquiry and discovery skills that enable us to acquire relevant factual information, to test hypotheses, and to weigh the likely consequences of alternative choices. Good science education promotes clarity, thoroughness, perseverance, respect for sound reasoning, impartiality, and open-mindedness–all valuable assets for morality as well.

On the other hand, moral education promotes values essential to good science education. Honesty, fairness, and cooperativeness are all virtues necessary for good scientific practice. Fellow scientists and others depend on the honest reporting of data. It is a matter of fairness (and honesty) to give proper credit to the work of others. Scientists typically work in teams, or at the very least depend on the work of others in furthering their own. Furthermore, questions need to be asked about appropriate or inappropriate scientific research and its resulting technological use. These questions concern biomedical research and treatment, experimentation on humans and animals, military research and the development of weaponry, the use of various forms of energy, environmental quality, and the entire range of scientific activity that can significantly affect public health, safety, and welfare—including the sustainability of various ways of living for future generations.

---

[8]This paragraph and the next are taken from my *Reasonable Children*, pp. 77–78.

Perhaps not all of this is appropriate for elementary school level science; but, suitably adjusted for their level of experience, much of it is; and, as Martin points out, the stakes may be high (Martin 1986, 107). If David Benjamin and Jeremy Scott are close to the mark, there is far too little critical thinking being encouraged in the schools; and if this is right, there is much more that can, and should, be done to foster the reasonableness of students. Including ethics as a part of science studies can help achieve this end.

References

Astington, J. W. (1993). *The child's discovery of the mind.* Cambridge, MA.: Harvard University Press.

Benjamin, D., & Scott, J. (1989). Review of Nagel's What does it all mean? *Thinking, 7*, 4, 29.

Black, M. (1972). Reasonableness. In R. F. Dearden, P. H. Hirst, & R. S. Peters (Eds.) *Education and the development of reason* (p. 202). Routledge and Kegan Paul: London.

Damon, W. (1988). *The moral child.* New York: Free Press.

Donaldson, M. (1979). *Children's minds.* New York: Norton.

Elgin, C. (2011). Science, ethics and education. *Theory and Research in Education, 9*(3), 251–263.

Goldfarb, T., & Pritchard, M. S. (1999). *Ethics in the science classroom: An instructional guide for secondary school science teachers.* Online Ethics in Science and Engineering. www.onlineethics.org/Resources/precollege/childrenreason.aspx.

Martin, M. (1986). Science education and moral education. *Journal of Moral Education, 15*(2), 99–108.

Matthews, G. (1980). *Philosophy and the young child.* Cambridge, Mass: Harvard University Press.

Matthews, G. (1995). *The philosophy of childhood.* Cambridge, Mass: Harvard University Press.

Northwest Association for Biomedical Research. *Ethics primer.* Online: NWABR.org.

Pritchard, M. S. (1985). *Philosophical adventures with children.* Lanham, Md: University Press of America.

Pritchard, M. S. (1996). *Reasonable children.* Lawrence, Kansas: University Press of Kansas.

Pritchard, M. S. (2005). Ethics in the science classroom: Science teachers as moral educators. In W. van Haaften, T. Wren, & A. Tellings (Eds.), *Moral sensibilities and education III: The adolescent* (pp. 113–132). London: Concorde Publishing House.

Splitter, L., & Sharp, A. M. (1996). *Teaching better thinking: The classroom Community of Inquiry.* Melbourne: Australian Center for Educational Research.

Sustainability Ethics Across the Curriculum

Randall Curren

Abstract This chapter identifies confusion about the normative dimensions of sustainability as an important obstacle to teaching the ethics of sustainability across the curriculum. It aims to overcome this obstacle by presenting a framework of sustainability ethics consisting of principles derived from the most basic commitments of common morality. Four rationales and related models for teaching ethics across the curriculum are identified, and an argument for infusing education at all levels with education in sustainability and sustainability ethics is framed on this basis. The prescribed model involves cross-curricular integration and collaborative public service projects, where possible.

Keywords Aggregate unsustainability · Common morality · Sustainable development · Virtuous character · Ethical domain · Complexity

Introduction

It is no exaggeration to say that the aggregate unsustainability of the ways we live now is diminishing the life prospects of the growing population we are collectively engendering, not to mention the prospects for countless other forms of life on this planet. This makes sustainability a matter of pervasive and fundamental importance, and it is one we would be well advised to teach systematically and at all levels of instruction. Systematic education in sustainability is a daunting undertaking, however, and it is one in which few countries have made much progress, despite UNESCO's declaration of a Decade of Education for Sustainable Development (2005–2014), the announcement of implementation frameworks, and a burgeoning scholarly literature on education and sustainability (UNESCO 2005; Rowe et al. 2015; Curren and Metzger 2017: 153–179). One of the many factors that have contributed to this lack of progress is the difficulties educators across the disciplines

R. Curren (✉)
University of Rochester, Rochester, USA
e-mail: randall.curren@rochester.edu

E. E. Englehardt and M. S. Pritchard (eds.), *Ethics Across the Curriculum—Pedagogical Perspectives*, https://doi.org/10.1007/978-3-319-78939-2_17

face in meaningfully engaging the ethical dimensions of sustainability. The most basic stumbling block is that sustainability is spoken of in many ways and its normative content has not been defined in a way that supports the articulation of an ethic of sustainability.[1] Principles of sustainability have been formulated with business, industrial, urban planning, and other applications in mind (IISD 2013; Condon 2010), but these are typically rules of practice for reducing the carbon footprint of facilities, products, and travel; reducing storm runoff; preventing the collapse of fisheries; and so on. There are normative ideas lurking in the background of such principles but not an articulated system of ethical principles.

The primary task of this chapter is to frame a general ethic of sustainability that can be taught across the curriculum and discussed in conjunction with cases, rules of practice, codes of corporate and professional conduct, and readings in ethical analysis. It begins in the idea that the moral heart of the concept of sustainability is living in ways that do not diminish opportunities to live well in the future. It frames several principles of sustainability ethics and offers guidance on applying them. The principles are presented as deriving from the most basic commitments of common morality, not as emanating from a conception of the value of nature. It is important to be clear at the outset that this reflects a strategic decision to elaborate the ethical dimensions of the idea of sustainability as such, rather than to elaborate comprehensive ethical guidance for acts that may impinge upon nature, the integrity of ecosystems, and other life forms.

It is also important to note that although sustainability is inherently concerned with intergenerational justice, inasmuch as it is concerned with the preservations of opportunity to live well into the future, it does not obviously concern other forms of justice. The idea of *sustainable development* does embrace another form of justice and it should not be equated with sustainability. The two are unfortunately often conflated, despite the fact that the term "sustainable development" arose as an uneasy geopolitical compromise between global cooperation to promote environmental sustainability and global cooperation to raise living standards through further economic growth. The health of ecosystems essential to economic activity has declined as the global economy has grown, and this suggests that raising the standards of living of those most in need will require greater equity and less growth. Progress toward sustainability will require taking this seriously, but the cause of progress is not well served by the assumption that the requirements of "social justice" are intuitively obvious. This assumption is often bundled into the discourse of sustainable development, and an effect of this assumption and the conflation of sustainable development and sustainability has been a good deal of controversy and uncertainty about whether and how sustainability should be taught.

---

[1]For a sociological survey of the language of sustainability, see Portney (2015). For some exploratory studies in the ethics of sustainability, see Moore and Nelson (2010), Raffaelle et al. (2010).

Rationales for Teaching Ethics Across the Curriculum

Instruction in ethics across the curriculum might be predicated on at least four different rationales. The most ancient and enduring of these rationales is that the overarching purpose of education is to promote wisdom or good judgment, not only within specific domains of inquiry and practice but to equip students for life (Curren 2014b). For self-determining individuals, heads of households, and citizens who share responsibility for the fate of a society, civilization, and planet, the difference between lacking good judgment and possessing good judgment is monumental. Good practical judgment is sensitive to the value of what is at stake in decisions, and it is in this sense inescapably ethical. It requires ethical perception and understanding no less than perception and understanding of other relevant aspects of situations and options that may be addressed by diverse sciences and other disciplines. The complexity of the decisions we face is largely a product of the complexity of the social and built environments that we have created for ourselves through the application of diverse forms of expertise. In these circumstances, an education for good practical judgement requires an integration of ethics with an open-ended array of disciplinary perspectives and awareness of diverse forms of expertise sufficient to at least discern who is a trustworthy expert on a subject and who is not.

This implies a form of ethics across the curriculum entailing curricular integration focused on matters of practical importance to individuals, society, and civilization itself. Within this form there could be many variants, including ones that involve students in group projects focused on the identification and solution of real problems. John Dewey and other Progressive Era educators in the United States promoted a form of "community civics" along these lines in the early twentieth century, and the Association of American Colleges and Universities (AAC&U) has recently advanced a globalized version of it for college students (Reuben 1997; U.S. Bureau of Education 1915; Hovland 2006; Curren and Dorn 2018). The integration of ethics with other curricular content might occur entirely in the context of the project work in which students engage or it might begin by infusing ethics in conventional courses spanning the curriculum.

A second rationale for ethics across the curriculum is similar to the first in its concern with practical judgment that is informed by relevant ethical and disciplinary considerations, but its focus is on professions as the relevant spheres of activity. On this model, ethics might be integrated into courses across the curricula of engineering, architecture, public health, law, agriculture, and other programs of study leading to professional employment.

A third rationale for infusing education with ethics is a perceived need for character education. The focus of public attention has been on primary and secondary education, but the reach of this rationale is illustrated by the introduction of collegiate service learning requirements, resurgence of colleges with strong religious identities, and references to character education in the professions. The extent to which this rationale is perceived to justify the teaching of ethics across the curriculum varies with different conceptions of what character education is. Views that take *phronesis* or good judgment to be essential to good or virtuous character should be open to instruction in ethics that would span different school subjects or courses within a program of pre-professional study (Annas 2011; Curren 2015, 2017; Kristjánsson 2015). A *phronesis*-based character education approach to ethics across the curriculum might differ from a judgement-focused model only by focusing on institutional ethos and learning through practice.

A fourth rationale for ethics across the curriculum takes a wider view of the ethical domain and its relationship to the curriculum. It revolves around the idea that fields of study are devoted to the goods of various practices and should not be conceived of as bodies of technique and knowledge having only instrumental value (Peters 2007; Strike 2003; Raz 2003; Elgin 2011). On this view, teaching with integrity involves a demonstrated commitment to the goods proper to the subject and initiation of students into a capable devotion to them. Such devotion would involve the requisite intellectual, creative, athletic, or other virtues. If ethics pertains to what has value and education is ethical to the extent that it communicates and teaches proper regard for what has value, then this is a conception of ethics across the curriculum.

Why Sustainability Ethics?

Where might sustainability ethics fit within this scheme of rationales for ethics across the curriculum? The short answer is that the first three rationales would support a diffusion of sustainability and its ethical aspects across a general or liberal arts curriculum, professional curricula, and all curricula, respectively. A fuller defense of this answer would involve spelling out the significance of sustainability and its ethics for personal, civic, and professional decisions—its significance for living well, for the survival and flourishing of sociopolitical systems, and for doing good work.

The fuller defense I have developed elsewhere begins with documentation of the deleterious impact of human activities on the natural systems on which the activities collectively depend, it identifies the factors that shape those activities, and it identifies norms of conduct conducive to sustainability (Curren and Metzger 2017). The net result of this is to demonstrate that the unsustainability of current human activities is a pervasive aspect of the contexts in which human decisions are made, and to provide basic ethical guidance that can yield individual and collective decisions compatible with sustainability. Students should be educated in ways that prepare them for the world that awaits them, and the unsustainability of human existence will be a dominating aspect of their world for the foreseeable future. Denying them knowledge of this fact would be unconscionable and denying them understanding of the ethical basis of cooperation that is essential to a better future would also be unconscionable.

The fourth rationale supports a form of ethics across the curriculum, but not the importing of ethical considerations and frameworks into subject domains where they do not already have a foothold. Taken at face value it appears ill-suited to justifying instruction in an ethic of sustainability. However, there is a subtle way in which ethical sensibilities associated with sustainability may be better cultivated in educational settings oriented to the internal goods of fields of study than in ones that capitulate to a narrowly instrumental view of the value of learning. Students who find non-instrumental value in what they study can find meaning in a competent pursuit of the goods they come to value. In doing so, they would come to value an aspect of human civilization and be pained by the thought of it being extinguished. The potential death of civilization is as close to the heart of sustainability as the multitudes of present and future persons whose opportunities to live well are inextricably dependent on the fate of civilization and the natural systems on which it depends.[2] Attachment to the goods of civilization encountered in one's field of study can be motivationally powerful, and meaning in life seems to require not just devotion to goods independent of oneself but a measure of competence in that devotion (Wolf 2010; Curren 2014c). If this is the case, then the teaching of diverse fields of inquiry and endeavor might properly touch on their historical trajectories and the wider context of their future prospects. It might touch on matters of sustainability and the ethical norms of cooperation essential to achieving it.

The upshot of these various rationales and related models for teaching sustainability ethics is that infusing sustainability ethics across the curriculum can be defended on several grounds and can involve conventional elements, such as case analysis, application of principles, examination of ethical arguments, debate, and elements of character education or cultivation of an ethical professional identity and commitments. A distinctive aspect of sustainability is that it is a problem that is too complicated to be addressed within the confines of any one or two fields of study. It does not follow that every discipline can contribute equally to understanding it, but it does follow that cross-curricular *integration* is essential. Faculty from different fields will need to collaborate with each other to provide students with an approximation of the education in sustainability that they need (Curren and Metzger 2017: 121–122, 153–179). Without this, instruction in the ethics of sustainability

[2]Note by way of contrast the emphasis environmental educators place on the love of nature acquired through environmental education that brings students into intimate contact with and understanding of nature. While conceding the ethical significance of such contributions to education in sustainability, my concern here is to point out that ethical learning of comparable significance for sustainability may occur across the curriculum (Ferkany 2018; Curren and Metzger 2018). Andrew Light gestured toward this quite literally, in a November 3rd, 2016 lecture on his shuttle diplomacy as the U.S. State Department special envoy to India, a labor of two intense years that was pivotal in persuading the government of Narendra Modi to sign on to the Paris Climate accord. Explaining why stabilizing Earth's climate was important enough to him to make the transition from only doing theoretical work to also doing policy work, he gestured toward the library and campus of the University of Rochester that surrounded him and said "All of this." A rough translation might be, "The aspects of civilization that are most worth saving."

may be impaired by an insufficient grasp of the context of decisions pertaining to sustainability.

A second important aspect of sustainability is that it is an unsolved problem of public life with countless moving parts that students can collaboratively engage in practical ways. I noted above that the AAC&U has proposed a model of undergraduate education designed to advance active global citizenship through collaborative, problem-focused learning. This is in many ways a global collegiate counterpart to the community-based, problem-solving, and service-oriented "community-civics" promoted by Progressive educators in the early decades of the twentieth century. Project-based collaborations are a vehicle for integrated learning heavily promoted by sustainability educators, and service-oriented projects of the kinds envisioned by the AAC&U could be valuable for the development of sustainability-related virtues (Curren and Dorn 2018).[3]

The Scope and Spheres of Sustainability Ethics

The ethic of sustainability I will present is an application of principles of common morality informed by the sustainability facts of life. Response to these facts is required by common morality, inasmuch as it regards us all as having a duty of care to make reasonable efforts to understand the circumstances in which we act and avoid doing harm. This is an approach sharply at odds with the common view that existing ethical frameworks are unhelpful in addressing matters of sustainability because they were created in and reflect a world of abundant frontiers. Many implications of the basic ethic I rely on have been ignored through long stretches of history, but they were not ignored by those who articulated this ethic most powerfully in the environmentally devastated world of ancient Greece to which I trace its origins (Curren and Metzger 2017: 69–72).

I said above that the preservation of opportunities to live well is the normative focus of concern for sustainability. Sustainability ethics may consequently be defined as the domain of ethics and ethical inquiry pertaining to every sphere and aspect of human activity as they bear on the capacity of natural systems to provide opportunities at least as good as the ones available now, indefinitely into the future. To say that sustainability ethics pertains to every sphere of human activity is to identify it as an aspect of universal personal ethics, social ethics, and political justice. *Personal ethics* pertains to everyone, everywhere, at all times, whatever they are doing. *Social ethics* concerns the norms of conduct of various institutions and forms of endeavor, and our roles within them, including professional roles. *Political justice* concerns the responsibilities of citizens and governments. Various

[3]The sections that follow incorporate material from Curren (2014a: 338–343) and the expanded and revised presentation of that material in Curren and Metzger (2017: 54–69). For a fuller presentation and related case studies, background, and references, see Curren and Metzger (2017).

roles, including those of professionals and of citizens, carry special responsibilities over and above those of common morality or personal ethics, as I use the term here. To say that sustainability ethics pertains to every aspect of human activity, as it bears on the preservation of opportunities to live well, is to imply that many aspects of our conduct make a difference to sustainability. It also assumes that many factors beyond personal character shape our conduct, such as institutional culture, policy context, and built environments. Personal virtues, basic commitments of common morality, principles derived from those basic commitments, codes of professional ethics, and ethical criticism of practices, institutions, structures, and acts of government all have roles to play in the ethics of sustainability.

Sustainability ethics is not confined to one sphere of activity, as business ethics is. Nor is it centered on concern with the value of the non-human natural world, as environmental ethics is. Sustainability ethics overlaps environmental ethics to the extent that the former concerns the life prospects of both humans and non-humans, and it may identify principles or responsibilities that should supplement those presently acknowledged in business ethics or other domains of professional ethics. Sustainability ethics overlaps the domain of environmental justice, inasmuch as sustainability is an issue of environmental justice, but not all issues of environmental justice are issues of sustainability. Environmental justice is concerned with fairness in the distribution of environmental burdens and benefits, but some (synchronic) unfairness in the distribution of environmental burdens and benefits may be consistent with sustainability, which concerns the (diachronic) preservation of opportunity into the future. Workers in solar panel factories might suffer unjust exposure to environmental toxins while doing work that contributes to a sustainable human footprint, for instance. Wider release of toxins could be a problem of both environmental justice and sustainability, of course, if it impairs the functional integrity of ecosystems, but locally it might be a problem of environmental justice and not of sustainability.

Principles of Sustainability Ethics

I begin from the idea of an ethic of respect for persons as rational beings and the duties entailed by such an ethic. These include the duty to take care not to cause harm and the duty of individuals who interact with one another to negotiate fair terms of cooperation governing their interactions, even if they interact only through the remote effects of their conduct. Violations of basic moral respect include acts that cause harm by diminishing opportunities to live well, failures to diligently investigate the potential of acts to diminish opportunity, efforts to deceive others about the possible destructiveness of acts or practices, and obstruction of efforts to negotiate or enforce fair terms of cooperation in pursuing sustainability.

Another basic starting point is that I will speak of the totality of practices of a human collectivity as *ecologically sustainable* just in case it is compatible with the long-term stability of the natural systems on which the practices rely. Sustainability of this kind is basic, because it pertains to the preservation of the biocapacity or renewable natural capital (RNC) that produces irreplaceable supporting, provisioning, and regulating "services" essential to human existence and well-being. A stable climate is essential to moderate reliable rainfall and other aspects of water availability, for instance, and there is no feasible replacement for naturally occurring fresh water on the scale of current human use. The first principle of sustainability ethics concerns ecological sustainability.

1. *Take care to ensure that the totality of human practices is ecologically sustainable.*

This is a straightforward implication of a basic duty to take care not to harm, given the fundamental role of RNC in human existence and well-being. Interpreting and applying this principle requires measures of sustainability and clarity about how it should guide the conduct of different kinds of decisions and actors. The most promising general framework presently available identifies ten "planetary boundaries" we should take care not to cross. These pertain to such things as atmospheric carbon, phosphorus runoff, land cover conversion, atmospheric aerosol loading and nitrogen removal, and pollution. With respect to the kinds of actors and decisions that might be involved, this principle identifies a responsibility of governments and government officials but also of individuals in their capacities as citizens and in their private and professional affairs to favor choices compatible with and conducive to the long-term stability of natural systems.

Although ecological sustainability is most basic, another form of sustainability that is significant for the preservation of opportunity is *throughput sustainability*, defined as follows: The totality of practices of some human collectivity is sustainable if and only if the material throughput on which it relies is compatible with the projected provisioning capacity of natural systems. Depletion of non-renewable accumulated products of ecosystems (NNC), such as aquifers, is a less fundamental concern than ecological unsustainability, so long as the total material throughput of human activities is less than a prudent maximum of RNC. If total human reliance on natural capital (NC) required only 30 percent of biocapacity, for instance, then ending the use of fossil water and fossil fuels might be manageable. Because it is presently about 150% of biocapacity and rising, humanity will find it impossible to sustain its current practices on the present scale. The economic throughput or flows of materials and energy on which opportunities depend will decline, and the opportunities themselves will necessarily change. The opportunities will not necessarily be worse on the whole, but the prospect of declining throughput is nevertheless a serious threat to the preservation of opportunity. A fundamental duty of care not to harm thus entails a second principle of sustainability ethics.

2. *Take care to ensure that the throughput requirements of human practices are compatible with the projected provisioning capacity of natural systems.*

This too is a principle that identifies a responsibility of governments and government officials but also of individuals in their capacity as citizens and in their private and professional affairs. The application of this principle would involve the same kind of measures of ecological stability as the first one does, insofar as it requires projections of future provisioning capacities of natural systems that may be overburdened, and it would require supply-side throughput measures as well.[4] I will refer to these first and second principles as *principles of opportunity preservation*.

An important implication of these first two principles being duties of care is that they have a precautionary aspect. Ecological and throughput sustainability require that natural limits not be exceeded, but *taking care* to ensure these limits are not exceeded would involve making investigations to thoroughly understand the circumstances of human actions and taking precautions to avoid doing harm, such as by holding in reserve some excess or redundant system capacity. It is also important to note that even when these aggregate limits are exceeded, the harm that can be anticipated is incremental. The endangering and diminution of opportunities to live well is not an all or nothing matter, so the underlying duty to take care to avoid harm would remain relevant. There is, in other words, no threshold effect that could justify saying it is "too late" to prevent disruptive planetary changes from occurring and we are all excused from making an effort to avert future misery.

A further important aspect of taking care to ensure that the totality of human practices is sustainable is that the relevant objects of choice include not only specific acts but—insofar as various actors have the power, authority, or influence to alter them—the attributes, norms, settings, structures, cultures, institutions, systems, and policies that are more or less conducive to ecological or throughput sustainability. This can be made explicit in a corollary to the first and second principles:

Corollary to 1 and 2: Take care to ensure that the human attributes, practices, institutions, systems, and policies within your control, authority, or influence are conducive to ecological and throughput sustainability.

The formulation *take care to ensure* could be glossed here, as in other contexts, as *make every reasonable effort to ensure*, where reasonableness must be judged in light of all that is at stake, the costs of exercising epistemic and preventive care, exerting influence, and so on.

As individuals, our actions contribute to an unsustainable human footprint incrementally, globally, and often without any identifiable victim, making it hard to know what would count as full or sufficient compliance with these principles. The principles call upon everyone to make reasonable efforts, but how much is enough? How much difference will the voluntary actions of individuals even make? One answer is that *harm is incremental*, so our personal increments of pollution and waste do matter. Another answer is that changing the way we individually live—and doing so voluntarily before adequate government policies are in place—is essential and can be efficacious in altering the political calculus of sustainability.

---

[4]The contrast with supply-side measures would be waste-side measures of such things as carbon emissions and phosphorus runoff.

It is nevertheless true that the translation of environmentally safe *collective* limits into publicly established and sanctioned limits on what *individuals* may do is the means by which we can ultimately determine what is and is not sufficient personal restraint. Our current situation with respect to sustainability is analogous to the early age of automobiles before there were speed limits and other traffic laws. A driver could slow down to reduce the risk of a collision, but how slow is safe enough? Democratically enacted speed limits and other traffic laws are the means by which we have collectively defined acceptable risk associated with automobile accidents. A carbon tax or permit (carbon cap and trade) system is one piece of what we need now to define the limits of acceptable environmental risk, and we need parallel policy interventions to fairly allocate the burdens of collective self-restraint with respect to other planetary boundaries. A system of user permits for NNC would also be required.

To say this is to affirm that even as environmentally conscientious individuals seek to live in ways that are compatible with sustainability, a fundamental burden we *all* bear—individual, institutional, and government actors alike—is to seek fair terms of cooperation in living sustainably. I noted above that seeking fair terms of cooperation is a basic requirement of interpersonal respect when individuals' acts have impacts on others, and it is undeniable that we are already interacting with others around the globe in ways that damage their interests. Our unsustainable levels of atmospheric carbon emissions are already causing catastrophic drought, storms, acidification and warming that undermines ocean ecosystems, and spreading disease vectors, all with present and growing costs in human lives and property. We in the U.S. are imposing these costs while having withdrawn from the Paris Climate agreement and failing to negotiate a mutually agreeable climate treaty and other environmental protection treaties. These are violations of basic norms of interpersonal respect that require facing those on whom we impose risk and harm and negotiating reasonable limits on those impositions. In practice, this implies that individuals have a responsibility to encourage government action to negotiate and enact binding agreements and to hold their elected officials accountable for failure to do so.

This leads to a third principle of sustainability ethics.

3. *Seek fair terms of cooperation conducive to sustainability. Actors whose actions affect each other have an obligation to cooperate in negotiating fair terms of cooperation in living in a manner that is collectively sustainable.*

Fair terms of cooperation would undoubtedly define not only the terms of participation in achieving a sustainable human footprint, but also what will constitute *wrongful* impositions of environmental risk on identifiable populations and jurisdictions. To that extent, the terms of cooperation in achieving sustainability would also address present matters of environmental justice, such as cross-border acid rain pollution that causes damage to forests in another country.

Fairness in cooperation requires that the terms of cooperation be known and clear to all parties and that acceptance of those terms not be predicated on one party being ignorant of what is known to another. People freely accept terms of agreement, believing that doing so serves their prudential and moral interests, but they often lack information others possess, giving those others (if they are selfish) an incentive to misrepresent facts that might make agreements less attractive to the other parties. Clear and accurate disclosure of information is thus an aspect of fairness or mutual respect in civic cooperation and other voluntary transactions, such as commercial transactions. It is often termed *transparency* or, in the political sphere, a *publicity* requirement—a requirement to make certain things public, such as documents submitted as evidence or records of testimony (Gutmann and Thompson 1996: 95ff). Transparency in this sense is essential to the legitimacy of agreements, but it often falls short of ensuring that the agreements and commercial transactions people agree to actually serve their interests.

The *efficacy* or efficiency of civic and economic relations in serving people's interests requires a stronger form of transparency. It requires not only that all parties to the terms of cooperation or transaction know the relevant facts known to other parties, but also that everyone knows what is actually at stake in the respects that matter to their legitimate prudential and moral interests—in the respects that matter to what they do and should care about. Achieving *full transactional transparency* of this kind requires coordinated efforts to discover and communicate the relevant facts. This may involve everything from the advancement of fundamental science, such as climatology, to the use of relevant sciences in investigating the probability of various costs and benefits of a proposed venture, as in preparing an environmental impact statement produced by an agency or business to justify a project that may have environmental risks (Gutmann and Thompson 1996: 98). Governments will need to make decisions regarding the allocation of responsibilities to undertake and report investigations into matters of public interest, such as the environmental impact of business ventures. Their duties to make these decisions would, however, be an aspect of justice in the pursuit of sustainability and as such a constitutional matter rather than a matter of common morality as such.

This having been said, we can derive a fourth principle of sustainability ethics from a basic ethic of interpersonal respect.

4. *Do not obstruct transparency and cooperation with regard to sustainability.*

Obstruction of transparency has taken a variety of forms, including corporate denials of matters of scientific consensus that the firms' own scientists accept, creating a false appearance of scientific controversy, orchestrating campaigns to discredit honest scientists doing valuable work, and misrepresenting the nature of science to cast specious doubt on entire fields of science (Oreskes and Conway 2010; Powell 2011; Negin 2015). Obstruction of cooperation may occur in any of these ways and more directly through lobbying; lawsuits; sponsoring politicians willing to cater to industry's short-term interests; and providing legislators with draft legislation that weakens environmental standards, enforcement, or both. Obstruction of the cooperation in which others might engage in fulfillment of their own ethical obligations may be considered a violation of an implicit universal moral duty associated with this third principle: a duty not to prevent others, by force or deception, from fulfilling their own moral obligations.

Obstruction of transparency about the environmental impact of products and business practices is not only objectionable as dishonest and a barrier to seeking fair terms of cooperation in pursuit of sustainability, but may also induce hazardous reliance on vulnerable systems, exposing people to risk of harm they have not voluntarily accepted and might otherwise choose to avoid.

Inducing hazardous reliance is a way of exposing people to unreasonable risk of harm, and thereby failing to take care not to harm them. Consider as an example that causing people to be at a party on a yacht at sea that cannot return them safely to port exposes them to unreasonable risk of harm and that it may occur in any number of ways. These might include not paying attention to the yacht's safe occupancy limit and inviting people in excess of that limit, setting sail despite warnings that violent storms are converging on its intended route, using so much fuel to make ice for the champagne that it cannot return to port ahead of the storms, and leaving the life vests behind to make room for the champagne. An aspect of the hazard associated with unsustainability is that excessive reliance on systems of limited capacity is a determinative causal factor in the growing probability and magnitude of harm likely to occur through ecological damage and depletion of natural capital.

The wrongness of putting people in harm's way by inducing or causing hazardous reliance can be captured in a fifth principle.

5. *Do not subject individuals or collectivities to detrimental reliance. Do not cause anyone to be in a position of fundamental reliance on hazardous or vulnerable systems or resources—systems or resources that cannot be relied on without exposure to unreasonable risk to their fundamental interests.*

This principle identifies imposition of risk per se as a form of wrong, and it focuses on the kinds of *systemic* risks that are at stake in discussions of sustainability: risks that ecosystems will collapse or that basic societal systems will suffer sharply declining capacity before sustainable alternatives can be developed and scaled up to replace them.[5] It addresses acts of *inducement*, in which people are induced to rely on something that is already hazardous or unreliable or will become so as a consequence of the reliance, and it addresses acts that cause something that is already fundamentally relied upon to become less reliable or adequate. An example of the former is the conduct of the fossil fuel industry in discouraging action to limit the use of their products. Any large-scale business that has the capacity to profitably develop and market products more conducive to global

[5]Readers familiar with legal doctrine will recognize this principle as a moral version of the legal doctrine of detrimental reliance or reliance-based estoppel ("equitable estoppel"), more general than the legal doctrine with respect to the pathway to reliance, more specific with respect to the character of reliance, and lacking the requirement of an actionable or compensable harm. Morally, it is sensible to treat imposition of unreasonable risk per se as a wrong, our concern being to capture what is ethically distinctive about unsustainability.

sustainability than its current products and fails to do so is arguably in violation of the principle of detrimental reliance. An example of the latter sort would be the poaching of fish in territorial waters, where islanders survive on fish caught by traditional methods within a few meters of shore.[6]

The principle may apply to the actions of specific individuals and collectively to the whole of a society or civilization. Regarding the latter, it says, in essence, that it is wrong to impose unreasonable risk on future generations by causing them to be in a position of fundamental dependence on systems that cannot be dependably relied on to provide them with adequate opportunities to live well. It is also designed to capture both the wrongness of inducing people to live unsustainably, putting them at risk, and the wrongness of putting future generations at risk through present unsustainable living. In general, it seems to capture what is morally troubling about unsustainability in a way other principles do not.

The principle that one should prevent harm that one can prevent at little cost to oneself has been invoked with good reason as justifying investments in reducing carbon emissions and stabilizing Earth's climate (Garvey 2008: 85). If there were no other reason for individuals and government and corporate actors to make the modest sacrifices that may be sufficient to avert catastrophic climate disruption, this principle would still provide one. However, because it applies independently of any causal contributions the actor may make to the prospect of harm, it does not capture salient aspects of the wrongness of sustainability-related acts that put others at risk. One should rescue a drowning child who is at risk through no fault of one's own, but failing to do so is a very different matter from inviting a child to a birthday party big enough to capsize the ill-equipped yacht on which it takes place and sailing off into storms.

The Dust Bowl of the 1930s provides an illustration of people being induced to live unsustainably and thereby subjected to unreasonable risk of harm. Homesteaders were induced to farm a region unsuitable for farming and largely destroyed the grasslands constituting North America's second-largest ecosystem. (Egan 2006). Named the Great American Desert in 1820, the high plains grasslands later rebranded the Great Plains were designated by surveyors as too dry for farming. Nevertheless, with encouragement from the railroads and prairie state senators, the Enlarged Homestead Act of 1909 promoted dryland farming by distributing parcels of undeveloped federal lands. Then, as homesteading peaked in 1914 and World War I began, the US government encouraged planting of more wheat in response to the exclusion of Russian wheat from global markets. For a few unusually moist years high plains wheat was profitable. Farmers expanded, taking on debt justified by a high price for wheat, and with the war's end and falling wheat prices they expanded again to cover their debt. At each step of the way, they plowed

[6]In its application to such cases, the *principle of detrimental reliance* adds pointed specificity to the *first principle's* prohibition against diminishing natural capital, subject to qualifications associated with the *third principle* or requirement to seek fair terms of cooperation. Those qualifications are met in the poaching case unless there is reason to think the protection of territorial waters is so politically unjust as to have no moral significance.

up the native grasses that anchored the soil and an ecosystem sustaining hundreds of species. In the drought of the 1930s that followed, the unanchored soil was gathered by winds into rolling mountains, ten-thousand feet high or more, blinding and suffocating cattle, obliterating roads, and dropping thousands of tons of dust on cities hundreds of miles away. Infants died of "dust pneumonia" and birth rates plummeted. A quarter of a million people who had been induced to settle and farm a region that had never supported more than a few Native American hunting camps and villages fled, leaving behind one hundred million acres in ruin.

Conclusion

Sustainability is a topic that lends itself to teaching across the curriculum, within existing courses and in team-taught interdisciplinary courses and projects, including service projects. The ethics of sustainability has an important role to play in this, and my hope for this chapter is that it will provide a helpful basis for advancing this aspect of teaching ethics across the curriculum.[7]

References

Annas, J. (2011). *Intelligent virtue*. Oxford: Oxford University Press.
Condon, P. (2010). *Seven rules for sustainable communities: Design strategies for the post-carbon world*. Washington, DC: Island Press.
Curren, R. (2014a). Defining sustainability ethics. In M. Boylan (Ed.), *Environmental ethics* (2nd ed., pp. 331–345). Oxford: Wiley-Blackwell.
Curren, R. (2014b). Judgment and the aims of education. *Social Philosophy and Policy, 31*(1), 36–59.
Curren, R. (2014c). Meaning, motivation, and the good. *Conference keynote and professorial inaugural lecture*. London: Royal Institute of Philosophy, January 24, http://www.youtube.com/watch?v=rhjZvbvpJYQ&feature=youtu.be.
Curren, R. (2015). Virtue ethics and moral education. In M. Slote & L. Besser-Jones (Eds.), *Routledge companion to virtue ethics* (pp. 459–470). London: Routledge.
Curren, R. (2017). *Why character education?*. London: Wiley-Blackwell.
Curren, R., & Dorn, C. (2018). *Patriotic education in a global age*. Chicago: University of Chicago Press.
Curren, R., & Metzger, E. (2017). *Living well now and in the future: Why sustainability matters*. Cambridge, MA: MIT Press.
Curren, R., & Metzger, E. (2018). The art of preserving opportunity: A response to Ahlberg, Ferkany, Macleod, and Ruitenberg. *Theory and Research in Education, 16*(1), 113–117.
Egan, T. (2006). *The worst hard time*. New York: Mariner Books.
Elgin, C. (2011). Science, ethics and education. *Theory and Research in Education, 9*(3), 251–263.

[7]For guidance on teaching sustainability ethics in the geoscience classroom in particular, see Metzger and Curren (2017).

Ferkany, M. (2018). Legitimizing education in sustainability. *Theory and Research in Education, 16*(1), 99–103.
Garvey, J. (2008). *The ethics of climate change*. London: Continuum.
Gutmann, A., & Thompson, D. (1996). *Democracy and disagreement*. Cambridge, Mass.: Harvard University Press.
Hovland, Kevin. (2006). *Shared futures: Global learning and liberal education*. Washington, D.C.: Association of American Colleges & Universities.
IISD. (2013). *The CERES principles*. http://www.iisd.org/business/tools/principles_ceres.aspx.
Kristjánsson, K. (2015). *Aristotelian character education*. London: Routledge.
Metzger, E., & Curren, R. (2017). Sustainability: Why the language and ethics of sustainability matter in the geoscience classroom. *Journal of Geoscience Education, 65*(2), 93–100. http://nagt-jge.org/toc/jgee/65/2.
Moore, K. D., & Nelson, M. P. (Eds.). (2010). *Moral ground: Ethical action for a planet in peril*. San Antonio: Trinity University Press.
Negin, E. (2015). Documenting fossil fuel companies' climate deception. *Catalyst,* 14(Summer), 8–11.
Oreskes, N., & Conway, E. (2010). *Merchants of doubt*. New York: Bloomsbury Press.
Peters, R. S. (2007). Education as initiation. In R. Curren (Ed.), *Philosophy of education: An anthology* (pp. 55–67). Oxford: Blackwell.
Portney, K. (2015). *Sustainability*. Cambridge, MA: MIT Press.
Powell, J. (2011). *The inquisition of climate science*. New York: Columbia University Press.
Raffaelle, R., Robison, W., & Selinger, E. (Eds.). (2010). *Sustainability ethics: 5 questions*. Copenhagen: Automatic Press.
Raz, J. (2003). *The practice of value*. Oxford: Oxford University Press.
Reuben, J. (1997). Beyond politics: Community civics and the redefinition of citizenship in the progressive era. *History of Education Quarterly, 37*(4), 406–416.
Rowe, D., Gentile, S. J., & Clevey, L. (2015). The US partnership for education for sustainable development: Progress and challenges ahead. *Applied Environmental Education & Communication, 14*(2), 112–120.
Strike, K. (2003). The ethics of teaching. In R. Curren (Ed.), *A companion to the philosophy of education* (pp. 509–524). Oxford: Blackwell.
UNESCO. (2005). *Education for sustainable development: United Nations Decade (2005–2014) website*. http://en.unesco.org/themes/education-sustainable-development.
U.S. Bureau of Education. (1915). "The teaching of community civics," U.S. Bureau of Education Bulletin, No. 23, Whole Number 650. Washington, D.C.: Government Printing Office.
Wolf, S. (2010). *Meaning in life and why it matters*. Princeton, NJ: Princeton University Press.

Ethics Bowl: An Approach to Implementing Ethics Across the Curriculum

Robert F. Ladenson

Abstract Ethics bowl is an academic competition combining a valuable and distinctive educational experience with the excitement and fun of a competitive game. Section "Format, Procedures, and Rules of the APPE-IEB" describes the format, procedures, and rules of the Association for Practical and Professional Ethics Intercollegiate Ethics Bowl (APPE-IEB). Section "Brief History of the APPE-IEB (Association for Practical and Professional Ethics Intercollegiate Ethics Bowl)" summarizes the APPE-IEB's origin and development. Section "Basic Educational Objectives and of the APPE-IEB" identifies and analyzes the APPE-IEB's basic educational objectives, making apparent the close coincidence of these objectives with those of ethics across the curriculum. Finally, Sect. "Is the APPE-IEB Too Competitive?" discusses the apparent tension between the APPE-IEB's basic educational objectives and its utilization of the desire for competitive success as a motivator.

Keywords Ethics · Bowl · Moral decisions and judgments · Democratic deliberation

Introduction

Ethics bowl is an academic competition combining a valuable and distinctive experience for students with the fun and excitement of a competitive team game. National ethics bowl competitions take place annually on both the college and the high school levels in which many students participate.[1] The first section of this chapter describes the format, procedures, and rules of the Association for Practical

[1]See Sect. "Brief History of the APPE-IEB (Association for Practical and Professional Ethics Intercollegiate Ethics Bowl)".

R. F. Ladenson (✉)
Illinois Institute of Technology, Chicago, USA
e-mail: ladenson@iit.edu

E. E. Englehardt and M. S. Pritchard (eds.), *Ethics Across the Curriculum—Pedagogical Perspectives*, https://doi.org/10.1007/978-3-319-78939-2_18

and Professional Ethics Intercollegiate Ethics Bowl (APPE-IEB).[2] The second section summarizes the APPE-IEB's origin and development. In section three I will develop an interpretive analysis of the basic educational objectives implicit in the APPE-IEB as it has developed and grown since its inception in 1997. The close coincidence between the basic educational objectives of the APPE-IEB, as I conceive of them, and those of ethics across the curriculum, I believe, will be apparent. Finally section four identifies and discusses an apparent tension between, on the one hand, the APPE-IEB's basic educational objectives, and on the other hand, its utilization of students' desire to achieve competitive success as a motivator.

Format, Procedures, and Rules of the APPE-IEB

In an ethics bowl match a moderator asks two teams of three to five students questions that pose ethical problems over areas such as: social/political topics (e.g. war and peace, freedom of expression, economic justice, health care, etc.); the classroom (e.g. plagiarism or other kinds of cheating in an academic context); personal relationships (e.g. dating, friendship, etc.); and professional ethics (e.g. engineering, business, medicine, law, etc.). Each of the two teams in an ethics bowl match is asked a different question, which each team answers according to the following format. The moderator poses a question to one of the teams (hereinafter 'team 1'). Team 1 then has two minutes to confer, after which it must state its answer (in no more than ten minutes). Team 1 does not respond cold, however. Six weeks prior to the competition all participants receive a set of cases that present ethical issues upon which the questions a team may be called upon to answer at the ethics bowl are based. When team 1 has completed the presentation of its answer the opposing team (hereinafter 'team 2') has one minute to confer, after which it comments (for no more than five minutes) upon team 1's response to the moderator's question. Team 1 then receives one minute to confer, after which it responds (for no more than five minutes) to team 2's commentary.

A panel of three judges then poses questions to team 1 to elicit the team's viewpoint on important aspects of the question, or to seek clarification in regard to the team's responses to points made by the commenting team. The judges, who are highly qualified persons in diverse fields, receive advance copies of the cases upon which questions are based at the same time as do the student participants, about six weeks before the ethics bowl competition. The same format then is repeated with the two teams exchanging roles. That is to say Team 2 becomes the presenting team and team 1 becomes the commenting team. Team 2's question is based upon a different case than the one upon which Team 1's question was based.

---

[2]The official Rules for the APPE-IEB National Championship Competition can be found at: http://appe.indiana.edu/ethics-bowl/cases-rules-and-guidelines/.

After the judges have concluded their questions to team 2 the moderator asks the three judges to indicate their respective evaluations of each team's presentation and commentary. Prior to the competition the judges have been instructed as to the criteria they are to apply in their evaluations, which are as follows:

Presenting Team's Response to Moderator's Question: (30 total points)

1. Was the presentation clear and systematic? Regardless of whether or not you agree with the conclusion, did the team give a coherent argument in a clear and succinct manner?
2. Did the team's presentation clearly identify and thoroughly discuss the central moral dimensions of the case, doing so in a way consistent with a collegial and thoughtful discussion?
3. Did the team's presentation indicate both an awareness and thoughtful consideration of, as well as a respect for, different viewpoints, including especially those that would loom large in the reasoning of individuals who disagree with the team's position?

Opposing Team's Commentary (10 total points)
Did the commenting team give a fair and accurate representation of the presenting team's argument and present a thoughtful commentary on the argument?

Presenting Team's Response to Commentary (10 total points)
Did the team thoroughly and respectfully respond to the commentary presented by the other team, addressing the relevant concerns raised by the commenting team?

Brief History of the APPE-IEB (Association for Practical and Professional Ethics Intercollegiate Ethics Bowl)

I first developed the idea of the Ethics Bowl in 1993, organizing in that year a small Intramural Ethics Bowl event at the Illinois Institute of Technology (Illinois Tech) with the assistance of colleagues and associates in Illinois Tech's Center for the Study of Ethics in the Professions (CSEP).[3] Two years later, in 1995 we also organized a small local ethics bowl to which we invited three teams from nearby universities to compete along with the winner of Illinois Tech's intramural ethics bowl competition.

[3]My CSEP colleagues during the period 1993–1995, which were the years before I took the lead in organizing and conducting the first APPE-IEB were: Michael Davis, Ellen Fox, William Pardue, and Vivian Weil.

In 1997 I took the lead in developing and organizing, for the first time, a nationwide intercollegiate ethics bowl.[4] This event, in which fourteen teams participated, representing colleges and universities throughout the United States took place in conjunction with the annual meeting of the Association for Practical and Professional Ethics (APPE). (The idea of holding an Intercollegiate Ethics Bowl associated with APPE occurred to me because the previous spring I had put on a brief demonstration of the Ethics Bowl at the dinner of the APPE annual meeting that was extraordinarily well received.)

The APPE-IEB, as the competition subsequently came to be known, was open to any college or university that wished to take part. From the outset it generated enormous enthusiasm on the part of all who participated—students, faculty who coached teams, and judges and moderators. In the years immediately following the first APPE-IEB many participating faculty spontaneously volunteered to help in any way they could with the planning and organizing of the event. Such was extremely fortunate because during this period The APPE-IEB experienced steadily increased participation, which, in turn, necessitated increasingly larger staffs of volunteers to organize and conduct the competition. Indeed, after six years the APPE-IEB became a victim of its own popularity. The field of teams expanded from fourteen to forty, which was the largest number that could be accommodated in a one-day event, taking place in a hotel. By 2003 there was a waiting list of twenty schools that wanted to be included. In some instances schools had to wait several years for an opportunity to participate.

To address this situation the (*ad hoc*) Coordinating Committee of the APPE-IEB and I, organized a small group to plan a reorganization and expansion of the competition, which ultimately developed into four sub-committees, in which twenty-five individuals were involved actively.[5] By way of background, prior to the reorganization/expansion planning effort, six regional ethics bowls had been established that functioned independently of the APPE-IEB, but which were modeled upon it closely.[6] The reorganization/expansion plan envisaged inviting the independent regional bowls to link up in a tiered APPE-IEB competition. Under the

---

[4]From 1997 through 2003 the APPE-IEB was supported by generous grants from Sears Roebuck & Co.'s Office of Business Practices. During the period 2000–2006 the APPE-IEB was supported as well by a generous matching grant from Robert Galvin and Robert Pritzker in connection with the Illinois Tech capital campaign. In 2003, the APPE-IEB was supported also by a grant from the AMS Corporation, and in 2004 it was supported in part by the Raj Soin College of Business of Wright State University.

Also, I would be gravely remiss not to acknowledge the unstinting, enthusiastic support and assistance of Brian Schrag, who was Executive Director of APPE from its founding in 1990 to 2011, and of Stuart Yoak, Brian's successor as APPE Executive Director.

[5]The members of the coordinating committee (which was an informal, ad hoc group of people from whom I requested help), were: Anthony Brinkman, Patrick, Croskery (who, at the time was the (again ad hoc) co-director of the APPE-IEB), Joanne Ladenson, and Sarah Pfatteicher.

[6]The six independent regional ethics bowls were the following: Indiana (later, Central States), Texas, Wasatch (Utah), Northeast, California, and Pacific Northwest.

plan the top scoring teams in the regional bowls would compete for the national championship at the APPE annual meeting.

A seventh and an eighth regional ethics bowl were created in 2006.[7] During the 2006–07 school year the reorganized and expanded APPE-IEB took place. Eight regional bowls were held in the fall. The thirty-two top scoring teams in the eight regional bowls then competed in the early spring for the APPE-IEB National Championship at the APPE annual meeting. Two more regional ethics bowls were created the following year, bringing the number of regions to ten.[8] In 2014 a national ethics bowl competition among community colleges was instituted. As of 2016, the winner of the national community college ethics bowl competition qualifies for the field of teams in the APPE-IEB National Championship Competition.

Since the first APPE-IEB took place in 1997 a vast number of other ethics bowl competitions have been organized and conducted that are not affiliated with the APPE-IEB but which utilize formats, rules, and procedures influenced by those of the APPE-IEB, but adapted and modified in diverse ways.[9] The largest of these competitions is the National High School Ethics Bowl (NHSEB). During the 2016–2017 school year 3924 high school students from 281 high schools throughout the United States took part.[10]

My focus of discussion in this chapter, however, is limited to the APPE-IEB, given the extent of my experience with and knowledge about it. Such seems reasonable to me, rather than attempting to describe particular details of other ethics bowl, much less to generalize about all of them.

Basic Educational Objectives and of the APPE-IEB

A. Relationship to Ethics Across the Curriculum

Ethics Across the Curriculum has two related ideas at its core concerning respectively educational objectives and teaching methods. The first idea is that education, both in college and in high school, can, and should, make a valuable, distinctive contribution to students' moral development in the following respect. It can (and should) help students to develop their abilities in regard to reasoning, deliberation, and judgment concerning important moral issues that are complex, controversial, highly viewpoint dependent, and difficult to resolve. The second core idea is that

[7] The Southeast and the Upper Midwest regional bowls.

[8] The Rocky Mountain and the Mid-Atlantic regional bowls.

[9] A Google Search on 'Ethics Bowl' will generate hundreds of items, indicating the amount and diversity of different of ethics bowl competitions at this time, which include, among others, competitions for medical students, graduate students in archaeology, undergraduate college students, high school students, and, most recently, middle school students.

[10] The News and Observer, Durham County (NC), April 11, 2017.

effective teaching methods to accomplish the above objective require innovative educational approaches to be implemented throughout an entire curriculum. That is to say, a single ethics course, even if mandated for every student in a given academic program, is inadequate.

This section analyzes in depth the educational objectives of the APPE-IEB. The close coincidence between these objectives and those of Ethics Across the Curriculum, I believe, will be made readily apparent in the analysis that follows. The APPE-IEB begins from the following premise. Many of the issues students face during both their high school and their undergraduate college experience, as well as many of those they will confront in their future life experience, are complex, controversial, highly viewpoint dependent, and difficult to resolve. Such applies in the case of topics such as cheating, plagiarism, personal relationships, gender inequality, and campus social or political controversies, as well as to issues of professional ethics, which may come into play in their future careers. The above premise applies also to matters of public importance upon which students will have to develop informed opinions to exercise their rights and to fulfill their responsibilities of democratic citizenship.

Education on both the levels of high school and college, can make the following three valuable and distinctive contributions to the development of a student's capabilities relative to ethical reasoning and judgment: (1) It can help students develop a framework of analysis for addressing ethical issues in an intellectually well-organized manner; (2) It can provide an opportunity for students to acquire valuable information relevant to arriving at judgments about ethical issues of special importance to them in light of their respective interests, concerns, and career aspirations; (3) It can foster the capacity of ethical understanding over a broad range of subjects.

The last of the above items, the capacity for ethical understanding, needs further words of elucidation. The positions people take on important, but complex, controversial, highly viewpoint dependent, and difficult to resolve ethical issues tend to depend strongly upon factors such as an individual's politics, personal values, gender, or religion. Even from the standpoints of ethical theories that posit objective general standards of ethical reasoning and judgment, such as Utilitarianism or Kantianism, no one reasonably may assume that his or her ethical judgments about such issues proceed from a neutral standpoint absolutely uninfluenced by the above kinds of factors. A question thus arises of what ethical understanding means relative to complex, controversial, highly viewpoint dependent, and difficult to resolve ethical issues. I believe that it consists largely of viewing from the inside other ethical positions besides those with which a person agrees. It involves not only awareness of the arguments advanced on behalf of those positions, but also understanding of the concerns that motivate the arguments, and even, to some extent, an appreciation of their force. Development of the capacity for moral understanding relative to complex, controversial, highly viewpoint-dependent, and difficult to resolve issues thus only can take place on a large scale in societies where a strong and effective right of free speech exists.

The existence of a strong and effective right of free speech in society, however, although necessary in this regard, is far from sufficient. In the case of the above mentioned kinds of issues, a person's capacity for ethical understanding tends to develop most readily, I think, in an environment with three special characteristics: (1) a person feels motivated strongly to state her views on the issues; (2) she feels also motivated strongly to listen carefully to what others have to say about their views on the issues; (3) everyone involved in discussing the issues is motivated strongly to do so in a way that brings out clearly the similarities and differences in outlook among the discussants' views.

The APPE-IEB, I believe is a form of experiential education that seeks to create a learning environment in which students can develop three crucial capabilities for attainment of ethical understanding in regard to complex, controversial, highly viewpoint dependent, and difficult to resolve ethical issues: open-mindedness; willingness and readiness to engage in meaningful conversation; and deliberative thoughtfulness.

(i) *Open-Mindedness*: As noted earlier, six weeks before an APPE–IEB regional or national competition takes place the participating teams receive a set of ethics cases. There are fifteen cases, each one of which is about a single spaced typed page. The teams are not given questions about the cases. Instead they have to identify key ethical issues themselves and then develop a team position for each case, which they then must be able to state clearly and to justify. All cases are complex, controversial, highly viewpoint dependent and difficult to resolve. No team ever can quickly reach full agreement among its members on most, let alone all the cases.

A team's challenge thus is to identify key ethical issues raised by the cases and then to work out positions on them that everyone on a team agrees are reasonable in the sense that a morally conscientious person could accept them after careful consideration. To reach this kind of agreement among themselves each team member must be able to listen to the others with on open mind. Each team member has to be able to consider seriously different views from his or her own, and to appreciate their force—not in the sense of being persuaded necessarily, but in recognizing why a morally responsible person could find them persuasive. So, the incentive to do well in the competition tends to motivate ethics bowl team members to listen open-mindedly.

(ii) *Willingness and Readiness to engage in meaningful communication:*

To recapitulate briefly the format of an ethics bow match, summarized in section (I.):

- The moderator poses a question about a case to one of the teams (Team 1);
- The Team 1 presents its answer to the moderator's question about the case;
- The opposing team (Team 2) comments on Team 1's answer;

- Team 1 responds to Team 2's commentary;
- The Judges then pose questions to Team 1;

  The above format is repeated with Team 2 responding to a different question and Team 1 commenting.

The above procedural format raises an apparent question. What happens if the commenting team agrees with the presenting team's answer to the moderator's question? Here is the answer: Ethics bowl cases, without fail, are conceptually deep, and factually complex. The likelihood is vanishingly small that a team's answer to a question posed by an APPE-IEB moderator will leave the opposing team utterly speechless, so that the presenting team need not say anything in response. To the contrary, a commenting team that agrees with the presenting team's answer may discuss why it finds problematic the team's justifying argument. Furthermore, even if it finds the presenting team's justifying argument persuasive it may wish to develop another justifying argument that highlights other ethically significant considerations in the case, because no ethical issue in an ethics bowl case ever can be completely closed (as contrasted with being settled, for immediate practical purposes). The mark of an excellent commentary is the contribution it makes to continuing, mutually beneficial, discussion.

(iii) *Deliberative thoughtfulness:* There are three evaluation criteria ethics bowl judges are required to apply. The first is clarity, and the second is thoroughness. The third concerns deliberative thoughtfulness, in a sense essential to ethical understanding in connection with the kinds of issues posed by ethics bowl cases. In this context ethical understanding thus consists largely of understanding the views of people with whom one disagrees. It means not simply awareness of what they've said or written, but calls as well for a serious effort to understand their views from the inside—to comprehend the key concerns motivating them, and, at least, to some extent, appreciating their force.

Prepping and training efforts for APPE-IEB judges thus emphasize the critical importance of posing questions to teams in a match that probe the team members' depth of ethical understanding of views different from the one's they've presented to regard to the cases they consider. Correspondingly, a key part of an ethics bowl team's preparation and coaching involves trying to anticipate questions the judges will ask in light of the team's answers to the moderator's questions. With respect to judges' questions, it becomes apparent when observing ethics bowl matches that most of the time the concerns underlying these questions were identified and discussed carefully by the teams in their preparations for the competition. Students attest, however, that on many occasions they have found themselves put hard to the test, with great educational benefit, by judges' questions, which the students themselves recognize, approached cases from different directions than those they considered, or that pursued the students' lines of reasoning to deeper levels than they reached in their discussions prior to the competition.

B. Virtue Ethics, Moral Community, and Democratic Deliberation

A discussion of the APPE-IEB's educational objectives and methods has to consider, and respond to, a criticism I shall develop immediately below. In order to understand the point of this criticism, one first needs to know, by way of background, that practical and professional ethics education has two predominant intellectual orientations—what one may call, for lack of better terms, the applied ethical theory and the virtue ethics approaches—which are perceived widely to pull in opposed directions.

The applied ethical theory approach considers major philosophical theories, of which Utilitarianism and Kantianism are prime examples, to have immense value as conceptual resources for reasoning and judgment in regard to complex, controversial, highly viewpoint dependent, and difficult to resolve ethical issues. According to this approach educational efforts apropos practical and professional ethics should familiarize students with major philosophical conceptions in this area and encourage students to work through the concrete ethical issues of concern (e.g. relative to business, medicine, engineering, social policy, etc.) in terms of these conceptions. The applied ethical theory approach values, and seeks to develop in students, attributes such as logical and analytical thought, intellectual rigor, and capacity for critical distance in connection with issues of practical and professional ethics.

In contrast to the applied ethics approach, the proponents of virtue theory view the fostering of moral virtue as, by far, the most important aim for courses in practical and professional ethics. Virtue ethics proponents would observe that the emphasis of the applied ethical theory approach upon analysis of ethically complex, controversial, highly viewpoint dependent, and difficult to resolve issues would make sense in a world where independent and creative thought about genuinely contestable ethical questions was stifled by an overwhelming consensus in society on interpretation and application of ethical principles and values. For virtue ethics theorists, however, this is not our world. In contrast, they consider one of the most troubling elements of contemporary life to be a deep decline in ethical consensus, and a resulting attenuation of the individual's sense of membership in a moral community. Virtue ethics theorists believe that practical and professional ethics education should aim to reinforce, or recover, this sense of membership in a moral community by fostering the tendency of students to identify with shared standards of ethical reasoning and judgment, and to help students develop important virtues associated with commitment to such standards.

Some virtue theorists thus might conclude that the APPE-IEB does not address the most important objective of practical and professional ethics education from their standpoint. Even worse, from the perspective of such theorists, the APPE-IEB might tend to work against this objective owing to its unavoidable, near-complete focus upon hard cases about which significant disagreement exists.

I agree with the virtue ethics theorists to the extent I think the APPE-IEB would be educationally unsuccessful if it led students to exaggerate the extent, and to misunderstand the nature, of the disagreement that exists relative to interpretation and application of important ethical principles and values. I do not believe this is the case, however. For the reasons developed immediately below, in my opinion, the APPE-IEB has key elements that should resonate strongly with proponents of virtue theory.

In this regard, it seems to me that the APPE-IEB not only is consistent with, but also reinforces, a view of ethical reasoning and judgment as activities conducted within a community whose members identify with shared standards of right and wrong, and better or worse, in connection with ethical questions. Students report that their discussions of cases in team meetings to prepare for the APPE-IEB tend to result either in significant narrowing of differences of opinion or else in a clearer mutual understanding of each other's viewpoints, which enables them to agree upon team positions relative to the cases. Furthermore, the identification of community members with shared standards of ethical reasoning is expressed strongly by the student's interactions with judges in their matches. As noted above, students often acknowledge the educational benefit of judges' questions, which they realize approached cases from directions different from the ones they considered, or that pursued the students' lines of reasoning to deeper levels than they reached in their pre-match preparation.

Some virtue theorists might concede that the aspects of the APPE-IEB described above harmonize with their general approach but insist that this does not go far enough. According to their view, education in practical and professional ethics not only should provide students a limited model (from their standpoint) of a moral community committed to shared standards of ethical reasoning and judgment, but also should help students develop important virtues essential to membership in such a community. I believe the APPE-IEB serves this purpose effectively. An explanation of why, I believe so, however, calls for a brief analysis set out immediately below, of an important idea of social and political ethics that has not yet figured in this discussion—democratic deliberation.

Democratic deliberation occupies a conceptual space between bargaining, on the one hand, and proselytizing, on the other. Bargaining, in the clearest cases, is an activity that precedes voluntary exchange in which parties motivated by the desire to further their respective interests try to gain as much as possible while giving up as little as they can. Democratic deliberation, again, in the clearest cases has a different focus. It involves matters for public discussion, implicating principles and values of political and social ethics that are shared within a community but about which significant disagreements exist concerning their interpretation, and hence about their applications in diverse circumstances.[11]

Seeking to persuade, in the context of democratic deliberation, however, needs to be distinguished from proselytizing. A proselytizer attempts to convert other persons—to "win them over"—so that their most important ethical principles and values coincide exactly with his/her own. In contrast, the idea of persuasion, relevant to democratic deliberation, takes (and leaves) people more or less as they are. Democratic deliberators do not aim at getting people who disagree with them upon a particular political issue to renounce their deeply held convictions about social-political ethics, and to embrace entirely different new ones. They aim instead to produce a change in outlook on the particular issue through logical argument and rhetorical appeals directed at finding a compromise all parties can consider beneficial in terms of their key interests consistently with their respective interpretations of core shared principles and values of political ethics.

Understood in the above way, one thus may view democratic deliberation as a practice with distinctive virtues internal to it that include as central components attitudes and dispositions, which the APPE-IEB aims to develop through experiential education.[12] Efforts at persuasion on highly controversial matters for public discussion can degenerate easily into bitter conflict, and, in the worst cases, even hatred, unless parties exemplify the basic civic virtues of tolerance, patience, and restraint.[13]Tolerance, however, in the sense relevant to democratic civic virtue, includes not only commitment to upholding the fundamental civil rights of every member of the democratic body politic but also deliberative thoughtfulness concerning the rationales for, and the applications in diverse particular circumstances, of fundamental democratic rights. The virtue of patience, in the democratic context, encompasses both open-mindedness and readiness to engage in meaningful communication with others whose opinions on public matters differ from one's own. Restraint, as a member of a democratic body politic, calls for avoidance of gestures or actions likely to convey a message of closed-mindedness on one's part, and to

---

[11]In this regard, Ronald Dworkin considers the conception of a "community of principle" intrinsic to constitutional democratic government. Under this conception, according to Dworkin,

> members of a genuine political community … accept that they are governed by common principles, not just by rules hammered out in a [mere] political compromise. Politics … for such people [in large part] is a theater of debate about which principles the community should adopt as a system, which view it should take of justice, fairness, and due process. Members of a community of principle accept that their political rights and duties are not exhausted by particular decsions their political institutions have reached, but accept, more generally the scheme of principles those decisions presuppose and endorse. So such a member accepts that others have rights and that he has duties following from that scheme, even though they have never beenformally identified or declared.) (*Law's Empire* (Cambridge, MA, 1986), 210).

[12]The idea of virtues as internal to practices is presented and developed in a form highly influential among academic philosophers by Alasdair MacIntyre in his seminal book *After Virtue* (Notre Dame, IN, 1984).

[13]See Ladenson (2011).

consider carefully in good faith the viewpoints of those with whom one disagrees deeply on public matters.

For the above reasons thus I believe the APPE-IEB has intrinsic features that make it an excellent educational activity not only from the standpoint of the applied ethical theory approach to education in practical and professional ethics, but also from the perspective of virtue theory. In this regard, some of the APPE-IEB's most ardent supporters over the years have had strong intellectual inclinations toward the virtue theory approach. The preceding analysis, I think, illuminates why such has been the case.

Is the APPE-IEB Too Competitive?

The basic educational objectives and methods of the APPE-IEB concern fostering of attitudes and dispositions that promote successful cooperative activity under circumstances that easily can degenerate into irreconcilable conflict. A principal motivating factor the APPE-IEB draws upon, however, is the desire of participating students for victory in an intense academic competition. An unavoidable tension thus exists between the APPE-IEB's inherently cooperative basic educational objectives, and on the one hand, and its strong reliance upon competitive success as a motivator, on the other hand.

In considering this tension I believe one must keep the following general points in mind. Cooperation and competition both are significant aspects of the most important forms of human social life, such as democratic politics, public affairs, the arts, sports, and (even) marriage, family relations, and/or personal friendships. The cooperative and competitive aspects of such forms of life, in all their diversity, must be negotiated and reconciled continually. At times this may not be simple. No formula exists to determine a successful resolution for every particular circumstance.

From its inception the people who became involved actively in organizing and conducting the APPE-IEB viewed it as embodying a distinctive spirit, combining egalitarianism, inclusiveness, and civility. I believe that over the past two decades, as the APPE-IEB has grown, such an attitude has both become more widespread and has deepened. The spirit of the APPE-IEB thus is now something, which, in my opinion, many persons who consider themselves stakeholders in regard to the APPE-IEB's continued flourishing are deeply concerned to sustain.[14] I realize, however, there are no research studies that provide "hard" data to confirm that the

[14]Expressive of such concern, in 2015 the APPE-IEB Executive Board created an award for the team at the National Championship Competition that exemplifies best the spirit of the APPE-IEB. (The award winner is selected by how many votes a team receives from the teams against which it competed in its three preliminary matches, as well as from the moderators and judges who interacted with the team.) I was honored deeply that the Executive Board named the award in my honor.

experience of having taken part in the APPE-IEB has a lasting positive effect, *after the competition ends*, upon how students tend to approach complex, controversial, highly viewpoint-dependent, and difficult to resolve ethical issues. Furthermore, design and implementation of an adequate research study in this regard, I realize also, surely would pose daunting challenges.[15]

I can, nonetheless, note the following. Since creating the Ethics Bowl in 1993 I have received vast numbers (probably hundreds) of communications from students and faculty participants that are overwhelmingly (indeed, almost unanimously) positive. Their most reinforcing aspect for me, however, has been the extent to which the comments of students concerning what they believe they have gained from participating in the APPE-IEB track closely the conception of its basic

---

[15]It seems to me, however, that, in this regard, a research study should be noted that was conducted by Diana E. Hess and Paula McAvoy, described in detail in Hess' and McAvoy's book, *The Political Classroom: Evidence and Ethics in Democratic Education* (New York, 2015).

The Hess McAvoy study involved over 1000 high school students and 35 high school teachers in three states—Illinois, Indiana, and Wisconsin. It took place over four years (2005-2009), and had both quantitative and qualitative components. The quantitative component was a statistical analysis of student responses to questionnaires given them both prior to and after taking courses in which controversial public issues were a significant part of the course subject matter. The qualitative component consisted of classroom observation and extensive interviews of both students and teachers.

A principal purpose of the study was to learn about students' perceptions of what they learned from their social studies classes that covered controversial public issues. In this regard Hess and McAvoy were interested especially in finding out about the extent to which students expressed the judgments that they became more interested, as a result of taking the classes, in understanding the viewpoints of people whose opinions on the issues differ from theirs', and in discussing the issues with them.

Hess and McAvoy distinguished three kinds of courses described below.

(i) Best Practice Discussion: Students in these classes engaged in discussion of controversial public issues more than 20% of the time. These discussions also involved students preparing in advance, significant student-to-student talk, and high levels of student participation.
(ii) Discussion: Students in these classes engaged in discussion 20% or more of the time, but fell short of Best Practice Discussion because most of the talk was student-to-teacher and not student-to-student. Students often were not expected to prepare for discussion and usually only a small number of students took part in the discussion.
(iii) Lecture: These classes did not meet the threshold of using some form of discussion at least 20% of the time. The dominant pedagogical strategy was teacher lecture.

Hess and McAvoy report that "the qualitative and quantitative data [produced by their research study] showed that students in Best Practice Discussion classrooms were more likely to report that they are more interested in politics as a result of taking the course, more likely to enjoy political talk, and more comfortable with disagreement." From the standpoint of the quantitative data the study generated, which was analyzed statistically via multiple linear regression modeling, when compared to all other predictors in the model, the Best Practice Discussion classroom was by far the strongest predictor" of the above mentioned reports.

It was encouraging for me to learn about the results of the study summarized above. In this regard, I believe it is apparent that the APPE-IEB is a prime example of Best Practice Discussion, as Hess and McAvoy define it.

educational objectives I have tried to articulate clearly in this chapter. Here, for example, are the words of a student from the University of Montana, Ms. Dixie Dishon, who was interviewed by Catherine Crier on the Fox Cable Network's *Crier Report* in 1997 several hours after her team won the 1997 APPE-IEB National Championship Competition:

> Catherine Crier: How much has the… [APPE-IEB]… affected you personally Dixie, your own approach to life?
>
> Dixie Dishon: Hmmm, I would say that my ethical stances are probably more reasonable and object[ive]. I can listen to other people's opinions now and kind of see their light, where before when I got into a debate I was very staunch in my beliefs and very particular with my own ethics basis for my argument; and doing these particular questions and especially having to concur with my group we found that doing research and coming into a discussion where two of us would want to go with the question in one way and say that it was justifiable, and two people wanted to say it was morally objectionable really forced us to try to come to a consensus on what we would do in those situations.

Conclusion

As noted earlier, ethics across the curriculum has two related ideas at its core. First, education, both on the undergraduate college and the high school levels, can, and should, make a valuable, distinctive contribution to students' abilities in regard to addressing thoughtfully important ethical issues that are complex, controversial, highly viewpoint dependent, and difficult to resolve. Second, effective teaching to accomplish these objectives calls for utilizing innovative approaches throughout an entire curriculum rather than limited to a single course. In this chapter I have sought to elucidate why, from the standpoint of these core ideas, ethics bowl qualifies as a valuable item for inclusion in a "toolkit" of approaches for implementation of ethics across the curriculum.

Reference

Ladenson, R. (2011). Civility as a democratic civic virtue. In D. S. Mower & W. L. Robison (Eds.) *Civility in politics and education* (pp. 201–220). New York.

Linking Academic Integrity and Ethics Across the Curriculum: Groundwork for Sustainability in Practical and Professional Ethics

Daniel E. Wueste

Abstract The piece argues that there is a connection between academic integrity (AI) and teaching ethics across the curriculum (EAC) that extends beyond shared terminology in a practical and purposeful way, i.e., in a way that is responsive to a challenge in practical and professional ethics. The twofold purpose of the essay is (a) to explain how linking AI and EAC is responsive to this challenge and (b) to make the case for the approach it involves. Two large questions are addressed. The first is about how EAC should be done, if it is connected to AI. The second (two-part) question is (a) What would success look like? and (b) How would we know that it had been achieved—how would it (success) be measured? The first concern receives the lion's share of attention and involves taking cues from Jonathan Haidt's social intuitionism and writing about logic and the law by Oliver Wendell Holmes Jr., and, in particular, John Dewey. Several considerations that argue for the connection are discussed before the two-part question about success is addressed in the conclusion.

Keywords Ethics across the curriculum · Academic integrity · Reasoning
Moral judgment · Social intuitionism · Framing · Behavioral ethics
Debiasing · John dewey · Logical method and law · Logics of inquiry
and exposition

Introduction

The title suggests a connection between academic integrity and teaching ethics across the curriculum. One may well wonder, however, what such a connection would come to beyond the fact that folks concerned with either use the term "integrity" and speak of values (e.g., fairness, honesty, trust, respect, responsibility,

---

D. E. Wueste (✉)
Clemson University, Clemson, USA
e-mail: ernest@clemson.edu

E. E. Englehardt and M. S. Pritchard (eds.), *Ethics Across the Curriculum—Pedagogical Perspectives*, https://doi.org/10.1007/978-3-319-78939-2_19

courage) (ICAI 2014). Here I will endeavor to show that the connection certainly extends beyond shared terminology, and this in a practical and purposeful way, i.e., in a way that is responsive to a challenge in practical and professional ethics.

Questions of practical and professional ethics (PPE) are prompted by challenges or issues that arise in context(s) of human interaction. The questions, like the challenges that prompt them, are not abstract or theoretical. They are concrete and practical. For example, it seems clear that PPE confronts a serious challenge in the arena of scientific research when one reads in the journal *Nature* that in "the early 2000s, only about 30 retraction notices appeared annually" but in 2011, though "the total number of papers published [had] risen by only 44%, the Web of Science [was] on track to index more than 400." It might have been comforting that this finding was followed by the observation that while "the number of retraction notices …shot up 10-fold…[it] is likely that only about half …[were] for researcher misconduct" (Van Noorden 2011). If so, however, that comfort would have been short-lived. In 2015, *The New York Times* ran a story on the "growing list" of retracted scientific studies. The prompt for the article was a retraction from the journal *Science* of a study on changing attitudes about gay marriage that received widespread attention. Noting that many retractions receive essentially no attention outside the scientific community, the article goes on to say that in "some instances, the studies that were clawed back made major waves in societal discussions of the issues they dealt with"; the *Times* article includes a list of such prominent retractions of studies that dealt with vaccines and autism; stem cell production; cloning and human stem cells; cancer in rats and herbicides; pesticides and estrogen; as well as "several dozen" psychological studies, "based on falsified data and faked experiments," that made claims that received much attention, for example, that eating meat makes people selfish (Roston 2015).

The existence of such an ethically suboptimal state of affairs here, in the arena of scientific research, is unsettling for many reasons, not least that it flies in the face of what we have been encouraged to believe about science and its practitioners, namely that they can and should be trusted. We are unhappy that such a state of affairs exists in business and politics, but (sadly) we aren't surprised. One explanation of this difference is suggested in Everett C. Hughes's observation that, so to say, the basic rules of play in these fields differ: With scientists and other professionals the rule is *credat emptor* (to trust) while in business (politics too?), it is *caveat emptor* (to beware) (Hughes 1965). In any case, the current state of affairs in each of these domains constitutes a challenge for practical and professional ethics.

In what follows my twofold purpose will be to (a) explain how a linking of academic integrity (AI) and ethics across the curriculum (EAC) is responsive to this challenge and (b) make the case for embracing the approach it involves. I will begin with a story.

A Short True Story

In December 2010, I was an invited presenter at a conference in Egypt. The conference, "Fostering Academic Integrity," sponsored by the American University in Cairo, was the "first conference of its kind in the Middle East and North Africa" (American University in Cairo 2010). The conference addressed many topics, including "Academic Integrity and its Impact on Business Ethics," for which I was the keynote presenter. With an eye to engaging not only the panelists charged with responding to my remarks but members of the audience as well, not long into my presentation I posed some questions. As prompts for engagement these questions were quite successful. Those present, in particular, members of the audience (academics and business leaders, among others), did not hesitate to speak up and to engage with each other. That, of course, is good. However, the discussion took a turn that was unwelcome. It began with the claim that, obviously, students need to be prepared to succeed in the real world. The prompt for the unwelcome turn was a "wake-up-and–smell-the coffee observation": the real world is rather more corrupt than it is ethical, which, in turn, led to the suggestion that instead of talking about ethics and integrity students should be taught strategies for real-world success. This would require familiarity with and mastery of techniques for success in the world beyond the classroom that college is supposed to prepare them for: "we should teach them about what is done and how to be successful in doing it." The point was far from subtle: success in the real world, like sausages and laws, involves a process that most would prefer not to see. Yet, desiring the end—whether success, sausage, or law—one must know and master the techniques for attaining it.

Things had gotten seriously off track. I interrupted, asking for their attention; and then, slowly, with conviction, I noted that the current state of affairs—the way things are at present—isn't particularly good; that, indeed, what we have is a far cry from how it ought to be. Our job, I said, isn't to help students acquire the wherewithal to succeed in the world as it is, troubled and corrupt as it may be; no, our job is to equip them to make things better. Ethically better. It's important that they, no less than I, saw the situation in business as less than ethically optimal. Where we differed is in our thinking about whether that might be changed by young people who, having been taught ethics in college, are equipped to do it. I affirmed this possibility, they seemed to deny it; it is in any case something they found doubtful.

Supposing that teaching ethics may be worthwhile in this way, we confront two large questions. The first is how this teaching should be done. While there is some controversy on this score, the good news is that so far as answers are concerned there are some strong competitors. The second (two-part) question presents a formidable challenge for whichever approach prevails in the competition. That question is (a) What would success look like? and (b) How would we know that it had been achieved—how would it (success) be measured?

The How Question: Some Competing Answers

In the EAC community answers to the "how question" are, I should say, of three types, two familiar and one rather new, namely, those that focus on theory, those that focus on character (*being* ethical), and those that, influenced by what has been called "behavioral ethics," focus on recognition and management of cognitive biases, or "debiasing." The latter two present themselves in contrast to and as alternatives to the first. Yet, with each, as noted above, the success question looms large. The assumption that the teaching of ethics is something to be done by philosophers, largely, if not exclusively, in philosophy courses, is strongest in the first and weakest in the last of these three types. The attractiveness of this assumption wanes with the commitment to the goal of changing behavior and/or a grounding in social science research. The ardor and depth of one's commitment to teaching ethics *across-the-curriculum* makes a difference as well.

In the Western tradition, and, in particular, in American colleges and universities, the default in ethics teaching is the teaching of ethical theories. This often includes the application of the theories in situations/cases in business, medicine, or engineering, for example. Either way, as Lisa Kretz writes, "[r]atiocination about ethics is the focus." Success consists in a student's "ability to articulate and, in some cases apply, ethical theories" (2015:151). Like advocates of debiasing, the other alternative to the tradition identified above, Kretz worries about a disconnect between "moral education and moral behavior." Or put another way, thinking, as most do, that "morality ultimately lies in action" (2015:152) and, accordingly, that "inspiring ethical behavior is.... obviously relevant to teaching ethics well," she thinks it very strange that "the current, dominant, pedagogy of Western ethics" focuses on moral reasoning rather than moral behavior; on being "able to articulate ethical positions and related arguments [instead of how to] *be* more ethical" (2015:152). Ethics teaching currently doesn't, though it should, focus on practical behavior; ethics teaching should "enhance the moral quality of lived practice," by, for example, nurturing an "engaged moral citizenry." Kretz would not have us jettison moral reasoning; rather, she advocates ethics teaching that combines it with a "more robust concern with education for moral action" (2015:166). That is essential if one embraces, as she does, the "goal of making the world a more ethical place" (2015:158). And one needn't have an especially penetrating view of the world as it is to see that Kretz is right in thinking that there is a profound need for such improvement.

One noteworthy feature of the focus Kretz recommends is that it points to a way of answering the second question above, about measuring success—how we would know that it had been achieved. She invites attention to a Taiwanese study by Shih-Jang Hsu that shows that and how providing an opportunity to effect change in the context of environmental education fostered "responsible environmental behavior…environmental responsibility, intention to act, perceived knowledge of environmental issues, and perceived knowledge of/skills in using environmental action strategies" (2015:163). Kretz believes that a comparable increase in responsible ethical behavior might be achieved, if students were "comparably empowered in ethics courses" (2015:163). While precisely what this idea of empowerment comes to is beyond the scope of this discussion, this much is clear. If we (a) accept the goal of fostering ethical action and (b) resolve to proceed on the basis of the best available evidence [which shows that (i) in moral judgment "There are two cognitive processes at work-reasoning and intuition-and the reasoning process has been overemphasized" and (ii) "moral action covaries with moral emotion more than with moral reasoning" (Haidt 2001:815; Kretz 2015:160)], then ethics teaching "must employ theories and practices that reflect how emotion functions in tandem with critical thinking" (Kretz 2015:159).

The assessment method in Hsu's study involved a survey instrument that was employed in a two-month follow-up as well as at the start and the end of the term. While Kretz quite rightly notes that self-reporting raises questions about validity, what she says next is, I believe, more important: "In circumstances where observable action must occur or fail to, tests of behavior are verifiable" (Kretz 2015:163). If such circumstances were to obtain within the academy, a more robust concern with education for moral action might have the advantage of occurring in a space where success could be reliably verified.

One important caveat: what Kretz recommends is not indoctrination, but rather "inspiration of moral behavior." Transparency with students about pedagogical goals is critical and, perhaps most important, "the moral behaviors to be encouraged are those that the students have identified as being desirable through their own moral contemplation" (2015:152). I have discussed this idea of "ownership" in the academy elsewhere (Wueste and Fishman 2010). I will return to the point about assessment.

Turning to the second alternative to the traditional approach, which I earlier called debiasing, it should be noted that it emerges from empirical work by, among others, social psychologists and behavioral economists, that purports to establish that traditional philosophical ethics has it the wrong way around: moral reasoning does not in fact precede or lead to judgment (these writers would speak of cause and effect here); rather it follows it, in the sense of "comes after," for the purpose of legitimating or justifying judgments that were made independent of the reasons now proffered by way of justification. There is a sense in which their characterization of moral judgment is rather like a description of a lawyer's process in developing a legal theory of a case. The lawyer knows the conclusion he wants; it will be a conclusion (judgment/holding) for the plaintiff, if plaintiff is his client, for example. The theory he builds and the arguments he constructs don't lead to the decision; they are constructed post hoc; their purpose is not discovery, it is justification (Dewey 1924:23). I will return to this; it's an important point.

Debiasing or behavioral ethics is motivated by an interest in finding a way effectively to address problems of moral wrongdoing (Prentice 2015). The work of behavioral economist Dan Ariely, for example, focuses attention on cheating behaviors which are found in academe, to be sure, but not only there, as well as dishonesty generally (Ariely 2017, 2012). According to Bazerman and Gino, behavioral ethics is the "study of systematic and predictable ways in which individuals make ethical decisions and judge the ethical decisions of others when these decisions are at odds with intuition and the benefits of the broader society." The distinctive feature of this approach is that it is descriptive rather than normative, and, according to its advocates, it is precisely because of this difference that "behavioral ethics is better suited than traditional approaches to addressing the increasing demand from society for a deeper understanding of what causes even good people to cross ethical boundaries." (Bazerman and Gino 2012).

That there is a problem here can scarcely be doubted. It's a general problem in the sense that it's not a problem solely for educators committed to teaching practical and professional ethics across the curriculum. The problem is not abstract or theoretical, it is concrete and practical; it is also pressing because its consequences are often profound and far-reaching. Examples are not far to seek and the interest in identifying the causes of the problem and developing an effective prophylactic is easy to understand.

Concrete Practical Problems, Behaviors and Biases: The Importance of How Things Are Framed

In April, 2017, American Airlines agreed to give raises to pilots and flight attendants. The reaction of analysts at some investment banks was bitter. For example, Kevin Crissy, an analyst at Citigroup, sent a note to the bank's clients voicing his frustration: "Labor is being paid first again. Shareholders get the leftovers." JP Morgan's Jamie Baker spoke to the matter as well: "We are troubled by AAL's wealth transfer of nearly $1billion to its labor groups." Sheelah Kolhatkar, in a piece for *The New Yorker*, writes that "similar sentiments are everywhere in the financial establishment. Both Costco and Whole Foods…have been criticized by Wall Street investors and analysts for years for, among other things, their habit of paying workers above the bare minimum" (2017). Kolhatkar records this observation by Peter Cappelli, professor and labor economist at the Wharton School: "The interesting thing is always to ask them, 'What's the value proposition for employees? Why should these people work only for the interest of the shareholders? How are you going to get people to work hard?…I don't think they have an answer." One explanation would be that such a question simply would not arise given the way the they have framed the matter (2017).

At the end of June 2014, the French bank BNP Paribas pleaded guilty and agreed to a landmark $8.9 billion settlement with the "U.S. Department of Justice, U.S. Attorney's Office for the Southern District of New York, the New York County District Attorney's Office, the Board of Governors of the U.S. Federal Reserve System (FED), the New York State Department of Financial Services (DFS), and the US Department of the Treasury's Office of Foreign Assets Control (OFAC)" (BNP Statement 2014). BNP had "breached U.S. sanctions against Sudan, Iran and other countries transferring billions of dollars there." *Forbes* characterized BNP's statement on the matter this way: "BNP, which pleaded guilty, runs through the detail from the settlement as if it'd just been found guilty of unpaid traffic tickets." Noting that such settlements seem to have become the norm since 2012, *Forbes* discusses JP Morgan Chase's $13 billion settlement (2013) "to resolve allegations that it sold bad mortgage securities," as well as HSBC's $1.9 billion settlement of money laundering charges, UBS's $1.5 billion settlement for wire fraud, both in 2012, and Credit Suisse's settlement of $2.6 billion along with a plea of guilty to charges of "conspiracy to aid and assist U.S. taxpayers in filing false returns," and ends the article with the observation that banks think of billion dollar settlements as "the cost of doing business" (Touryalai 2014). Here too, it seems that the way in which the situation is framed is critical.

Seven years ago, *Newsweek* magazine posed and addressed the question whether fines work to change corporate behavior. At that time there already was a lot of ink on paper as well as words in digital form addressing this and related questions such as who, exactly, is punished when fines are imposed and what sort of culpability is involved when the fines are paid but there is no admission of wrongdoing, and/or the temporal distance between the offense and the penalty is, say, 8 years (which was the case with Johnson and Johnson fined in 2013 for things done in 1999–2005) (Gillen and Hodge 2014)? It would be nice if these questions, and more important, the actions that prompt them, were less frequent and less concerning now than they were then. Sadly, as the discussion above shows, this is not the case.

The *Newsweek* piece offers a very succinct answer to the question whether fines change corporate behavior: "Not when the penalties add up to less than 1 percent of annual revenue. At scales like this, corporations can treat regulatory action as just another cost" (Newsweek 2010; Otten 2010). Drawing on several sources, notably the Department of Justice, Labor Department, Securities and Exchange Commission, and the Minerals Management Service—which was replaced, not long after the BP disaster in the Gulf of Mexico, by the Bureau of Ocean Energy Management, Regulation and Enforcement—*Newsweek* reports some remarkable numbers that certainly seem to support its negative answer to the question about the efficacy of fines in changing corporate behavior. For example, in 2007, the year in which, until the 2010 "spill" in the Gulf, BP was fined the most, "it settled charges of illegally manipulating energy markets, breaking environmental laws, and anticompetitive practices," and paid fines of $391 million. Its revenues that year (2007) were $284 *billion*. So, the fines it paid came to "just 0.14 percent" of its revenue. Goldman Sachs was accused of securities fraud in April 2009. In July 2010 the SEC announced Goldman's penalty, which, at that time, was "the largest in SEC history"; the article notes that this penalty—$550 million (Harris 2013)—amounts to "four days of revenue for the bank." Massey Energy, owner of the Upper Big Branch mine in Raleigh, West Virginia, where 29 mineworkers were killed on April 5, 2010, racked up "65,832 [safety or health] violations" in the space of ten years (2000–2010). Many of the violations carried no or a very low penalty and Massey contested the costlier fines. According to the *Newsweek* piece, if Massey were to pay all of the fines associated with cited violations in 2009, that is, $13 million, it would be paying fines amounting to "less than half of one percent of [its $2.7 billion] annual revenue" in 2009.

Once again, *Newsweek* answers the question whether fines change corporate behavior in these words: "Not when the penalties add up to less than 1 percent of annual revenue. At scales like this, corporations can treat regulatory action as just another cost." It's the last point—an echo of points made earlier—that interests me, namely, that businesses, at least big ones, treat fines for regulatory violations as "just another cost." Of course, unless we are bewitched by legal fictions, we know that businesses can't *do* anything. People within a business organization, on the other hand, can do any number of things, which could include treating fines for violations of, say, safety and health regulations, as nothing more than a cost of doing business. In this respect, what's going on is rather like what's going on when a driver decides to take a chance and treat the parking ticket and fine he is likely to get, if those responsible for enforcing parking regulations get to his car before he can move it, as the cost of parking in a space close to where he wants or needs to be. Not every driver does this, of course. This suggests that those who don't respond to parking regulations merely as indicators useful for predicting the likelihood of sanctions (i.e., of costs they may have to deal with), understand the situation differently than those who do see the regulations in this way. Put another way, they frame the situation differently, or, perhaps it would be better to say that the situation is framed differently for them. I will return to this. The same might be said of those who don't and those who do text while driving, even though it is against the law—as is the case in South Carolina, for example. You will have noticed that the stakes are higher in the latter example. While it's unlikely that anyone will be harmed as a result of a parking violation, the situation is altogether different with texting while driving.

Texting While Driving Causes:

1. 1,600,000 accidents per year—National Safety Council
2. 330,000 injuries per year—Harvard Center for Risk Analysis Study
3. 11 teen deaths EVERY DAY—Ins. Institute for Hwy Safety Fatality Facts
4. Nearly 25% of ALL car accidents
   http://www.textinganddrivingsafety.com/texting-and-driving-stats
   https://www.fcc.gov/consumers/guides/dangers-texting-while-driving

Bans and fines: The details of laws banning texting while driving and fines for violations vary widely. California's fine is the lowest: $20.00 http://www.motherjones.com/media/2013/10/numbers-texting-and-driving/

Yet—and we're getting closer to what I think needs to be investigated—here too, there are (at least) two ways in which the situation is framed and they differ from one another in the same respect, namely, whether the regulations are taken to be nothing more than indicators useful for predicting the likelihood of monetary sanctions, or not.

Returning to the weighty case of corporate behavior, one wonders how people at BP, Massey, or Goldman Sachs, for example, come to regard a fine for the violation of regulations as a cost of doing business. Put another way, how does it come to pass that for these people the situation is framed in such a way that safety and health regulations, for example, appear on their radar as nothing more than indicators useful for predicting the likelihood of monetary sanctions? One reason to spend some time puzzling over this is that, with Massey and BP, seeing safety regulations in this way appears to be a key element in the story of the deaths of 29 mine workers and 11 workers on the Deepwater Horizon, as well as environmental degradation in the Gulf of Mexico. Another reason is that while one would have thought consideration of evidence from similar incidents in the past would have made it clear that much more is at stake with safety and health regulations than can be captured in an understanding of them as indicators of costs that may be incurred, if regulators take action after citing the company for a violation, that didn't happen. That is to say, the decision makers continued to think as they had in the past; the evidence did not sway them. One wonders why.

There are two questions here, which together constitute a problem with respect to the teaching of ethics that deserve attention, particularly if, like those who embrace debiasing, one is interested in developing an effective prophylactic for moral wrongdoing. One question is about the way in which cases such as these are framed. The other, put in general terms, is about the role and effectiveness of evidence and reasoned argument in decision-making. Let's begin with the second question. Since the most recent examples come from business, I think it would be well to stay with business for a bit, though of course, the problem here is quite general in scope.

The Role and Effectiveness of Evidence and Reasoned Argument

Jeffrey Pfeffer and Robert Sutton provide some help with the second question in their book, *Hard Facts, Dangerous Half-Truths, & Total Nonsense*. All too often, they suggest, "belief [trumps] evidence" (2009:10). They point to "the use and defense of stock options as a compensation strategy" to illustrate the point. While many executives believe that stock options and other incentive pay arrangements lead to enhanced performance, they write, "[t]here is little evidence that equity incentives of any kind" do this. They go on:

> One review of more than 220 studies concluded that equity ownership has no consistent effects on financial performance. Another massive study and review of research on executive compensation published by the National Bureau of Economic Research reported that most schemes designed to align managerial and shareholder interests failed to do so; instead, executive compensation practices just operated as devices to enrich senior managers, who usually received most of the stock options (2009:11).

Nevertheless, executives continue to insist that such pay strategies are not merely helpful, but critical elements of a business plan and they work hard to block governmental action that would lead to their having to treat stock options as an expense in the composition of income statements. Pfeffer and Sutton put the key point this way: "[b]eliefs rooted in ideology or cultural values are quite 'sticky'—they resist disconfirming evidence and persist in affecting judgments and choice, regardless of whether or not they are true" (2009:12). John Dewey, writing about logical method and the law, explains a comparable point made by Oliver Wendell Homes Jr. about judicial reasoning, which too often, and regrettably, applies concepts "irrespective of the consequences of their application to concrete matters-of-fact." Continuing, Dewey explains that "concepts once developed have a kind of intrinsic inertia on their own account; once developed the law of habit applies to them. It is practically economical to use a concept ready at hand rather than to take time and trouble and effort to change it or to devise a new one" (1924:20).

One finds a similar idea in the work of moral psychologists such as Jonathan Haidt. Championing a "social intuitionist model" of moral judgment as an alternative to "rationalist models," which are championed by some psychologists, but mostly by philosophers, including some very heavy hitters—Haidt mentions Kant, Hare, and Rawls, for example—Haidt's model casts light on the stickiness Pfeffer and Sutton point to in trying to explain the inability of reason, evidence and argument to carry the day. The expectation that reason, evidence and argument would prevail, familiar from rationalist models, is out of place with flesh-and-blood judgments made in real time, in the real world, with real consequences for real people. In a tantalizing line at the end of a provocative paper, "The Emotional Dog and Its Rational Tail: A Social Intuitionist Approach to Moral Judgment," Haidt reveals his inclination to side with the philosopher that Kant is at pains to answer. Haidt writes, "[t]he time may be right to take another look at Hume's perverse thesis: that moral emotions and intuitions drive moral reasoning, just as surely as a dog wags its tail" (2001:830). As Haidt has it, the rationalist model has it the wrong way around.

The order of events isn't in the direction of reasoning causing moral judgments—the causal language is Haidt's. Haidt argues, on the basis of multiple studies, that moral reasoning is largely, if not entirely, an ex post affair. According to his social intuitionist model, moral judgments arise from (or are caused by) intuitions; one reasons about judgments after they have been made, perhaps because a psychologist doing a study, or a philosopher in a classroom, asks for reasons, likely supposing, wrongly, Haidt would say, that what is proffered will answer the question of how the participant in the study or student arrived at the judgment (what caused it).

According to Haidt, "[m]oral intuition is…the psychological process that the Scottish philosophers talked about, a process akin to aesthetic judgment: One sees or hears about a social event and one instantly feels approval or disapproval" (2001:818, 2006:21). Fleshing out the similarity to aesthetic judgment, Haidt says, when you see a painting, for example, your liking it (or not) is instant and automatic; reasoning has nothing to do with it. What happens if someone asks why you like it? According to Haidt,

> You search for a plausible reason…and you latch on to the first reason that makes sense (maybe something vague about color, or light, or the reflection of the painter in the clown's shiny nose). Moral arguments are much the same: Two people feel strongly about an issue, their feelings come first, and their reasons are invented on the fly, to throw at each other (2006:21).

Perhaps moral disagreements do sometimes involve reasons invented on the fly and thrown back and forth. One might hope that, if so, the situation is not altogether different than that with reasoning that involves a formal fallacy such as denying the antecedent or affirming the consequent. Just as the latter can be diagnosed and effectively addressed in the teaching of logic, the former can be effectively addressed in the teaching of ethics. While the quotation above might lead one to think Haidt would have no patience with such a suggestion, I turn now to showing that, rightly understood, and with a bit of help from John Dewey, what Haidt calls social intuitionism is quite compatible with this idea and suggests an answer to a question posed earlier, namely, how we ought to teach ethics across the curriculum, if our aim is to equip our students to make things ethically better than they are at present rather than helping them acquire the wherewithal to succeed in the world as it is, however troubled and corrupt it is.

Framing, Social Intuitionism, and the Logics of Inquiry and Exposition

As noted earlier, according to Haidt's social intuitionism moral judgments are the product of intuition. The philosopher's account to the contrary notwithstanding, moral reasoning is an ex post affair. What Haidt calls intuition is a variety of cognition that contrasts with reasoning in the following way: "intuition occurs quickly, effortlessly, and automatically, such that the outcome but not the process is accessible to consciousness, whereas reasoning occurs slowly, requires some effort, and involves at least some steps that are accessible to consciousness" (2001:818). As Haidt has it, while it may be the case, at least sometimes, that reasoning generates moral judgments, intuition is the default.

This contrast of intuition and reason is an element of mainstream psychology (since the 1990s). In another place, self-identifying with the mainstream view, Haidt says that "there are really two processing systems at work in the mind at all times: controlled processes and automatic processes" (2006:13). The former are rational, the latter emotional. He developed a metaphor, which has been widely and happily embraced (Heath and Heath 2010), that captures both this mainstream idea and what Hume himself recognized as a "somewhat extraordinary" view of the relationship between reason and emotion. Hume's view, expressed in his *Treatise of Human Nature*, is well known.

> We speak not strictly and philosophically when we talk of the combat of passion and of reason. Reason is, and ought only to be the slave of the passions, and can never pretend to any other office than to serve and obey them (Hume 1739:Bk II Sec. 3).

In Haidt's metaphor of the elephant and rider, the automatic (emotional) processes are represented by the elephant and the controlled (rational) processes are represented by the rider:

> the rider is an advisor or servant; not a king, president, or charioteer with a firm grip on the reins. The rider is...conscious, controlled thought. The elephant, in contrast, is everything else. The elephant includes the gut feelings, visceral reactions, emotions, and intuitions that comprise much of the automatic system (2006:17).

With Haidt's metaphor it's not difficult to see, with the mind's eye, as it were, the lopsided power relationship Hume spoke about. Dan Heath and Chip Heath provide a straightforward description.

> Perched atop the Elephant, the Rider holds the reins and seems to be the leader. But the Rider's control is precarious because the Rider is so small relative to the Elephant. Anytime the six-ton Elephant and the Rider disagree about which direction to go, the Rider is going to lose. He's completely overmatched (2010:6–7).

For a proper understanding of moral judgment attention should be focused on the automatic processes of the human mind, and in particular, what Haidt calls intuitions (Haidt and Joseph 2004:64–65). Intuitions, Haidt tells us, are rooted in our biology and evolution, which includes forces of nurture as well as nature: "moral intuitions derive from innate psychological mechanisms that coevolved with cultural institutions and practices" (Graham et al. 2009:1030). Accordingly,—and this is important—although they are innate, they are modifiable. Following cognitive scientist Gary Marcus (2002, 2004), Haidt holds that "innate 'does not mean unmalleable; it means organized in advance of experience'" (1031). These mechanisms, which he also calls foundations,[1] are something like Velcro for the teachings of parents and others within a social group about virtues, vices, and the interlocking expectations of social practices. They are, Haidt suggests, again following Marcus, key elements of a first draft of the moral mind that has been written by evolutionary forces and will be edited by experience (1031).

It is very important that this editing is largely, if not entirely "an interpersonal process"; *that* is what makes Haidt's model of moral judgment a *social* intuitionist model. I will return to this in a moment. But first, I need to make good on a promise made earlier to return to the matter of framing and the two ways I put the point about framing differences in talking about parking regulations. Earlier I said,—and hopefully it's not so long ago that you've forgotten—some folks do and some folks do not respond to parking regulations merely as indicators useful for predicting the likelihood of sanctions (i.e., of costs they may have to deal with). One way to talk

---

[1]According to Haidt, Graham, Nosek, and Joseph's moral foundations theory there are five foundations: Harm/care; Fairness/reciprocity (the first two are "individualizing foundations); Ingroup/loyalty; authority/respect; Purity/sanctity (the last three are "binding foundations). The five foundations have been called out in efforts to explain how liberals and conservatives differ. The story that's told, which is based on four psychological studies, is that liberals more vigorously embrace and employ the first two, harm/care and fairness/reciprocity, than conservatives who embrace and employ all five in a rather even-handed way (2009). Haidt's discussion of the moral foundations in *The Righteous Mind* (2012), which is based on further research, is richer in that, for example, it includes a sixth foundation, namely, liberty/oppression, which is grouped with what were earlier called individualizing foundations, and includes an interesting change: fairness is contrasted with cheating rather than reciprocity.

about this would be to say that they frame the situation differently; alternatively, one might say that the situation is framed differently for them. The second description differs from the first roughly in the way intuition differs from reasoning for Haidt and other social intuitionists. You will recall that for social intuitionists, "intuition occurs quickly, effortlessly, and automatically, such that the outcome but not the process is accessible to consciousness, whereas reasoning occurs slowly, requires some effort, and involves at least some steps that are accessible to consciousness." As I read Haidt, the elements of the first draft of the moral mind—the foundations—are frame setters for the processes of intuition that generate moral judgments. "They are," he says, "the psychological systems that [give rise to] feelings and intuitions that make local stories, practices, and moral arguments more or less appealing during the editing process" (Graham et al. 2009:1031).

With this we have something more than Pfeffer and Sutton's notion of stickiness in reckoning with the neglect of or rejection of evidence and argument that, one would have thought, would lead managers to different places, where, for example, stock options are no longer thought to be essential elements of a compensation scheme for executives, or managers would see that much more is at stake with safety and health regulations than can be captured in an understanding of them as indicators of costs that may be incurred, if regulators take action after citing the company for a violation. I submit that social intuitionism offers something of value if we are searching for a way to deal with morally impactful frames not only prophylactically but constructively, which is what we'd be up to if our aim were to create and sustain a culture of integrity on a college campus, or in a business or non-profit organization, for example.

Social intuitionists hold that moral reasoning is largely, if not entirely, an ex post affair. Reasoning doesn't generate moral judgments; rather, moral judgments arise from (or are caused by) intuitions; in general, one reasons about judgments after they have been made. But intuitions are "developed and shaped" through imitation and participation in social practices (Haidt 2001:828). The development of intuitions, like moral judgment itself, is an interpersonal process (2001:814). To be sure, there is no causal link between my reasoning ex post and my moral judgment, *however, my ex post facto reasoning can be causally linked to the intuitions of others and thus to their judgments.* One suggestion that we can ferret out here is that we should focus attention on, so to speak, tuning up intuitions; in the context of college teaching, one ought to focus on tuning up intuitions rather than teaching moral reasoning. After all, as Haidt notes, "attempts to directly teach thinking and reasoning in a classroom setting generally show little transfer to activities outside of the classroom" (2001:829). Pointing to Kohlberg's "Just Community" schools, Haidt suggests that a community in which "moral talk [is] ubiquitous and…adults model good moral thinking" can, among other things, help develop another tool for honing intuitions, namely "private reflection" (2001:829) which would instantiate on the individual level what can occur on an interpersonal level when one person's ex post reasoning affects the intuitions of another person. I should say that developing such a community, where moral talk is ubiquitous and good moral thinking is modeled, is an attractive and plausible strategy for dealing with morally impactful frames, not only prophylactically but constructively, that plainly fits squarely and comfortably within a responsible effort to promote academic integrity on a college or university campus. When the focus is on *promoting* academic *integrity*, rather than coupling increased policing with swifter and heavier sanctions, the door is opened to fruitful collaborative work in teaching ethics across the curriculum. This is true for several reasons (Wueste 2008, 2014) most important, perhaps, is this: academic integrity, in the broad sense in which it is about much more than student cheating, comprises every discipline, activity, and person engaged in the academic enterprise.

Haidt's Elephant and Rider

The Bugbear of Legal Reasoning: Reason and Fiat

There is more here, however, and as we turn our attention to it the way(s) in which the promotion of academic integrity can be linked to teaching ethics across the curriculum will emerge. But we need to step back for a moment before we forge ahead.

One problem with Haidt's image of the elephant and the rider is that it seems to presuppose that etiologically an ethical judgment is the product either of reasoning or emotion/intuition. In this respect, it is reminiscent of the familiar but misleading "either-or" in discussions of judicial reasoning, where, it is claimed, a judge's decision is either logically compelled in correct application of the law (reason) or it is the product of unfettered judicial discretion (fiat). This is interesting for a number of reasons, not least that when softening somewhat his description of the relationship between the elephant and the rider, Haidt suggests that, in the context of a community like the one described above, the rider "becomes a lawyer":

> Gut feelings, intuitions, and snap judgments happen constantly and automatically…but only the rider can string sentences together and create arguments to give to other people. In moral arguments, the rider goes beyond being just an advisor to the elephant; he becomes a lawyer, fighting in the court of public opinion… (2006:22)

That's how the rider establishes a causal link between his ex post facto reasoning and the intuitions of others (and thereby to their judgments); in other words, it is by means of moral argument that the rider participates in the development and shaping of intuitions, which, the reader may recall, is accomplished by imitation and participation in social practices.

The prompt for moral argument, indeed the prompt for the development of logic and argument in general, is social. As John Dewey says,

> It is quite conceivable that if no one had ever had to account to others for his decisions, logical operations would never have developed, but men would use exclusively methods of inarticulate intuition and impression, feeling; so that only after considerable experience in accounting for their decisions to others who demanded a reason or exculpation, and were not satisfied till they go it, did men begin to give an account to themselves of the process of reaching a conclusion in a justified way (1924:24).

Dewey, who makes this observation in the context of a discussion of logical method and law, suggests that, even if his observation about the emergence of logic is not well taken, it's clear, in the legal context, that a reasoned statement that articulates a basis and points to logical connections is the only alternative to a reliance on "arbitrary dicta," accepted solely on the basis of the "authority or prestige of the judge" (1924:24). "It is at this point," Dewey continues, "that the chief stimulus and temptation to mechanical logic and abstract use of formal concepts come in." Why? Because, although it is impossible to deny there is a personal element in such judgments, it's necessary that, so far as possible, when they emerge, they do so in an impersonal, objective, rational form. We are, as he puts it, strongly tempted to "surrender the vital logic which has actually yielded the conclusion and to substitute for it forms of speech which are rigorous in appearance and which give an illusion of certitude" (1924:24). I should say that Dewey's point, though made about logic and law, is no less apt in discussion of logic and morality. In any case, given the importance of moral argument in Haidt's *social* intuitionist model and his invitation to understand the form such argument takes as lawyerly, or akin to legal reasoning, it will be well, as we return to the problem with Haidt's image of the elephant and rider identified earlier, to continue the discussion in light of Dewey's observations. The problem is that this image seems to presuppose a false dichotomy (reason or emotion) very much like the misleading "either-or" in discussions of judicial reasoning, where, it is claimed, a judge's decision is either (a) logically compelled in correct application of the law (reason) or (b) the product of unfettered judicial discretion (fiat).

Inveighing against what Roscoe Pound would later call mechanical jurisprudence (which tracks with (a) in the disjunction above) Oliver Wendell Holmes Jr. wrote,

> The language of judicial decision is mainly the language of logic. And the logical method and form flatter that longing for certainty and for repose which is in every human mind. But certainty generally is illusion, and repose is not the destiny of man. Behind the logical form lies a judgment as to the relative worth and importance of competing legislative grounds, often an inarticulate and unconscious judgment, it is true, and yet the very root and nerve of the whole proceeding. You can give any conclusion a logical form (Holmes 1920:181).

A more famous quotation from Holmes helps to clarify what he has in mind.

> The actual life of the law has not been logic: it has been experience. The felt necessities of the times, the prevalent moral and political theories, intuitions of public policy, avowed or unconscious, even the prejudices which judges share with their fellow-men, have had a good deal more to do than the syllogism in determining the rules by which men should be governed (Holmes 1881:1).

Here Holmes is equating logic with the syllogism, and, thus, it is not surprising that he finds that experience (or good sense) stands in opposition to logic. Dewey is helpful in understanding Holmes' thinking. He writes, "the philosophy embodied in the formal theory of the syllogism asserted that thought or reason has fixed forms of its own, anterior to and independent of concrete subject-matters, and to which the latter have to be adapted whether or no." This is a problem, to be sure. But the larger problem is "that while the syllogism sets forth the *results* of thinking, it has nothing to do with the *operation* of thinking" (1924:22) Dewey uses the case of Socrates' trial to illustrate his point. The issue, he says, was not whether Socrates was mortal, but rather whether his mortality should be concretely established on a specific date by a specific means. But—here's the point—*that* "does not and cannot follow from a general principle," the major premise of a syllogism. As Justice Holmes said in his rightly famous dissent in Lochner v. New York, "General propositions do not decide concrete cases" (Lochner 1905). Dewey explicates this famous aphorism in these words: "No concrete proposition, that is to say one with material dated in time and placed in space, follows from any general statements or from any connection between them" (1924:22).

A couple of ideas quoted earlier, one from Dewey and one from Holmes, integrated below in one statement (Dewey in italics), are indicative of where this takes us. As Holmes rightly reminds us, any conclusion can be given a logical form. However,

> Behind the logical form lies *the vital logic which has actually yielded* a judgment as to the relative worth and importance of competing legislative grounds, often an inarticulate and unconscious judgment, it is true, and yet the very root and nerve of the whole proceeding.

The vital logic Dewey refers to is not the logic of the syllogism, but "another kind of logic" (1924:21) that addresses a different challenge. As Dewey has it, the logic of the syllogism is a logic of "exposition" that is deployed to meet the challenge of showing that an already determined judgment is justified. The challenge addressed by the "vital logic" that stands behind the syllogistic form of the logic of exposition "is not to draw a conclusion from given premises; that can best be done by a piece of inanimate machinery by fingering a keyboard. The problem is to *find* statements, of general principle and of particular fact, which are worthy to serve as premises" (1924:23). Accordingly, Dewey calls this "a logic of search and inquiry" (1924:24). It is responsive, he says, to the need of another kind of logic that "reduce[s] the influence of habit, and… facilitate[s] the use of good sense regarding matters of social consequence" (1924:21). As Dewey has it, and I think we should agree, the influence of habit is unwelcome insofar as it is associated with "feelings of ease and stability-feelings, which have little to do with the actual facts of the case" (1924:21).

According to Gerald Postema, Dewey's two logics—the logic of inquiry and the logic of exposition—"are complementary, especially in law." But he argues that there is more to this complementarity than Dewey explicitly acknowledges. He begins with an account of Dewey's thinking.

> Dewey described a process of tentatively testing conclusions, searching for more general principles to warrant them, and reconsidering both of them if the most appealing general principle does not yield the intuitively most attractive conclusion. This process presumably is carried on until some more or less rational equilibrium is established. (Postema 2011:97)

Taking it that this process (searching for equilibrium) is not public, Postema suggests that the search for rational equilibrium can be extended to include the process of justification, which is public. The key point is that just like the search process, the justificatory process will include adjustments in the pursuit of equilibrium. After all, it can and does happen that in fulfilling the expectation that judges will provide a justification for a decision in terms of available law, either of these two things may happen. A judge may discover that the conclusion that prompted the search process cannot be justified by the law and that she is thus "forced to reconsider the result of [her] "inquiry" and to search for some other conclusion." It might also happen that "the existing law is inadequate and so the fact that a justifiable decision formulated in its terms is not available counts against the law and not the decision" (2011:97). According to Dewey, this is what Holmes was getting at when he wrote that

> the whole outline of the law is the resultant of a conflict at every point between logic and good sense —the one striving to work fiction out to consistent results, the other restraining and at last overcoming that effort when the results become too manifestly unjust (Holmes 1920:50; Dewey 1924:50).

So, again, Postema argues that the search for what he calls rational equilibrium can be extended to include the process of justification. And that, in turn, leads to a blurring of the line between the two logics. Instead of two processes with different purposes and audiences, the two logics constitute two parts of one "more complex reasoning process." This process, he says, "is neither merely deductive argument nor merely a matter of being caused to come to one conclusion rather than another, but something in that possibly large territory between them" (2011:97–98).

If we were drawing a map, this territory would be located between reason and fiat. In any case, its existence is precisely what we were hoping to find, having claimed that Haidt's image of the elephant and rider presents the false dilemma—reason or fiat—in the moral context that is so familiar in discussion of legal reasoning.

Teaching Ethics Across the Curriculum in the Territory Between Reason and Fiat

The work of teaching ethics across the curriculum should focus on this territory between reason and fiat. This work needs to be done in a social context that, in key respects, resembles settings that students will work in after graduation while at the same time being "where they are" so that they have lived experience that resonates in the present and is translatable. A college or university campus environment where efforts are quite consciously directed at promoting academic integrity (hereafter PAI setting) satisfies both requirements.

When we discuss linking the teaching of ethics across the curriculum and efforts to promote academic integrity, what we (or at any rate what I) have in mind is practical and professional ethics, rather than personal morality, which involves norms that, for example, simply and straightforwardly proscribe and unambiguously disapprove a man's bragging about being able to get away with grabbing a woman by her genitals (Fahrenthold 2016) let alone doing such a thing, regardless of his position or role as, for example, a public figure or elected official. As noted at the outset, the challenges and questions that prompt this discussion are challenges and questions in practical and professional ethics, which, I think, is best understood as *a set of normative constraints that are present* in something larger than oneself that one has, by choice, become part of; *an enterprise, undertaking, or organization, for example,* that has or should have its own integrity, which should be understood in the same way as individual integrity should be, as an achievement without closure, as an ongoing project, a task, that entails sustained effort, if what has been achieved is to be maintained.

The Milieu of Academic Integrity

One way of understanding academic integrity is to think in terms of the interlocking responsibilities of those engaged in teaching, learning, and research, collectively, the academic enterprise. Any set of interlocking responsibilities, say those of professionals and their patients or clients, will occasion disputes about whether or how they have been carried out, and, in particular, malfeasance. We considered some examples in the arena of scientific research at the outset in a discussion of challenges in practical and professional ethics. Such disputes involve, among other things, motivations of personal gain, pressure from peers or powerful superiors, expectations of loyalty, or a belief that a governing norm is inapposite in the current situation; they also involve accusations, investigations, hearings that resemble judicial proceedings in some ways (for example, in raising questions of interpretation in the application of rules) and controversies about the imposition of sanctions (whether they may or must be imposed; whether they are too harsh, too lenient, fair/just, and effective). The same can be said of the situation in which a student is accused of plagiarism, falsifying or fabricating data, or cheating on an examination, for example. Here, then, we see two key respects in which the PAI setting resembles settings that students will enter after graduation. Also, and plainly, we see how this resemblance implicates argument and reasoning of the sort we have been discussing, i.e., lawyerly reasoning, in which, following Dewey, the task "is not to draw a conclusion from given premises, [but rather] to *find* statements, of general principle and of particular fact, which are worthy to serve as premises" (1924:23). Premises identified in the logical process of search and discovery are deployed in the complementary logic of exposition/justification, which, in the social intuitionist's understanding of moral reasoning, is an ex post affair, the purpose of which is to justify a conclusion which preceded it, a conclusion/judgment, that in Haidt's terms, arose from (was caused by) intuition. Accepting the assistance an embrace of Dewey's two logics provides, advocates of teaching ethics across the curriculum needn't worry that Haidt's insistence that moral judgments are not caused by reasoning impugns the project. From the vantage point Dewey provides, accepting this claim costs us nothing; in fact, it both amplifies a critical fact, namely, that so far as a moral judgment is concerned, what caused it is much less important than the strength of its justification, and focuses attention in ethics teaching on the challenge of achieving what Postema calls rational equilibrium.

Questions about and issues of academic integrity are fraught, though as with the law, there are many who are quite sure that any question that arises has a straightforward answer readily grasped by any right-thinking person, i.e., *the* right answer. Holmes rejects this mindset in speaking of judicial dissent: seen in this way it must be, though it surely is not the case that, dissenters "were not doing their sums right, and if they would take more trouble, agreement inevitably would come" (Holmes 1897). In the PAI setting, no less than in the legal setting, the conclusions people start with likely will differ (and this for a variety of reasons, in both cases). But then, as with lawyers for plaintiff and defendant, attention is properly focused on the arguments that emerge from the logical process of search and discovery for statements of general principle and particular fact to justify the conclusions with which they started. The question to be answered in the legal context as well as the PAI setting, is about the strength of competing arguments. Answering that question may be difficult, but difficult is importantly different from impossible.[2] And since this sort of question is in fact the sort of question one confronts in dealing with genuine ethical dilemmas (more on this shortly), we have another significant respect in which the PAI setting resembles the settings students will enter after graduation.

The challenges of practical and professional ethics are often genuinely dilemmatic. That is to say, they involve situations in which the choice one faces is not one between acting rightly or acting wrongly, but rather, a choice between options each of which has a claim to be being right. There are two noteworthy things here. First, a genuinely dilemmatic choice maps onto the lawyerly conundrum discussed above, where the critical question is which of the competing arguments is stronger. Second, that question will direct one's attention to the values that underpin and argue for the rightness of the choices, as the dilemma likely arises because one is committed to values that point in different directions as one stands at the proverbial fork in the road (one may be committed to the value of loyalty *and* the value of truth, for example, in a situation where they argue for doing different things). The values that underlie a genuine dilemma may be grounded in one's role as a professional, for example, or one's engagement in an enterprise, say, the academic enterprise, where, according to the International Center for Academic Integrity, the fundamental values are honesty, trust, fairness, respect, responsibility, and courage (ICAI 2014). As one thinks on this (and, of course, one would hope that students would find themselves doing that) it is likely to emerge that in addition to values such as these, which are bound up with an enterprise, value commitments also come with one's occupying a role, say that of parent, spouse, or friend. Of course, one may occupy several roles, the three just mentioned, for example; in fact, it seems

---

[2]Although it is conceivable, it's not likely that the arguments are equally strong. Here I follow Ronald Dworkin, who makes this point about competing legal arguments (Dworkin 1982, 1986; see Wueste 1999).

that one's occupying several roles is inevitable. And thus, here, with some help from teachers committed to teaching ethics across the curriculum, students can see three things: (a) one's role responsibilities can generate genuine ethical dilemmas, (b) because being part of something larger than oneself (e.g., the academic enterprise) involves value commitments, it can also be a source of genuine ethical dilemmas, by, for example, involving value commitments that point in a different direction than those associated with one's being a friend, and (c) the stakes in such situations can be very high because, among other things, they may involve individual as well as professional or organizational integrity. This is most salutary because it makes plain something that is often not fully appreciated, namely, that ethical challenges are frequently quite complicated and what should be done in such cases is neither something one learns and masters at mother's knee nor decidable through the application of an algorithm (which, as we saw earlier, is the idea that, following Pound, Dewey refers to as mechanical jurisprudence).

Conclusion

The purpose of this discussion was to explicate a connection between academic integrity and teaching ethics across the curriculum that extends beyond shared terminology in a way that is responsive to a challenge in practical and professional ethics. I told a story that revealed a disagreement about whether teaching ethics can prepare young people to make things better—ethically better—and thus be preferable to teaching them how to succeed in a troubled and corrupt world, i.e., the world as it is. Believing that teaching ethics may be worthwhile in this way, I undertook to answer two large questions entailed by this belief; in both cases, the answer involves a connection between academic integrity and teaching ethics across the curriculum. The first question is about how this teaching should be done. The second (two-part) question presents a formidable challenge for the approach I recommend as well as those currently being employed. That question is (a) What would success look like? and (b) How would we know that it had been achieved—how would it (success) be measured?

Have we arrived at the announced destination? A summary of our travels, and responses to two questions posed at the start will, I hope, provide what is needed to answer that question. To wit:

A brief discussion of competing approaches to teaching ethics, in particular, behavioral ethics, suggested that attention should be focused on finding a way to effectively address problems of moral wrongdoing such as cheating, which Dan Ariely has shown to be a problem not only in academe but generally (Mazar et al. 2008). And that, in turn, led to a discussion of some examples and a question about causes and effective prophylactics. It emerged that framing was key and that Jonathan Haidt's social intuitionism was especially apt and promising, though his disparaging the philosopher's focus on moral reasoning was somewhat disconcerting. The first takeaway from this discussion of Haidt's work was that attention should be directed to honing the intuitions social intuitionism identifies as the causes of moral judgments. Embracing Haidt's suggestion that this is best done in a community where moral talk is ubiquitous and good moral thinking is modeled, it appeared that (a) this is an attractive and plausible strategy for dealing with morally impactful frames, not only prophylactically but constructively, and (b) this strategy is an excellent fit with a responsible effort to promote academic integrity on a college or university campus, and especially so, if it were to be combined with the teaching of ethics across the curriculum.

The second takeaway from the discussion of Haidt's work emerged from consideration of a false dilemma suggested by his well-known image of the elephant and rider. Since the problem here is quite like that posed by the familiar but misleading "either-or" in discussions of judicial reasoning, and because Haidt tells us that the rider develops and shapes others' intuitions (in social practices) by "becom[ing] a lawyer," we drew on work in legal philosophy and embraced (a) John Dewey's distinction between the logic of search and inquiry and the logic of exposition/justification, and (b) Gerald Postema's extension of the process of rational equilibrium, which, he argues, is implicit in the complementarity of Dewey's two logics. These moves led to the suggestion that in teaching ethics across the curriculum we should focus on the intellectual territory between reason ("merely deductive argument") and fiat ("merely…being caused to come to one conclusion rather than another"); the task highlighted in the logic of search and inquiry ("find[ing] statements of general principle and of particular fact…worthy to serve as premises"); and the aim of achieving rational equilibrium. This would be done in the social setting (PAI setting) of the interlocking responsibilities of the people engaged in teaching, learning, and research, collectively, the academic enterprise.

Teaching ethics across the curriculum in the PAI setting has several things to recommend it. Because it maps very well onto Dewey's two logics, we have no quarrel with Haidt's research finding that moral judgments are not caused by reasoning. We're also in a position to defuse the unwelcome implication of his elephant and rider image that etiologically an ethical judgment is the product either of reasoning or emotion/intuition. Why? Because attention is focused, as it should be, on the challenge of achieving what Postema calls rational equilibrium and the strength of the justificatory argument(s) it produces. In addition, students will have the opportunity, in the context of lived experience, to learn about the sources and character of genuine ethical dilemmas and more fully understand what's at stake (on the individual and/or organizational level) when one confronts an ethical dilemma. Further, insofar as the PAI setting (a) includes key features of situations students will enter after graduation and (b) is a model of how things ought to be in an organizational setting, students will be prepared to deal with the ethical dilemmas they will confront and make the world a better place. Finally, this approach will make teaching ethics across the curriculum attractive to a larger and more diverse group of faculty because (i) academic integrity is a concern shared by all and (ii) we're not looking to them to teach ethical theory (both because it's not necessary and because ethical theory should be taught by faculty with the requisite expertise).

What of the two questions about success—what would it look like and how would we know that it had been achieved?

If such an effort were successful, we would see a more robust commitment to academic integrity and, consequently, fewer violations of the ethical norms governing academic work. We should also find that students have what they need to effectively work through the ethical dilemmas of practical and professional ethics.

Measuring success in the teaching of ethics has been vexatious. It would be presumptuous in the extreme to claim to have a solution. However, it's not unreasonable to claim that some progress could be made, if we were to move forward in the manner proposed here. We might, for example, as suggested by Lisa Kretz, develop an assessment protocol based on the model provided by Shih-Jang Hsu's study. Hsu's assessment method made use of a survey administered at the beginning and end of the academic term and a two-month follow-up. Perhaps a survey of students when they matriculate and as they graduate could work in roughly the same way as Hsu's survey did. With respect to the incidence of violations of academic integrity norms, in short, the incidence of cheating, colleges might begin by using the instrument developed by the International Center for Academic Integrity—ICAI—(for which there are more than two decades of longitudinal data) to assess the current situation on campus and then follow up with periodic administrations of that instrument as well as a follow up along the lines suggested in Hsu's study. A more robust follow up, which would require a good bit of work, would involve a survey within organizations in which alumni are working, say 5–7 years after a cohort graduates (the development of the survey is the largest part of the work that would have be done) to ascertain whether the behaviors found in the earlier survey continue (which could be a positive or negative, of course). Assessing the ability of students to effectively work their way through ethical challenges is a harder nut to crack. Perhaps the best that can be done here would be to get a sense of where students are when they come to university and when they depart by having them write in response to a case study. In the second writing, we would be looking to find evidence of enhanced awareness of the ethical dimensions of the situation (e.g., in the identification of stakeholders and clarity about what's at stake for them), and, in the context of a thoughtful discussion, evidence of a good faith effort in undertaking the complementary tasks of the two logics (i.e., evidence of their working to achieve rational equilibrium in their judgement about what ought to be done). Unfortunately, because reading and evaluating student essays is a labor-intensive undertaking, it is highly unlikely that a university would have the wherewithal to do this, except perhaps on a reasonably sized representative sample.

References

American University in Cairo. (2010). Academic integrity and its impact on business ethics. In *Fostering Academic Integrity Conference*, American University in Cairo, Egypt, December 5, 2010. AUC site: http://conf.aucegypt.edu/FAI10. Accessed September 25, 2017.

Ariely, D. (2012). Why we lie. *The Wall Street Journal*. The Saturday Essay, updated May 26, 2012, WSJ Online, Accessed May 28, 2012.

Ariely, D. (2017). Ethical Systems.org. http://www.ethicalsystems.org/content/dan-ariely. Accessed September 25, 2017.

Bazerman, M., & Gino, F. (2012). Behavioral ethics: Toward a deeper understanding of moral judgment and dishonesty. *Annual Review of Law and Social Science 8*, 85–104 (Volume publication date December 2012) First published online as a Review in Advance on July 12, 2012. https://doi.org/10.1146/annurev-lawsocsci-102811-173815. Accessed September 25, 2017.
BNP Statement. (2014). *BNP Paribas announces a comprehensive settlement regarding the review of certain USD transactions by US authorities.* https://group.bnpparibas/en/press-release/bnp-paribas-announces-comprehensive-settlement-review-usd-transactions-authorities.
Dewey, J. (1924). Logical method and law. *Cornell Law Review, 10*, 17–27. http://scholarship.law.cornell.edu/clr/vol10/iss1/2/. Accessed September 25, 2017.
Dworkin, R. (1982). 'Natural' law revisited. *University of Florida Law Review, 34,* 165–188.
Dworkin, R. (1986). *Law's Empire*. Cambridge: Belknap/Harvard University Press.
Fahrenthold, D. (2016). Trump recorded having extremely lewd conversation about women in 2005. *The Washington Post*. https://www.washingtonpost.com/politics/trump-recorded-having-extremely-lewd-conversation-about-women-in-2005/2016/10/07/3b9ce776-8cb4-11e6-bf8a3d26847eeed4_story.html?postshare=2491475870527101&tid=ss_tw&utm_term=.562469574057. Accessed September 25, 2017.
Gillen, D., & Hodge, C. (2014). Business behaving badly. *The New York Times*. https://www.nytimes.com/video/business/100000002614034/business-behaving-badly.html?playlistId=100000002585027. Accessed September 25, 2017.
Graham, J., Haidt, J., & Nosek, B. (2009). Liberals and conservatives rely on different sets of moral foundations. *Journal of Personality and Social Psychology, 96,* 1029–1046.
Haidt, J. (2001). The emotional dog and its rational tail: A social intuitionist approach to moral judgment. *Psychological Review, 108,* 814–834.
Haidt, J., & Joseph, C. (2004). Intuitive ethics: How innately prepared intuitions generate culturally variable virtues. *Dædalus* (Fall).
Haidt, J. (2006). *The happiness hypothesis*. New York: Basic Books.
Harris, C. (2013). Banks have been fined nearly $3 billion by the SEC and have spent $66 billion on litigation, but one major bank hasn't paid a dime and expects to pay $0 in litigation this year. *International Business Times*. http://www.ibtimes.com/banks-have-been-fined-nearly-3-billion-sec-have-spent-66-billion-litigation-one-major-bank-hasnt. Accessed September 25, 2017.
Heath, D., & Heath, C. (2010). *Switch: How to change things when change is hard*. New York: Broadway Books.
Holmes, O. W., Jr. (1881). *The common law*. Boston: Little, Brown & Co.
Holmes, O. W., Jr. (1897). The path of the law. *Harvard Law Review, 10*, 457. http://www.constitution.org/lrev/owh/path_law.htm. Accessed September 25, 2017.
Holmes, O. W., Jr. (1920). *Collected Legal Papers*. New York: Harcourt Brace and Howe (Dewey quotes Holmes at 1924 20).
Hughes, E. C. (1965). Professions. In K. S. Lynn et al. (Eds.), *The Professions in America. Daedalus* (pp. 1–3). Boston: Houghton Mifflin Company.
Hume, D. (1739). *A Treatise of human nature*. https://people.rit.edu/wlrgsh/Treatise.pdf. Accessed September 25, 2017.
International Center for Academic Integrity (ICAI). (2014). Fundamental values. http://www.academicintegrity.org/icai/assets/revised_fv_2014.pdf. Accessed September 25, 2017.
Kolhatkar, S. (2017). The financial page no more mister nice guy. *The New Yorker*.
Kretz, L. (2015). Teaching being ethical. *Teaching Ethics, 15,* 151–172.
Lochner v. New York, 198 U.S. 45, 76 (1905) (Holmes, J., dissenting).
Marcus, G. E. (2002). *The sentimental citizen: Emotion in democratic politics*. State College: Penn State Press.
Marcus, G. (2004). *The birth of the mind*. New York: Basic Books.
Mazar, N., Amir, O., & Ariely, D. (2008). The dishonesty of honest people: A theory of self-concept maintenance. *Journal of Marketing Research, XLV,* 633–644. (633).
Newsweek. (2010). Do fines ever make corporations change? September 13, 2010, p. 56.
Otten, L. (2010). *Does the punishment fit the crime*? https://www.lasallenonprofitcenter.org/does-the-punishment-fit-the-crime/. Accessed 25.

Pfeffer, J., & Sutton, R. (2009). *Hard facts, dangerous half-truths, & total nonsense*. Cambridge: Harvard Business Press.

Postema, G. (2011). *Legal philosophy in the twentieth century: The common law world a treatise of legal philosophy and general jurisprudence* (Vol. 11). New York: Springer.

Prentice, R. (2015). Teaching behavioral ethics. *Journal of Legal Studies Education* (February 2, 2014). Available at SSRN: https://ssrn.com/abstract=2389689.

Roston, M. (2015). Retracted scientific studies: A growing list. *The New York Times*, May 28, 2015. https://www.nytimes.com/interactive/2015/05/28/science/retractions-scientific-studies.html?_r=0. Accessed September 25, 2017.

Touryalai, H. (2014). BNP is just fine after $9B penalty. Are billion dollar settlements effective? *Forbes*. July 1, 2014. https://www.forbes.com/sites/halahtouryalai/2014/07/01/bnp-is-just-fine-after-9b-penalty-are-billion-dollar-settlements-effective/#4b8a1eca28d7. Accessed September 25, 2017.

Van Noorden, R. (2011). The trouble with retractions. *Nature*, *478*, 26–28. http://www.nature.com/news/2011/111005/pdf/478026a.pdf. Accessed September 25, 2017.

Wueste, D. (1999). The realist's challenge in professional ethics: Taking some cues from legal philosophy. *Professional Ethics, 7,* 3–22.

Wueste, D. (2008). Unintended consequences and responsibility. *Teaching Ethics, 9,* 13–24.

Wueste, D. (2014). Promoting integrity integritively: Avoiding the scylla and charybdis of abdication and zealotry. *Achieving Ethical Excellence Research in Ethical Issues in Organizations, 12,* 5–27.

Wueste, D., & Fishman, T. (2010). The customer isn't always right: Limitations of "customer service approaches to education, or why higher education is not Burger King. *International Journal for Educational Integrity*, *6*(1), https://www.ojs.unisa.edu.au/index.php/IJEI/article/view/672/499. Accessed September 25, 2017.

Part IV
Institutional Programs

Ethics Across the Curriculum at Utah Valley University

Elaine E. Englehardt

Abstract Utah Valley University (UVU) is one of the pioneer universities in designing and implementing an Ethics Across the Curriculum program (EAC). It began in 1986 as a general education core humanities course in Ethics and Values. Soon it was supported by funding from the National Endowment for the Humanities (NEH). The program received funding again in 1992 from the U.S. Department of Education's FIPSE, which deemed the program highly innovative. The program strengthened the overall curriculum at UVU as well as EAC efforts elsewhere. FIPSE project officers found the EAC project to be one of the most influential programs it funded. In 2001, the program received the Theodore M. Hesburgh Award for Leadership in Higher Education. The ease in replicating and disseminating the program to other institutions led FIPSE to award two additional grants to continue development and dissemination of the program over the next eight years. This article will detail the history of the program's implementation, design and theoretical underpinnings.

Keywords Ethics across the curriculum · Justification · Dissemination Evaluation · Scholars · Faculty participation · Students

The Beginnings

The Ethics Across the Curriculum (EAC) program at Utah Valley University has a long, multifaceted history. It involves recognizing the need for ethics to act as a vehicle in enriching the curriculum in multi-layered programs. The program started in 1986 with a general education core humanities course in *Ethics and Values,* facilitated by funding from the National Endowment for the Humanities (NEH). After considerable planning and funding, two distribution courses were added in the history of science and the history of civilization. By 1992, the courses had

E. E. Englehardt (✉)
Utah Valley University, Orem, UT, USA
e-mail: Elainee@uvu.edu

E. E. Englehardt and M. S. Pritchard (eds.), *Ethics Across the Curriculum—Pedagogical Perspectives*, https://doi.org/10.1007/978-3-319-78939-2_20

developed into an EAC program that would engage all departments in ethics programming. The EAC program was built on a continuous, firm foundation of faculty participation, administrative support, student acceptance and funding from the U.S. Department of Education that continued to 2004. This fluid program led to the establishment of a national organization, The Society for Ethics Across the Curriculum (SEAC), its journal *Teaching Ethics* and a variety of scholarly and practical educational programs.

From its humble beginnings at, what was then Utah Technical College, with 2500 students in 1986, the ethics program has grown as the institution matured to become Utah Valley State College, and now Utah Valley University, the largest university in the state with over 37,000 students in 2018. The ethics program remains central to the University's mission and has a strong presence throughout the campus.

Ethics and Values Course

Initially, faculty members were concerned that students did not understand the importance of the humanities, of being a well-educated person, or the value of being exposed to a variety of moral and social perspectives. A united humanities department initiated several programs to determine how students could better understand the importance of "how to *be* someone," as much as "how to *do* something." Furthermore, students seemed disposed to take the easiest class at the easiest time from the easiest teacher. In order to strengthen appreciation for the humanities, the interdisciplinary faculty held a variety of workshops with scholars of national reputation to design an improvement in the education of students. After three years of discussion, and a variety of trial workshops, in 1984 Professor Elaine Englehardt submitted a grant proposal to the National Endowment for the Humanities (NEH) to support an interdisciplinary humanities ethics course as the core general education requirement in humanities at her institution. All humanities faculty members had read the grant proposal and had written letters of support for it. Numerous administrators also supported the grant proposal.

The elements of the grant included the initiation of the course as well as an annual summer seminar for ten faculty participants, and ethics forums for the college and the community. Library acquisitions in interdisciplinary ethics areas were also part of the project. Success in receiving the NEH funding required three tries by Englehardt over a three-year period. In its first two rejection notifications, NEH indicated they were doubtful that a technical college could implement an interdisciplinary humanities core course. Reviewers noted little evidence of a strong commitment to the humanities at the college. The winning argument for funding explained that the college, with its limited offerings, was in greater need of funding to advance the humanities when compared with universities with established, prestigious programs. During the time between the first and third submissions, a core, interdisciplinary *Ethics and Values* course was designed and accepted by the general faculty at the institution. The college would have initiated the course whether or not NEH funding became available because all general education faculty members believed it was an important change for the institution. Having this broad support from the humanities department persuaded NEH of the college's commitment. In 1986, NEH awarded a three-year grant to Professor Englehardt to implement the ethics program.

Having NEH behind the new ethics program added immeasurable credibility to the undertaking. Additionally, strong administrative support signaled a desire to make the project a permanent addition to the college culture. The administration not only embraced the course but also all future ethics programs as the project unfolded over the years. Administrators added matching resources throughout the grant period and also fully funded the project as the grant support concluded.

A strong relationship with the National Endowment for the Humanities continued, and in 1989 Englehardt received funding for a second three-year project in history and the sciences. The grant funded interdisciplinary courses in the history of science and the history of civilization. With the assistance of Biology Professor Jim Harris, twenty-five faculty members became part of a program to strengthen interdisciplinary scholarship in history and the sciences. Once again, most project participants and administrators carefully read the grant application and wrote letters of support for the funding. This program also included faculty seminars, a speakers series and workshops. Ethics education continued as a component in the development of the courses. All three NEH sponsored courses remain successful, active, and highly enrolled. Administrative financial support has maintained all grant activities after the funding periods were complete.

The ethics course has had a strong impact on the humanities department and the college. When the course began, there was no philosophy department. In 2018, there is a flourishing Philosophy/Humanities department with 14 full time faculty members, a large number of adjunct faculty and about 250 majors. This progress is a direct result of the addition of the Ethics and Values course.

Ethics Across the Curriculum

In 1990, Englehardt approached NEH with the idea of an Ethics Across the Curriculum program for the institution. During the period of the successful implementation of these programs and courses, NEH had asked Englehardt to mentor over twenty colleges and universities in the USA in their developing stronger humanities programs, and also to emphasize the importance of ethics. *Advancing the Humanities*, was designed by NEH to help two-year colleges structure and adopt a commitment to the humanities. NEH program officer Judy Jeffrey Howard was a principal mentor in all three ventures. She became a strong advocate for UVU's interdisciplinary ethics programming as well as for hundreds of two-year colleges that were interested in strengthening the humanities from a variety of approaches.

However, when Englehardt approached NEH about submitting an Ethics Across the Curriculum grant, she was discouraged from making an application. Program officers believed that the "humanities could be too watered down" in an EAC program that would have interdisciplinary ethics as a vehicle to explore professional and practical aspects of course work. Howard recommended exploring funding from FIPSE.

FIPSE reviewers and administrators were enthusiastic about the scope and depth of the project, and in 1992 Englehardt was awarded a three-year grant for Ethics Across the Curriculum to incorporate professional and practical ethics into all departments and disciplines including: humanities, sciences, social sciences, communication, business, nursing, trades, technologies, vocations, and liberal arts at the institution. That year, with over 4000 FIPSE applications submitted, the EAC proposal was among the 42 programs selected for funding. The vision for EAC included a robust, multidisciplinary plan that involved ethics education for all interested faculty at the college, enhancements for courses, and a variety of outreach programs for the community. Art Instructor Barbra Wardle became a co-director to assist in evaluation and assessment of the program.

EAC Specifics

Historically, the Ethics Across the Curriculum program at UVU gradually developed from the humanities core course, to two interdisciplinary history courses, to a significant cross-curricular program. Many faculty members from across disciplines were eager to join in the EAC project. Initially, FIPSE funded 35 slots for faculty participants. In year one, administrators at UVU advanced funds for an additional fifteen faculty to join the three-year project. President Kerry Romesburg, Dean Veonne Howlett, Department Chair Sharon Staples and Vice President Lucille Stoddard all advanced funds to support the projects. This type of support lasted throughout the funding period and continues today.

The EAC at UVU has as its goal teaching faculty, and thus students, to recognize and resolve ethical issues within their disciplines. To achieve these ends, the program involves activities similar to the first two projects, but also some significant differences. These activities include: (a) a summer seminar with a scholar of national reputation; (b) monthly meetings with scholars in ethics and the professions; (c) rewriting syllabi to include ethics components; (d) the creation of case studies by faculty; (e) workshops with faculty and students in individual disciplines as well as interdisciplinary settings; (f) engaging advisory boards in professional areas to assist in developing ethics modules and cases that would be realistic for students entering the workforce; and (g) the cultivation of an ethics resource center for faculty and students. By supporting and sponsoring all the activities listed above, the ethics resource center (now named Center for the Study of Ethics) serves as an umbrella for the entire EAC program. Permanent funding for the Center allows the program to progress smoothly and evolve over many years to come.

Faculty members assist student understanding of ethics through the following processes:

- Expanding the ethical sensitivity of students by helping them examine their own ethical assumptions;
- Understanding potential inconsistencies in their value framework;
- Carefully examining relevant facts and developing decision-making strategies for resolving ethical issues;
- Reinforcing the notion that ethical values are not merely "subjective opinions";
- Reinforcing critical thinking throughout all discussions;
- Increasing understanding of current ethical problems within the professions and society generally.

Additional objectives for faculty are:

- To increase the inclusion of ethics in courses and to integrate the discussion of ethical issues with standard subject matter;
- To become skillful in discussing ethical issues with their students;
- To provide mentoring support and networking opportunities for each other.

This shared study of ethics builds bridges and fosters understanding across the disciplines.

Case study writing and analysis is an important feature of the program. In writing and analyzing the cases, Englehardt and colleagues have promoted utilizing some form of a step-analysis model (Englehardt 2010, p. xxiv).

- Identify personal responses to the case;
- Review the facts of the case;
- Consult relevant policies and professional ethics codes;
- Identify and anallyze key ethical considerations;
- Identify legal considerations;
- Identify a response grounded in the ethical, professional, and legal considerations.

The successful implementation of this program was heralded by Joan Straumanis and David Johnson, FIPSE project officers, as one of the most influential programs funded by the agency in strengthening the overall curriculum. The EAC project at UVU was ideal for dissemination. It is easy to replicate in a variety of formats and to disseminate to other institutions. A justification was proposed for the continuation of the program. The program was funded twice for dissemination.

EAC Intellectual Justification

Practitioners in every profession and trade encounter ethical problems. When the students we are training in these disciplines leave their institutions of higher learning, they are generally ill-prepared to think critically about the ethical issues they will encounter as professionals. For example, initially engineers designed air bags that caused harm to those too distant in weight or height from the "Ideal Male" model chosen by those engineers as the norm. Yet the choice was made apparently without noting the ethical implications involved. This kind of example can be found in every discipline.

The failure of higher education to prepare students to recognize ethical problems in their professions, and thus the failure to prepare them to think critically about determining what ought to be done regarding such ethical incidents, presents a national problem of massive proportions. We have only to read our daily newspapers —with their reports of malfeasance among financial advisors, the harms caused to various people by the products they use, and so on—to know how widespread a problem we have. Responding to this problem requires changing the very way we teach those who will become the professionals upon whom we depend.

The EAC program is significant in several ways:

(1) It fosters an increased knowledge and understanding of ethical problems. While many of the problems facing those in the professions and technology may look to be wholly quantitative, they are not. As the example of the air bags indicates, such problems have value implications and so have an ethical component. Our program requires and allows for a new way of understanding the kinds of issues that those in the professions and technologies face. Ethics education takes place in the classroom, in scholarly seminars, in workshops and in concert with the community in forums. Scholarly guidance for the program is provided by UVU philosophy and humanities faculty, outside scholars of national reputation and community advisory boards. The project appeals to faculty from a variety of disciplines. Project faculty attend workshops and seminars to strengthen their understanding of ethics; moreover, they are encouraged to change their syllabi to reflect ethics content.

(2) Ethics Across the Curriculum involves the development of case studies written by faculty, students, and area professionals and practitioners. Students and faculty work through both the technical and the ethical components using these cases. After ethical grounding from seminars and workshops, this program makes use of the cases to present both the material of a course and ethical issues within a discipline. EAC provides a new paradigm about how to teach ethics within the disciplines that make up the professions and trades. The strategy is exciting in that faculty and students can appreciate how ethics permeates the disciplines, both forming the very practical issues practitioners face day-to-day as well as analyzing the implications for society at large. Students and faculty analyze how making what may appear to be merely a practical decision—such as picking a norm for air bags—is an instance of ethical decision making, one that is supported by the sort of training in ethics the program achieves.

(3) EAC produces a variety of results, especially for improvements in teaching and student achievement. Students understand that experience within a discipline consists of solving various problems with ethical implications. With the increased reliance on technology and the increased inability of all of us to keep up with changes in technology, it is essential that those designing,

manufacturing, and maintaining the technological infrastructure of our society do so with an awareness of what harms could occur with breakdowns in communication, mechanical failures, and the like. We hope to see the impact of our success by noting a decline in kinds and number of ethical problems that occur in the professions, technologies and trades.

Elements of the Program

The essentials of the three-year initial project included an annual one-week seminar for faculty members from a variety of disciplines. Notable scholars from universities, professions, and the wider community were invited to share their expertise in professional and practical ethics, pedagogy in teaching the topics, and instruction in writing and analyzing case studies. In addition to teaching one week, intense, ethics seminars, the scholars spent time assisting faculty members with curricular and extracurricular projects. The scholars assisted faculty in writing and assessing case studies, rewriting syllabi, discussing teaching strategies, and reflecting on best practices in implementing applied ethics in classes. Some of these included Leslie P. Francis, Robert C. Solomon, Kathleen Higgins, James Sterba, Janet Kourany, Robert Lawry, Margaret Battin and Deni Elliott. The scholars have remained long-term friends of the EAC program. Hundreds of nationally recognized individuals have now participated in the UVU ethics program.

Watching the progress of faculty and students in the study of ethics has been a source of pride to all participants. All scholars say they were impressed with the depth of the faculty commitment to the students in the program. Robert Lawry, Professor of Law at Case Western Reserve University noted:

> Whether talking with a nursing or business instructor, the faculty were excited to discuss the topics in ethics that were or could be important to their students. Some of my favorite faculty members were from the trade areas such as auto repair or welding or refrigeration. Some of these faculty members had students asking them about ethical concerns at their current job. We could discuss and analyze the cases as the faculty were submitting them.

The following experience was detailed by Instructor Todd Low from the automotive program:

> This project has been an excellent way to increase communication with students, and to help them learn more about our profession. Each semester, I have students asking when we are going to talk about ethics. I don't have the ethics module planned until the end of the semester, but generally we start talking about ethics daily after about the first three weeks of class. Many of the students have already discussed ethics concepts in other courses. They are anxious for the discussions in my classes. I find the students want to discuss an individual problem or issue they have encountered. They may bring up subjects involving ethical problems during class time. I also have students approach me or other students outside of class to discuss an ethical problem, or ask for advice in a situation. Before we began including ethics problems in class, it never occurred to students to talk about ethical issues or morally problematic situations. They thought that class was just for learning about their profession, not how to deal with customers and other employees.

An important feature of Low's comment is that it suggests that the ethics emphasis in other disciplines at UVU encourages students to want to discuss ethics issues in all of their courses. Insofar as this happens, the EAC program encourages students to make connections across the curriculum. To further support this, the faculty members were asked to write four case studies tied to their disciplines yearly. They were then to provide analysis of these cases to assist in the teaching of the cases. The cases were compiled and shared on the UVU website. The following case was written by a UVU professor Dale Olsen in the collision repair program:

> As an employee in an auto body shop, Ron was instructed by his boss to do cosmetic repairs on an old vehicle. The owner's instructions were, make this car look great; I'm going to sell it next week. When driving the car into his bay, Ron noticed an exhaust smell. Upon further inspection, he found the exhaust was able to leak into the car because of rust damage. Ron is a new employee and mentions the problem to his boss, as he sees the damage as life-threatening to any driver of the car. His boss says to ignore the problem and just do the cosmetic repair job. What should Ron do?

Olsen and his students discussed the case at length. They determined the repairs needed to be made, even if Ron wasn't paid for doing the work. They recommended that Ron look for a new position as soon as possible as he shouldn't need to work for a firm that lacks an understanding of ethical behavior. Because this sort of problem can come up in virtually any disciplinary area it has been used across the curriculum as an effective case.

Community ethics boards were established by each department to enable the discussion of ethical problems that students entering the work place would likely face. Board members were often asked to analyze cases written by the faculty members throughout the course of the academic year. The above case was discussed with owners of collision repair shops and other automobile related businesses. All board members agreed they would have never accepted the individual as a customer. They also discussed the need for an ethics policy in their businesses and professions.

Nursing Division Chair Karen Swendsen explained that board members were impressed with the forward thinking that incorporated ethics cases and discussions into their curriculum. "Our advisory board was impressed with the development and analysis of cases by our faculty members. The faculty members voluntarily joined the EAC program and spent considerable time learning how to help students understand ethical problems they would most likely encounter in a health care setting," Swendsen said. She added that the nursing accreditation association also found the work commendable.

Additionally, a dual speakers program was held on a monthly basis. The first speaker was designated specifically for the EAC participating faculty members. Often speakers were selected from a discipline's advisory board. For example, the School of Business invited a patent attorney from a nationally ranked corporation to discuss intellectual property theft of software. These discussions not only resulted in case studies for students but also scholarly publications by faculty members. The speakers were an inspiration to faculty scholarship as well as pedagogical enhancement.

The second invited monthly talk was open to students and the community. Topics often kept pace with issues happening in the state, country, and world. Many faculty required students to attend the talks because they found them so valuable. "We found the talks to be a highlight for the semester for our students. When could they listen to the CEO of a company or a decorated FBI agent? These students were at our college and we could provide them with a quality supplement to their education," said Larry Holt, Professor of Digital Technology.

Center for the Study of Ethics

To keep pace with the programming of speakers, workshops and seminars, a Center for the Study of Ethics opened in 1993. The Center was designed to act as an umbrella for the EAC program. Englehardt was the founder and initial director of the Center for the Study of Ethics (CSE). Immediately, the Center initiated an Ethics Awareness Week (EAW), during which all disciplines were to host programs in professional and practical ethics for their students. As part of EAW, the center sponsored an "Excellence in Ethics" award, inviting the awardee to be the featured speaker for the week.

The Center for the Study of Ethics continues to thrive today. Philosophy Professor David Keller was Director of the Center for 14 years until his untimely death. His successor is Professor Brian Birch. These directors have also stressed EAW. It is held the last week of September and features five full days of programs from all colleges, schools and departments on campus. Many faculty members design their course syllabi around EAW events sponsored by the center. The topics cover all UVU disciplines, and scholars/speakers/guests are often selected in coordination with academic departments and deans.

Other EAC/CSE activities include providing regional leadership in supporting high school and intercollegiate ethics bowls for our students. Led by Dr. Karen Mizell, UVU students have competed across the nation debating ethical case studies. The Center also awards faculty fellowships annually to sponsor and sustain interdisciplinary ethics scholarship for faculty at UVU. A competition is held and two or three faculty members are awarded a stipend to conduct a scholarly proposal. These faculty members are featured speakers during the next Ethics Awareness Week. In addition, a Student Ethics Symposium is sponsored annually by the Center to promote critical and academic inquiry related to ethical issues of all disciplines. This daylong event features talks, panel discussions and workshops by students in their ethics specialty areas. The Center sponsors a robust speaker's program throughout the academic year. Students are strongly encouraged to attend these events either as a requirement of a course or as an opportunity to obtain extra credit.

The mission for the CSE has changed little over the years. The following is the current statement:

> The mission of the Center for the Study of Ethics (CSE) is to promote and facilitate the interdisciplinary exploration of ethical theory and practice. The Center administers a variety of programs designed to advance ethics education at Utah Valley University, within our region and in national and international contexts. The CSE is a non-advocacy organization committed to the inclusion of diverse perspectives and practices…
>
> CSE programming is designed to advance the study and application of ethics in variety of disciplines and contexts. We promote the disciplined study of ethical theory and practice and explore ways in which ethics may be integrated into professional and civic life (www.uvu.edu/ethics).

Assessment and Dissemination

Numerous successful and innovative assessment measures were developed during the externally funded project. The most successful teaching, learning, and evaluation innovation was a modified version of the Critical Incident Technique (CIT), which asks students to locate an ethical problem in a film, text, or TV program and then in two pages to describe the problem, analyze its ethical dimensions, and indicate possible responses. In 90% of our tests using the CIT, students could successfully define and analyze an ethical problem. Advising and assessing from FIPSE were officers Joan Straumanis and David Johnson. Assessment advisor Dora Markus worked closely with UVU faculty member Barbra Wardle in writing and analyzing assessment measures.

The external evaluator for the project was Donald Schmeltekopf, Provost of Baylor University. He evaluated the Ethics Across the Curriculum pre and post project. His evaluation included numerous interviews with faculty, students and administration during the three years of the project. He evaluated all cases written by faculty and discussed analyses of the cases. He also scrutinized the assessment tools. His overall review of the EAC project was one of excellence in all areas. He was willing to mentor and assist throughout the project as well as after the project was completed.

In 1998, the program at UVU was deemed exceptional by FIPSE program officers, and Englehardt was awarded a two-year grant to disseminate the Ethics Across the Curriculum project to four colleges/universities nationally. Each university was guided by a strong leader to develop and implement the program at their institution. David Ozar led the work at Loyola of Chicago (a private, Catholic University). Wade Robison headed the program at Rochester Institute of Technology, NY (a private technical university). Maura O'Neill directed the program at Chaffey College, CA (a two-year community college). Leslie Francis guided the project at the University of Utah (a tier one state flagship university.) Each institution needed to implement EAC in its own way. Each program needed to be designed by and for the faculty, students, administrators, and the community of the target institution. EAC is a vehicle that can be incorporated in all disciplines. The main requirement for success is funding for faculty to help them develop an

EAC program that works for the institution. Each of these colleges and universities developed different perspectives and strengths in the EAC experience. All faculty directors found their programs to be successful. After assessing the various institutions and EAC programs, David Johnson, FIPSE project officer, determined the dissemination was outstanding at all four institutions.

Based on Johnson's and Straumanis' recommendations, in 1999–2004, Englehardt received an additional 5-year FIPSE dissemination award. The institutions that joined the EAC dissemination plan were SUNY Plattsburg (a graduate state university) and Miami Dade Community College (a large 2 year college). Under this award, UVU assisted these institutions in developing and implementing EAC programs. Funding was also awarded to start a national organization—Society for Ethics Across the Curriculum (SEAC), and a journal entitled *Teaching Ethics*. The journal and society had the goal of enhancing ethics pedagogy nationally and globally. The idea for the Society for Ethics Across the Curriculum originated with Wade Robison of RIT. The EAC program was successful at RIT; however, Robison convinced Englehardt that a national society was needed to strengthen and encourage the interdisciplinary teaching of ethics at an affordable price. The first SEAC conference took place in 1999 at RIT. SEAC continues to meet annually. SEAC was officially formed the next year under the sponsorship of UVU in Salt Lake City with adoption of a constitution, the selection of a board of directors, the election of a president, and the naming of an editor for the new journal *Teaching Ethics*. Board members included Wade Robison, President; Elaine Englehardt, Gabriel Palmer Fernandez, Michael Pritchard, Susan Martenelli Fernandez, David Keller, Lisa Newton, and Steven Scales. Daniel Wueste was selected as the second president of SEAC, and the current president is Deborah Mower. Stephen Scales, Donna Werner and Alan Preti have served as treasurers for the organization.

The first editor of its journal was Brian Birch; followed by David Keller. Next Michael Pritchard and Englehardt were awarded the editing of the journal. Current journal editors are Alan Tomhave and Mark Vopat. An article on the journal by these authors is in this volume.

The Society for Ethics Across the Curriculum continues to be an outstanding arena for ethics discussions in practical, theoretical, applied and pedagogical areas. "The purpose of the Society for Ethics Across the Curriculum is to stimulate scholarship on ethics and the teaching of ethics in all academic disciplines and to afford an opportunity for the exchange of research." http://www.rit.edu/~w-ethics/seac/

Recognition and Growth

In February 2001, the Utah Valley State College was awarded the nationally acclaimed Theodore M. Hesburgh Award for Leadership Excellence in Higher Education. The award was presented to Elaine Englehardt and College President Kerry Romesburg for the Ethics Across the Curriculum program being the best faculty development program in the United States. Most noteworthy is the impact the EAC program discussed in this article has had on an institution, its students and its faculty members. It has brought change to thousands of student's lives, and it has great potential to improve society long into the future. Moreover, this brief history of the UVU EAC program shows that its success of EAC was not just about the vision and determination of one or a few people, but also about the unwavering commitment of the entire UVU community. It also speaks to the value of support for superior educational programs at the national level for the betterment of higher education across the nation. Notably, the university backing continues to the present in all areas of the UVU ethics programming.

Nationally, UVU actively participates in two interdisciplinary organizations, SEAC and the Association for Practical and Professional Ethics. Both organizations assist membership in Ethics Across the Curriculum programs through workshops, seminars and special speakers. The annual conferences provide time to share ideas through a variety of methods.

One of the highlights of APPE is the student Ethics Bowl. UVU students yearly compete in this regional and national discussion of ethics. The competition includes the study of more than 12 cases in regional and national settings. Teams of five students then discuss individual cases in front of three judges who determine which team makes the superior arguments. Karen Mizell and Jeff Nelson spend hundreds of hours yearly training these students. The Ethics Bowl is considered part of the EAC program at UVU. Karen Mizell has also been a national leader in high school ethics bowl. These programs are all part of UVU's future commitment to EAC. The Ethics Bowl is discussed in detail in this volume by Robert Ladenson.

The original universities that were part of the dissemination program keep in contact with one another. All of the universities continue to have strong programs and have blossomed in many directions in the areas of EAC. Most of these universities are active in SEAC and APPE and enjoy the workshops, seminars and Ethics Bowl competition.

Conclusion

At UVU, the goals we have for students and faculty members are goals common to students and educators everywhere. If they are to learn to understand ethical issues within their chosen discipline, they must learn how to recognize and analyze these ethical issues. The methods we have adopted to achieve those goals are fully replicable in any institution of higher education, most commonly through seminars, workshops, panel discussions, class room exercises and so on. EAC fosters an increased knowledge and understanding of problems. It requires and allows for a new way of understanding the kinds of issues that those in the professions and technologies face. Through EAC, ethics education takes place in the classroom, in scholarly seminars, in workshops and in concert with the community in forums. EAC provides a new paradigm about how to teach ethics within the disciplines that make up the professions and trades. The EAC strategy is exciting in that faculty and students can appreciate how ethics permeates the disciplines, both forming the very practical issues practitioners face day-to-day as well as analyzing the implications for their personal lives as well as chosen fields. Students, faculty and the professions can grow together utilizing ethics as a vehicle to study the important facets of their daily choices.

Reference

Englehardt, E., Pritchard, M., Romesburg, K., & Schrag, B. (2010). *Ethics in higher education administration*. Netherlands: Springer.

Designing an EAC Program for the School of Life Sciences at Arizona State University: Early Initiatives and Lessons from the Literature

Karin D. Ellison, Challie Facemire and Joseph R. Herkert

Abstract In Fall 2015 the School of Life Sciences at Arizona State University began a systematic expansion of an ethics across the curriculum program as part of its Life Science Ethics Program. This essay presents the initial elements of that program, reviews the literatures on ethics across the curriculum programs and responsible conduct of research education programs, and concludes with lessons from the literature for ethics education program development. While ethics across the curriculum programs, such as ASU's Life Science Ethics Program, are likely to be eclectic in specific activities and teaching approaches, they can benefit from (a) having a coordinating body, (b) attending to development of ethics and ethics education competencies across faculty in the community, and (c) using assessment to illuminate opportunities for improvement.

Keywords Ethics education · Ethics across the curriculum · Arizona State University · Life sciences

Introduction

Formation of ethical professionals is an enduring goal and ever-present challenge in science, technology, engineering, and math (STEM) education. The School of Life Sciences (SOLS) at Arizona State University (ASU) has recently committed to engage this important task through expansion and coordination of ethics offerings in the Life Science Ethics Program. This essay describes the early initiatives and

K. D. Ellison (✉) · C. Facemire
School of Life Sciences, Arizona State University, Tempe, Arizona, USA
e-mail: karin.ellison@asu.edu

J. R. Herkert
Genetic Engineering & Society Center, North Carolina State University, Raleigh, NC, USA

E. E. Englehardt and M. S. Pritchard (eds.), *Ethics Across the Curriculum—Pedagogical Perspectives*, https://doi.org/10.1007/978-3-319-78939-2_21

challenges of the SOLS ethics program, reviews trends in ethics across the curriculum (EAC) programs more broadly, and reflects on these trends to present lessons learned.

An interdisciplinary school within a growing and innovative institution, the large student body and diversity of degree offerings in SOLS are both a challenge and opportunity for an ethics program. SOLS was created in 2003 through the merger of the Departments of Biology, Plant Biology, and Microbiology and faculty from related units, including a cohort who study history, philosophy, and ethics of the life sciences. By the fall of 2015, the school had approximately 2730 undergraduates, 100 masters students, 220 doctoral students, and 80 tenured or tenure-track faculty members (School of Life Sciences 2017). The over 3000 students study in 25 different degree programs. SOLS offers ten undergraduate, six masters, and nine doctoral degree programs—including several degree concentrations, which have additional requirements beyond those of the main degree they fall under (See Table 1).

The large and racially diverse study body, drawn to ASU for its commitments to both access and excellence, add further aspects of diversity to the SOLS student body. In 2014, ASU had 83,000 students enrolled of whom 34% were minority students and 11% were international students. Institutional goals, including to "[m] aintain the fundamental principle of accessibility to all students qualified to study at a research university" and to "[m]aintain university accessibility to match Arizona's socioeconomic diversity, with undifferentiated outcomes for success" (Arizona State University ND) insure students enter with scholastic preparation that varies as widely as that of high school graduates in the state. A large—enrollment of 6900 in fall 2016 (Arizona State University 2016)—excellent, and highly-rated honors college draws a significant pool of outstanding students who would be competitive at any institution in the US (Arizona State University 2014).

Colleagues across campus engaged in developing ethics programs and other like-minded activities at a variety of levels create a supportive environment for developing an ethics program in SOLS. The Lincoln Center for Applied Ethics, founded in 1998 and currently directed by bioethicist and philosopher of science Dr. Jason Robert, supports a university-wide applied ethics program. The School of the Future of Innovation in Society (SFIS), founded in 2015, grew out of a group of science policy and science studies scholars. The ambition of the school is to make innovation more interdisciplinary, anticipatory, and democratic. New joint appointments between SFIS and the Fulton Schools of Engineering—four as of Spring 2017 but anticipated to grow to twelve—are explicitly charged with developing responsible research and innovation across the engineering schools.

ASU and SOLS thus provide a context for developing an EAC program that poses both significant opportunities and challenges.

Table 1 SOLS programs (Arizona State University 2018a, b)

Program	Main Area	Concentration(s)
BS	Biological Sciences	General
BS	Biological Sciences	Biology and Society
BS	Biological Sciences	Biomedical Sciences
BS	Biological Sciences	Conservation Biology and Ecology
BS	Biological Sciences	Genetics, Cell and Developmental Biology
BS	Biological Sciences	Neurobiology, Physiology and Behavior
BS	Microbiology	General
BS	Microbiology	Medical Microbiology
BS	Molecular Biosciences and Biotechnology	NA
BS	Neuroscience	NA
UG Minor	Biological Sciences	NA
UG Cert.	History and Philosophy of Science	NA
MS	Biology	General
MS	Biology	Biology and Society
MS	Biomimicry	NA
MS	Microbiology	NA
MS	Molecular and Cellular Biology	NA
MS	Plant Biology and Conservation	NA
PhD	Animal Behavior	NA
PhD	Biology	General
PhD	Biology	Biology and Society
PhD	Environmental Life Sciences	NA
PhD	Evolutionary Biology	NA
PhD	History and Philosophy of Science	NA
PhD	Microbiology	NA
PhD	Molecular/Cellular Biology	NA
PhD	Neuroscience	NA
Grad. Cert.	Biomimicry	NA
Grad. Cert.	Environmental Communicationand Leadership	NA
Grad. Cert.	Nonfiction Writing and Publishing	NA
Grad. Cert.	Scientific Teaching in HigherEducation Certificate	NA

BS = Bachelor of Science
UG = Undergraduate
Cert. = Certificate
MS = Master of Science
PhD = Doctor of Philosophy
Grad. = Graduate
NA = not applicable

Ethics in SOLS at ASU

From Fall 2016 to Spring 2021, SOLS and its Center for Biology and Society (CBS) have strategic initiative funding from the university to develop an EAC program in the school under the direction of one of the authors (Dr. Karin Ellison, Associate Director of CBS). This current undertaking, known as The Life Science Ethics (LSE) Program, builds on substantial strengths in ethics within the school including faculty, courses, and degree programs. To connect to recognized initiatives in biology education, the program takes the mandate of the ethics program to be very broad; it encompasses a range of approaches to critical thinking concerning the relationship between knowledge and innovation in the life sciences and social outcomes. The first stage of the ethics program has been a loosely coordinated set of initiatives including ethics content embedded in coursework and other school programs, novel ethics courses, discussion series, and events.

The LSE Program is building on substantial strengths among faculty who teach and conduct research in environmental ethics, bioethics, and research ethics as well as natural science colleagues active in ethics, policy, and compliance activities. Dr. Ben Minteer is an environmental ethicist and Arizona Zoological Society Endowed Chair in SOLS at ASU. His work explores historical and contemporary ethical obligations to species and wild lands and how these kinds of natural resources can be maintained. Bioethicist Dr. Jason Robert, director of the university's ethics center, teaches graduate bioethics courses and has an active research program. Dr. Ben Hurlbut works in the areas of history of science, science studies, and bioethics with a special focus on biological sciences, biotechnologies, and governance. Dr. Karin Ellison teaches responsible conduct of research to both undergraduates and graduate students and conducts STEM ethics education research.

Beyond faculty with primary interests in ethics, SOLS has a number of life science faculty active in ethics, policy, and compliance. For example, the school's director, Dr. Bertram Jacobs, also chairs ASU's Institutional Biosafety Committee and serves as a member of ASU's Intellectual Property and Institutional Review Committee. Dr. Dale DeNardo, ASU's Animal Care Program Director and Attending Veterinarian, is a SOLS faculty member. Evolutionary ecologist Dr. James Collins has a substantial publication record in ecological ethics and co-chaired the recent National Academies study, *Gene Drives on the Horizon: Advancing Science, Navigating Uncertainty, and Aligning Research with Public Values* (National Academies of Sciences, Engineering, and Medicine 2016).

Also established is the ability of the students to engage deeply with ethics through a full course of study. In the Biology and Society degree programs, students in SOLS can focus their studies on bioethics and environmental ethics. Options include an undergraduate major in Biological Sciences with a concentration in Biology and Society and M.S. and Ph.D. degrees in Biology with concentrations in Biology and Society. All three degrees are research-oriented with a capstone project, thesis, and dissertation, respectively.

Another strength in SOLS important for the direction of the developing ethics program is engagement with biology education. Led by faculty members who specialize in discipline-based education research, the school has a number of initiatives to strengthen undergraduate teaching including substantial training and assessment of teaching assistants, ongoing development of undergraduate courses using best practices in STEM education, a STEM education speaker series, and a newly designed, mediated, flipped classroom space.

By viewing ethics as integral to understanding the relationship between science and society, the LSE Program connects to and is enriched by broader unit promotion of excellence in undergraduate teaching. In 2011 the American Association for Advancement of Science completed the report *Vision and Change in Undergraduate Biology Education: A Call to Action* (Bauerle et al. 2011). This study recognizes that the concepts and disciplines of biology, as well as biology-based emerging technologies, are vital parts of modern society with complex impacts and ethical implications. It calls for biology education to promote understanding of these dimensions of life sciences. Specifically, the report champions the "ability to understand the relationship between science and society," and asserts "[b]iologists need to evaluate the impact of scientific discoveries on society, as well as the ethical implications of biological research." The LSE Program likewise emphasizes these learning goals.

Starting from courses, faculty, and degree programs focused on ethics, as well as a perspective that the school's ethics program should cultivate an understanding of relationships between biology and society, the initial phase of the LSE Program has been an expansion of activities in three main areas: embedding content in biology courses, incorporating ethics in co-curricular activities, and developing new special initiatives.

Initial Experiments

Embedding ethics material in large enrollment courses required for multiple majors within SOLS is a key strategy for reaching all of the over 2700 undergraduates in the school several times during their degree programs. The ultimate goal is to embed ethics content in all introductory biology courses and also high-enrollment upper division courses. Further, the program aims to coordinate science and society learning objectives across the courses. This work has begun with embedding content in BIO 189, Life Sciences Career Paths, and BIO 282, Conceptual Approaches to Biology for Majors. Life Sciences Career Paths is a massive undertaking; it had 800 freshmen enrolled in fall 2016. The one-credit course introduces students to areas in the life sciences, faculty in the school, and basic study skills. The ethics learning objective is to identify ethical issues related to the interplay between life sciences and society. In an online assignment, the students view an introductory video on bioethics, read a brief article on environmental ethics, and write a reflection on two ethics topics related to their majors that they'd like to learn more about. Embedded content has also been piloted in BIO 282—one of three versions of introductory biology that students select among. Embedded content in introductory biology will continue to develop students' abilities to recognize ethical issues and begin to introduce knowledge of relevant ethical norms and standards as well as ethical problem solving. In BIO 282, one lecture and one recitation session have been redesigned to focus on ethics. The lecture uses the tragedy of the commons as a framework for thinking about climate change while the recitation section focuses on the ethics of research with animals, introducing students to the 3Rs (reduction, refinement, and replacement).

Embedding ethics content in science courses across degree programs will complement three undergraduate majors and several graduate programs that give students a deep dive into life science ethics by requiring ethics courses. The concentration in Biology and Society within the undergraduate major in Biological Sciences combines core education in biology with substantial coursework on the social and ethical dimensions of the life sciences. The concentrations in Biomedical Sciences within the undergraduate majors in Biological Sciences and Microbiology require students to take either Bioethics or Biomedical Research Ethics. The Ph.D. programs in Neuroscience and in Cellular and Molecular Biology require students to take BIO 610, Introduction to Responsible Conduct in Research.

Co-curricular activities provide myriad opportunities for the ethics program to establish and reinforce that understanding ethics and relationships between science and society are integral to biology and not mere side issues. Two examples come from beginning of the year activities. Incoming freshmen in SOLS participate in Camp Ignite—a two-day retreat. With upper division undergraduates as peer mentors, students rotate through a series of sessions designed to build relationships and confidence before students start their study in SOLS. One session, "Difficult Decisions," has students discuss situations with ethical dimensions that undergraduates may find themselves in, such as interacting with a drunken roommate and seeing other students cheat on an exam. For graduate students, ethics is embedded in orientation week in teaching assistant training through a session on minimizing and responding to student cheating.

Special initiatives allow students and faculty in the school to engage more deeply in reflection on ethics. For example, the program has created "ethics labs." One-credit ethics labs parallel 300- or 400-level courses that do not have a traditional lab or recitation. Labs provide an opportunity for undergraduates to engage ethics issues related to life sciences in depth but still in the context of natural science courses. In addition, graduate students design and teach the labs with guidance from the ethics program director so the program is also fostering ethics education skills and experience among the school's graduate students. In 2016–7, Biology and Society Ph.D. students offered ethics labs paired with courses on neuroscience and conservation.

In 2016 and 2017, other special initiatives have included a monthly faculty discussion group on genome editing and a visiting scholars program. For example, in spring of 2016, Diane B. Paul, Professor Emerita, University of Massachusetts Boston, and Associate, Museum of Comparative Zoology, Harvard University, spent a week at ASU as a LSE Program Distinguished Scholar in Residence. Dr. Paul's research principally focuses on the histories of evolution and genetics, especially as they relate to eugenics and the nature-nurture debate, and policy-oriented work on issues in contemporary prenatal and neonatal genetic testing.

To recap, SOLS has embarked on a new phase in integrating perspectives on science and society and ethics into its educational activities. The school has well-established faculty, courses, and degree programs in biology and society, which include bioethics, environmental ethics, and research ethics. Expansion is envisioned primarily in the area of embedding content in science courses and co-curricular activities so that all the school's students develop competency in biology and society and life science ethics. Next, we review major trends in applied ethics education—particularly in EAC approaches in engineering ethics and RCR. We conclude by discussing how these approaches can inform ongoing development of the SOLS LSE Program.

What Is EAC?

According to the Society for Ethics Across the Curriculum (SEAC) (2000), EAC refers to "the teaching of ethics in all academic disciplines." Like Writing Across the Curriculum, which posits that writing should be taught in all disciplines not just composition courses, EAC is based on the principle that ethics instruction should not be limited to one or two courses in the Philosophy or Religion department but rather be spread throughout the curriculum. Matchett (2008) argues that ethics is, in fact, taught in all disciplines, deliberately or not, and that the goal of EAC should be to "make explicit attempts to coordinate or integrate the various ethical lessons their students might be learning."

The EAC concept has been found to be particularly well suited to professional disciplines such as engineering (Cruz and Frey 2003; Mitcham and Englehardt 2016). Cruz and Frey (2003) describe EAC as "a holistic and interdisciplinary approach to integrating ethical concerns throughout the university academic program." Moreover, they argue that EAC "proceeds by ethically empowering faculty and students." Newberry (2004) adds, "ethics-across-the-curriculum has the potential advantage of helping students understand that ethical and societal considerations are integral parts of engineering and not simply add-on material."

As will be discussed later, the goals of and approaches to EAC differ depending on the institutional and disciplinary contexts, but the core concept generally encompasses the points raised above: EAC is a coordinated, integrated, interdisciplinary approach that enables faculty and students to explicitly engage with ethical issues within the context of their academic programs or disciplines.

Elements of EAC Programs

Goals

When integrating ethics throughout a curriculum instructors and institutions typically have several goals. In engineering ethics, Davis articulated a widely used set of four teaching goals. These categories are: increased ethical sensitivity, increased knowledge of relevant standards of conduct, improved ethical judgment, and improved ethical will-power (Davis 1999b). Davis's categories reiterate many of the goals set out by Callahan and Bok in 1980 in their overview of the landmark Hastings Center's project on the teaching of applied ethics in higher education (Mitcham and Englehardt 2016; Callahan and Bok 1980). Any activity in an ethics program can have one or more of these goals as a learning objective. In addition, programs can strategically set up learning objectives for different activities to address all four goals in sequence.

Increased ethical sensitivity embodies what Harris et al. call "professional morality" (2016). Professional morality focuses on what the practice (be it engineering or life sciences) owes to society. In particular, ethical sensitivity means recognizing that professional choices can have positive and negative consequences for society.

Increased knowledge of relevant standards of conduct focuses on the required ethical guidelines for STEM fields. These norms and standards come from a number of sources. For some topics, the relevant standard has legal force and national scope, such as federal regulations concerning research with human and animal subjects. For other issues, professional organizations, employers, journals, or even local research teams may establish best practices. Generally engineering ethics focuses on issues canonized in professional codes, such as the IEEE code of ethics, and other issues presented in widely used engineering ethics text books, such as Martin and Schinzinger (2004) and Harris et al. (2013). Common topics include public safety and welfare, risk and the principle of informed consent, conflicts of interest, whistleblowing, trade secrets, and accepting gifts (Kline 2001). Similarly, for the life sciences, National Institutes of Health (NIH) and National Science Foundation (NSF) require ethics education for trainees (undergraduates, graduates, and postdocs) (Benya et al. 2013). The nine core areas of responsible conduct of research, which were articulated in a NIH policy announced in December 2000 but suspended shortly after, have become the default curriculum. The areas are: "(1) data acquisition, management, sharing, and ownership, (2) mentor/trainee responsibilities, (3) publication practices and responsible authorship, (4) peer review, (5) collaborative science, (6) human subjects, (7) research involving animals, (8) research misconduct, and (9) conflict of interest and commitment" (Steneck and Bulger 2007). The widely used RCR textbooks by Shamoo and Resnik (2015) and Macrina (2014) follow this approach by basically dividing the material they present into chapters on these topics.

The third and fourth categories are higher order learning goals—improved ethical judgment (moral reasoning) and improved ethical will power. Student understanding and application of concepts is key to ethics programs. Students need to be able to apply the concepts, not just reiterate them. Moral reasoning can be difficult to teach STEM students because there are not always simple or binary answers to whether something is good or bad. Applied ethics courses often use the case method to teach moral reasoning and introduce ethical problem-solving methodology that minimizes the use of formal moral theory (e.g. Bebeau et al. 1995; Bulger et al. 2002; Davis 1997; Elliot and Stern 1997).

Those methods help students to improve their ethical judgment and improve their ethical will power. The difficulty is that, while it is possible to assess a student's ethical judgment (see Assessment section below), the testing of ethical will power can only happen in the field (Smith et al. 2007). Therefore, the program can give the students all the tools to make the ethical choices and to recognize ethically fraught situations, but realization of the last goal comes down to the students' behavior on the job. This dynamic leads to contention over ethical will power as a learning objective, because it is unclear how to measure success. Still, improved ethical will power is also an underlying goal of federal regulations that require RCR education. They see ethics education as a way to reduce scientific misconduct.

Curricular change in the University of Michigan's College of Engineering in 1996 is a good example of intentionally staging learning goals across a curriculum. Steneck reports that the college defined goals in three areas across four kinds of courses. They focused on developing understanding of the relationship between technology and society, engineering professional responsibilities (engineering and society), and ethical reasoning. These goals would be pursued at progressively sophisticated levels across the four-year undergraduate programs. First, in Engineering 100, a required general introduction course, students establish fundamental understanding of technology and society and engineering and society issues through studying history of technology and engineering. Second, introductory courses to particular fields explore these issues for the specific field. Third, advanced engineering courses illustrate the concepts in more detail for a single engineering project. Finally, in design courses students analyze technology and society and engineering and society dimensions of their own projects (Steneck 1999). Zaikowski and Garrett (2004) propose a similar approach for undergraduate biology major's study of bioethics. They suggest three tiers. First, students are introduced to science and society issues and ethical reasoning in a general required course as a freshman or sophomore. Second, in the introductory biology sequence, students explore bioethics issues related to the course content. Finally, in capstone courses or research students further examine specific science and society issues or are exposed to the fundamentals of research ethics.

Pedagogical Methods

EAC programs use many methods to integrate ethics into curricula. Likely all the common methods for teaching humanities material—and many novel ones too—have been tried with ethics. Case studies are a commonly used method; students discuss ethical problems in historical or hypothetical cases. Instructors can bring in guest speakers into their classes, or to co-teach classes with them. Students can go through online training or service learning. Instructors can use films or film series. Lecture, discussion, and active learning exercises are all classic approaches. Finally, analysis, application, and writing professional codes of ethics can teach ethics to students (Barry and Herkert 2014; Mitcham and Englehardt 2016). With 25 exemplars from different institutions, the National Academy of Engineering's study "Infusing Ethics into the Development of Engineers: Exemplary Education Activities and Programs" illustrates this diversity (National Academy of Engineering 2016).

Adding to the potential complexity of an EAC program, these methods can be applied in many formats. Instructors can run full semester or quarter courses or labs on ethics (standalone courses), offer intensive workshops running from a half day to several weeks, incorporate one or several lessons or units into STEM courses, or fully integrate ethics and STEM content. Students can join undergraduate organizations or create them. For example, they can participate in Engineers without Borders or similar organizations (Eisen and Parker 2004; Barry and Herkert 2014; Mitcham and Englehardt 2016).

Arguably the most important method for teaching applied ethics is cases. Leading scholars have called for the use of cases in teaching applied ethics for at least 35 years (Callahan and Bok, 1980; Elliott 1995; Harris et al. 1996; Davis 1997). In a meta-analysis of studies that assess the effectiveness of ethics education in STEM, Antes et al. (2009) conclude case-based instruction focused on development of moral reasoning is more effective than other approaches. In 2016 Davis et al. reported on a survey of faculty at the Illinois Institute of Technology (IIT) about what ethics topics should be taught and how (The survey was one approach to assessing the impact of 25 years of running an ethics center and EAC program at IIT.). In open-ended responses, faculty most often reported lecture, discussion, and mixed methods. Of specific approaches mentioned, case discussion was by far the most common. In their overview of engineering ethics, Barry and Herkert (2014) highlight the use and integration of cases in the curriculum.

Among the many methods used in EAC programs, micro-insertion is worth noting for the potential to seamlessly integrate ethics and traditional STEM topics. Davis (1999b), a strong proponent of micro-insertion, explains: "…the strategy, 'micro-insertion,' involves small units of ethics ('ethics minutes' here and there rather than even an 'ethics hour')." Micro-insertion takes place on the instruction design level and appears through integration in the course of modified, small-scale technical problems (Riley et al. 2009). According to Davis (1999b), there are a variety of potential benefits of micro-insertion:

> Micro-insertion requires neither new courses nor radical changes in existing courses. It is both continuous with what professors of engineering, science, and even mathematics already do and… something their students are likely to appreciate.

Program Design

Within the substantial literatures on teaching engineering ethics and RCR, comparatively little describes department, school, or institution-wide programs. In research ethics, the Council of Graduate Schools (CGS) has sponsored three rounds of university-wide RCR/academic integrity program development grants involving 22 institutions, which culminated in three reports on best practices (Tate et al. 2006; Carlin et al. 2008; Council of Graduate Schools 2012; Kent 2013). For engineering ethics, Steneck described the early stages of an EAC initiative at University of Michigan's College of Engineering in 1996 (Steneck 1999). Recently, Mitcham and Englehardt related the development of an EAC program at Utah Valley University (UVU) and efforts to develop one at the Colorado School of Mines (CSM) (Mitcham and Englehardt 2016).

At the program level, ethics programs are eclectic; they combine multiple elements to achieve their goals. To provide an overview of the eight institutional programs supported by CGS's second program building effort, Carlin et al. report on online training, courses, seminars (which includes dinner series and workshops), and "other." At UVU, the EAC program started with an ethics and values core course in the mid-1980s and grew into a rich set of offerings. In 1992, UVU created the Center for the Study of Ethics to coordinate activities

> …including speaker series, awards, evaluations, and summer seminars. Associated activities that have lasted longer than 10 years include an Environmental Ethics Week; Ethics Awareness Week; the Kirk Englehardt Business Ethics Award, and a faculty summer seminar (30 years). (Mitcham and Englehardt 2016)

Activities at CSM have been similarly diverse including a core seminar (LAIS 100 Nature and Human Values), an ethics module embedded in a broadly required engineering course, ethics electives, a humanitarian engineering minor, an honors program with a strong ethics emphasis, a graduate RCR course, an experimental online forum for graduate students to discuss professional ethics, and an ethics bowl team (Mitcham and Englehardt 2016).

The CGS reports and Mitcham and Englehardt all emphasize the importance to ethics programs of leadership and coordination, through an advisory board or ethics center, as well as assessment. In Tate et al. (2006) the discussion of best practices in RCR education starts with appointing an advisory board. Support of upper administration and an advisory board also appear prominently in the Carlin et al. "Checklist for building RCR programs" (2008). Similarly, Mitcham details the efforts to establish and sustain an advisory board at CSM. UVU and IIT both rely on campus ethics centers for coordination. Calls for assessment, discussed below, are similarly prominent in discussions of programs. Leadership and assessment reinforce each other in promoting program level change goals. They can demonstrate an institution's commitment to ethics, which should help cultivate an ethical culture within institutions. Mitcham and Englehardt propose:

> The strong EAC vision is not just to have a collection of activities that could be gathered under an EAC umbrella, but to interpret, analyze, and assess how their interactions are related and contribute to meeting some common goal or goals. Ideally the goal would be to cultivate an institutional culture of ethical science, technology, engineering, and mathematics (STEM), as outlined in a 2014 NSF program that has sought to go beyond the development and teaching of individual courses (Cultivating Cultures for Ethical STEM) (2016).

Train the Trainers Programs

A critical part of integrating ethics into a curriculum is training the faculty and other instructors who do not specialize in ethics to teach in the area. Train the trainer programs show faculty how to integrate ethics into courses by teaching them how to use the different tactics described in the methods section. They may also cover relevant basic and professional ethics content, explore the rational for teaching professional ethics, and introduce participants to the need for assessment. In engineering ethics, RCR, and ethics centers, workshops have become a standard for teaching faculty and graduate students how to bring ethical content and reasoning to their classrooms.

In engineering ethics, Davis established an early train the trainers program of "summer workshops for IIT faculty in how to teach professional ethics in technical courses" (1999a). These workshops focused on teaching faculty how to refine engineering problems to incorporate ethical issues. According to Riley et al. (2009) "Davis has taught faculty how to revise ordinary technical problems in science and engineering to bring out the ethical issues underlying such problems." The workshops were short engagements, usually a few days to a week long, but they were intensive. The end goal of the workshop was not just to teach faculty how to teach ethics, but to "write their own micro-insertions or construct their own original problems [which] allows them to carry out micro-insertions and to teach others to do the same" (Riley et al. 2009). When these workshops were run at IIT the organizers "did not define the background of the faculty" (meaning that workshops were open to faculty from all disciplines), because this allowed multiple perspectives to come to the table over these ethical problems. The approach promoted creative problem solving. The combined knowledge of myriad specialties and backgrounds could be brought into the classroom of each of the attendees (Davis 1993, 1999a; Weil 2003). Train the trainers workshops have been found broadly useful and similar programs have been documented for University of Puerto Rico, Mayaguez (Cruz and Frey 2003), Pennsylvania State University (Litzinger et al. 2003), and Arizona State University, Polytechnic Campus (Herkert 2011).

Clemson University's Rutland Institute for Ethics has used an interesting variant of a train the trainers program. In connection with developing embedded ethics content for the laboratory portion of Introduction to Genetics, they needed to train teaching assistants quickly. To meet this need, they used microsessions in their training workshops (Smith et al. 2007). Smith et al. (2007) explain why they took this microsession approach:

> …the Rutland approach to ethics training is somewhat unique. Often when philosophers are asked to train others in ethics, they design something very similar to what they would do in a traditional ethics class, for example, reading and discussing primary literature. Early on, we at the Rutland Institute decided against this approach. The reason is, first, it takes far too long. Even in a week, there is a severe limitation on how much material one can convey. Second, and more important, it is an approach ill-suited to such a practical goal.

The Rutland approach focuses on building from the bottom up, starting with answering the question of what ethics is. It builds from there, giving instructors a better grasp of ethics while actively showing them how to grow their own programs from the ground up. They discuss basic techniques, how to start ethical discussions in the classroom, and how ethics in the classroom can be justified. Finally, they give the participants tools to move from the initial feelings of many students that ethics has little application and no universal truth to an understanding of how to engage with ethics and promote ethical reasoning. When teaching labs, teaching assistants can directly apply this same format (though usually the student classes are even shorter).

Similarly, several institutions have run prominent and influential Responsible Conduct of Research train the trainer programs. Kenneth Pimple ran the Teaching Research Ethics weeklong summer workshop at University of Indiana. Michael Zigmond and Beth Fischer ran a professional survival skills program at University of Pittsburgh and a companion summer workshop in Snowmass, CO (Pimple 1996; Steneck and Bulger 2007). Michael Kalichman, Francis Macrina, Dena Plemmons, and others have occasionally offered daylong workshops in conjunction with major national meetings such as Public Responsibility in Medicine and Research (PRIM&R) and American Association for the Advancement of Science (AAAS) (van der Burght and Kalichman 2017). These workshops introduce participants to the need and rational for teaching ethics, the RCR topical areas, common ethics education methods, such as case study discussion, and assessment.

Assessment

Assessment is an important, but controversial, factor in the development and implementation of ethics courses and EAC programs in science, technology, and engineering. For example, Antes et al. (2009) conducted a meta-analysis of 26 ethics program assessments and found a "modest" level of effectiveness in ethics instruction. Interestingly, they also found that the more effective programs featured instruction in stand-alone seminars as opposed to EAC programs. Plemmons et al. (2006) found in a study of student perceptions of RCR that students perceived "(1) a wide variety of positive outcomes for research ethics courses, but that (2) the impact on knowledge was greater than that for changes in skills or attitudes."

One assessment model used in engineering ethics education is to key assessment to some of the goals for engineering ethics education described by Davis (1999b): (1) ethical sensitivity, (2) knowledge of professional norms, (3) moral judgment, and (4) ethical will power. For example, Canary et al. (2012) developed study-specific measures to assess goals 1 and 2 and used existing moral judgment instruments to assess goal 3, with mixed results in terms of statistical significance.

Davis (2016) has proposed a method for assessing goals 1 and 2 based upon ratios of student scores on course-specific pre- and post-tests. Davis argues that this simple method can be used to compare assessment across courses and programs, and even across institutions.

While several tests have been developed and used to measure moral reasoning, the most frequently employed in an engineering and science context is the Defining Issues Test (DIT), based on Kohlberg's stages of moral development, and a refined version of the DIT known as the Defining Issues Test-2 (DIT-2) (Rest et al. 1999). Antes et al. (2009), however, caution that "the DIT, a measure of moral development, may be limited in its ability to address all potential, and desired, outcomes of instruction."

As this brief review indicates, there is much work to be done in implementing appropriate and effective assessment of ethics courses and EAC programs. Nevertheless, assessment remains a critical component for developing and justifying such programs and best assessment practices can be identified. For example, based on a review of the literature and their own experiences designing assessment protocols for ethics education research projects and a faculty development seminar for an EAC program, Canary and Herkert (2013) offered the following practical guidance for assessing ethics programs and centers:

- Take instructional design seriously. Incorporate multiple methods of instruction and varied learning activities, as appropriate for the particular content and context.
- Consider what goals are appropriate for each instructional endeavor and clearly articulate them with center/program faculty.
- Use the content, context, and goals to determine assessment strategies and design assessment tools. For many programs/centers, this will involve the use of multiple strategies to capture multiple foci, contexts, and goals.
- When possible, make use of project workshops and contacts with experts to fine-tune assessment methods. Build such workshops and consultations into program/center budgets.
- Make use of informal assessments as well. Assessment need not always be expensive or time consuming. Informal assessments of center projects can provide valuable information to faculty and administrators.
- When appropriate, use existing resources developed in previous studies.

Conclusions: Lessons Learned for SOLS

Revisiting EAC scholarship provides valuable insights for a developing EAC program. For the SOLS program, the literature validates some of the initial decisions and points to important directions for growth particularly in the areas of program design, training trainers, and assessment.

In the LSE Program, development of activities has begun to articulate pedagogical goals that combine examination of biology and society topics with applied ethics, cover at least the first three of Davis's four teaching goals, and approach these aims in a staged fashion. For undergraduates, the program aims to introduce ethics first in the SOLS "freshman experience" activities—Camp Ignite, Bio 189, Life Sciences Career Paths, and introductory biology. Expanding ethics content across all introductory biology courses and key upper division courses (e.g. genetics and evolution) will be an opportunity to assure all students are initially exposed to and then explore more deeply biology and society issues, ethical norms and standards, and ethical problem solving in the context of the life sciences. One-credit "ethics labs" paired with upper-division science courses and full three-credit biology and society courses, including bioethics, environmental ethics, and biomedical research ethics, give students the opportunity to explore one or more specialized topics in depth. Unlike engineering degree programs, most SOLS degree programs lack a capstone design experience where students can synthesize their ethics and technical learning. However, students who participate in undergraduate research programs, graduate from the honors college (which requires a thesis), or major in biology and society (which requires a capstone project) do have such a substantial culminating experience and could be encouraged to use the project for similar advanced ethics engagement.

The effort to expand ethics content across all introductory biology courses and key upper division courses (e.g. genetics and evolution) can learn from the train the trainer literature. Experimenting with embedding ethics content in individual courses achieves valuable goals but is not efficient. A smarter approach would be to recruit faculty teaching introductory biology courses into one or two cohorts for a train the trainers program like the ones outlined above. With graduate student teaching assistants running labs and recitation sections for many of the large courses in SOLS, developing an accelerated train the trainers program similar to the one used at Clemson and described in Smith et al. (2007) may also be beneficial.

On the program level, the SOLS LSE Program is clearly following the eclectic, mixed methods approach with standalone courses, embedded materials, and co-curricular activities. This heterogeneous mix of initiatives would likely benefit from an advisory board made up of faculty from the school. Rather than being divided into disciplinary departments, SOLS is made up of six interdisciplinary faculty groups. An advisory board that could connect the ethics program to each of these faculty groups might improve program visibility and buy-in through engaging faculty across the school.

Although authors Herkert and Ellison led two NSF Ethics Education in Science and Engineering (EESE) projects with substantial assessment components, the SOLS LSE Program has not yet taken on program-level assessment. Clearly assessment would be valuable. In the over 35 years since the Hastings Center's project on the teaching of applied ethics in higher education was published in 1980 the field has developed substantially. Activities and good intentions are not enough. We need to show the impact of EAC activities on students and science. Indeed, the first question from Ellison's colleagues at a recent departmental seminar presentation on the SOLS LSE Program was "How will we know if we've succeeded?"

In summary, this study has suggested a number of inter-related paths for continued development of the SOLS LSE Program:

(1) Institutionalize clear, measurable, goals for the program;
(2) Continue to develop a variety of curricular and extra-curricular experiences aimed at reaching all SOLS students;
(3) Establish a train the trainers program for faculty and graduate students who lack the experience and confidence to embed ethics materials in their courses or to deliver embedded content;
(4) Establish an advisory board comprised of SOLS faculty; and
(5) Implement assessment strategies that measure and monitor effectiveness of ethics instruction at the unit, course, and program level.

References

Arizona State University. (2018a) *Graduate | school of life sciences*. Accessed April 11, 2018. https://sols.asu.edu/degree-programs/graduate.

Arizona State University. (2018b). *Undergraduate degrees | school of life sciences*. Accessed April 11, 2018. https://sols.asu.edu/degree-programs/undergraduate-degrees.

Antes, A. L., Murphy, S. T., Waples, E. P., Mumford, M. D., Brown, R. P., Connelly, S., et al. (2009). A meta-analysis of ethics instruction effectiveness in the sciences. *Ethics and Behavior, 19*(5), 379–402.

Arizona State University. (ND). ASU charter and goals. https://newamericanuniversity.asu.edu/node/25.

Arizona State University (2014) Arizona State University achievements 2002–2014. http://president.asu.edu/sites/default/files/ASU_Achievements_Booklet_Final_032615.pdf.

Arizona State University (ND) Barrette facts and figures—fall 2016. https://barretthonors.asu.edu/about/facts.

Barry, B., & Herkert, J. (2014) Engineering ethics. In *Cambridge handbook of engineering education research*. Cambridge: Cambridge University Press.

Bauerle, C. M., American Association for the Advancement of Science, National Science Foundation (U.S.), Division of Undergraduate Education, National Science Foundation (U.S.), & Directorate for Biological Sciences (Eds.). (2011). *Vision and change in undergraduate biology education: A call to action: Final report of a national conference*. Washington, DC: American Association for the Advancement of Science.

Bebeau, M. J., Pimple, K. D., Mustavitch, K. M. T., Borden, S. L., & Smith, D. H. (1995). *Moral reasoning in scientific research: Cases for teaching and assessment*. Bloomington: Indiana University Press.

Benya, F., Fletcher, C., & Hollander, R. (Eds.) (2013). *Practical guidance on science and engineering ethics education for instructors and administrators: Papers presented at and summary from a workshop on December 12, 2012*. Washington, DC: National Academies Press.

Bulger, R. E., Heitman, E., & Reiser, S. J. (Eds.). (2002). *The ethical dimensions of the biological and health sciences* (2nd ed.). New York: Cambridge University Press.

Callahan, D., & Bok, S. (Eds.). (1980). *Ethics teaching in higher education*. New York: Plenum Press.

Canary, H., Herkert, J. R., Ellison, K., & Wetmore, J. (2012). Microethics and macroethics in graduate education for scientists and engineers: Developing and assessing instructional models. In American Society for Engineering Education (Ed.), *2012 ASEE annual conference program: final conference program and proceedings, June 10-13, 2012, San Antonio Convention Center, San Antonio, TX*. Washington, DC: American Society for Engineering Education.

Canary, H. E., & Herkert, J. R. (2013). Assessing ethics education in programs and centers: Challenges and strategies. In: F. Benya, C. Fletcher, & R. Hollander (Eds.), *Practical guidance on science and engineering ethics education for instructors and administrators: Papers and summary from a workshop December 12, 2012* (pp. 38–43). Washington, DC: National Academies Press.

Carlin, D. B., Denecke, D. D., & Council of Graduate Schools in the United States. (2008). *Best practices in graduate education for the responsible conduct of research*. Washington, DC: Council of Graduate Schools.

Committee on Gene Drive Research in Non-Human Organisms: Recommendations for Responsible Conduct, Board on Life Sciences, Division on Earth and Life Studies, & National Academies of Sciences, Engineering, and Medicine. (2016). Gene drives on the horizon: Advancing science, navigating uncertainty, and aligning research with public values. Washington, D.C.: National Academies Press.

Council of Graduate Schools. (2012). Research and scholarly integrity in graduate education: A comprehensive approach. Washington DC.: Council of Graduate Schools. http://cgsnet.org/research-and-scholarly-integrity-graduate-education-comprehensive-approach-2.

Cruz, J. A., & Frey, W. J. (2003). An effective strategy for integrating ethics across the curriculum in engineering: An ABET 2000 challenge. *Science and Engineering Ethics, 9*(4), 543–568.

Davis, M. (1993). Ethics across the curriculum: Teaching professional responsibility in technical courses. *Teaching Philosophy, 16*(3), 205–235.

Davis, M. (1997). Developing and using cases to teach practical ethics. *Teaching Philosophy, 20*(4), 353–385.

Davis, M. (1999a). Ethics across the curriculum. In *Ethics and the university*. (pp. 111–142). London: Routledge.

Davis, M. (1999b). Teaching ethics across the engineering curriculum. Retrieved April 13, 2018, from http://www.onlineethics.org/Resources/instructessays/curriculum.aspx.

Davis, M. (2016). From practice to research: A plan for cross-course assessment of instruction in professional ethics. *Teaching Ethics, 16*(2), 251–260.

Davis M, Hildt E, Laas K, Society for Ethics Across the Curriculum. (2016). Twenty-five years of ethics across the curriculum: An assessment. *Teaching Ethics, 16*(1), 55–74.

Elliott, D. (1995). Case studies for teaching research ethics. *Professional Ethics: A Multidisciplinary Journal, 4*(3–4), 179–198.

Elliott, D., & Stern, J. E. (Eds.). (1997). *Research ethics: a reader*. Hanover: University Press of New England.

Eisen, A., & Parker, K. P. (2004). A model for teaching research ethics. *Science and Engineering Ethics, 10*(4), 693–704.

Harris, C. E., Davis, M., Pritchard, M. S., & Rabins, M. J. (1996). Engineering ethics: What? why? how? and when? *Journal of Engineering Education-Washington, 85,* 93–96.

Harris, C. E., Jr., Pritchard, M. S., Rabins, M. J., James, R., & Englehardt, E. (2013). *Engineering ethics: Concepts and cases* (5th ed.). Boston, MA: Wadsworth–Cengage.

Herkert, J. R. (2011). The Lincoln teaching fellows program at the ASU Polytechnic Campus. *Teaching Ethics, 11*(2), 1–5.

Kent, J. D. (2013). Institutional strategies for effective research ethics education: A report from the council of graduate schools. In: F. Benya, C. Fletcher, & R. Hollander (Eds.), *Practical guidance on science and engineering ethics education for instructors and administrators*. Papers and summary from a workshop December 12, 2012 (pp. 47–54). Washington, DC: National Academies Press.

Kline, R. R. (2001). Using history and sociology to teach engineering ethics. *IEEE Technology and Society Magazine, 20*(4), 13–20.

Litzinger, T., Christman, J., Lau, A., Tuana, N., & Wise, J. (2003). Learning and teaching ethics in engineering: Preparing engineering faculty to teach ethics. *Proceedings of the 2003 American Society for Engineering Education Annual Conference & Exposition*, 8.810.1-8.810.7.

Macrina, F. L. (2014). *Scientific integrity: Text and cases in responsible conduct of research* (4th ed.). Washington DC: ASM Press.

Martin, W. M., & Schinzinger, R. (2004). *Ethics in engineering* (4th ed.). New York, NY: McGraw-Hill Education.

Matchett, N. (2008). Ethics across the curriculum. *New Directions for Higher Education, 2008* (142), 25–38.

Mitcham, C., & Englehardt, E. E. (2016). Ethics across the curriculum: Prospects for broader (and deeper) teaching and learning in research and engineering ethics. *Science and Engineering Ethics*,. https://doi.org/10.1007/s11948-016-9797-7.

National Academy of Engineering (Ed.). (2016). Infusing ethics into the development of engineers: Exemplary education activities and programs. Washington DC: National Academies Press.

Newberry, B. (2004). The dilemma of ethics in engineering education. *Science and Engineering Ethics, 10*(2), 343–351.

Pimple, K. D. (1996). Teaching research ethics. A Workshop at Indiana University. Fund for the Improvement of Postsecondary Education (ED), Washington, DC. Retrieved from http://eric.ed.gov/?id=ED413834.

Plemmons, D. K., Brody, S. A., & Kalichman, M. W. (2006). Student perceptions of the effectiveness of education in the responsible conduct of research. *Science and Engineering Ethics, 12*(3), 571–582.

Rest, J. R., Narvaez, D., Thoma, S. J., & Bebeau, M. J. (1999). DIT2: Devising and testing a revised instrument of moral judgment. *Journal of Educational Psychology, 91*(4), 644.

Riley, K., Davis, M., Jackson, A. C., & Maciukenas, J. (2009). 'Ethics in the details': Communicating engineering ethics via micro-insertion. *IEEE Transactions on Professional Communication, 52*(1), 95–108. https://doi.org/10.1109/TPC.2008.2012286.

School of Life Science. (2017). Academic program review self-study report. Arizona State University.

Shamoo, A. E., & Resnik, B. R. (2015). *Responsible conduct of research* (3rd ed.). Oxford: Oxford University Press.

Smith, K., Wueste, D., & Frugoli, J. (2007). Using "ethics labs" to set a framework for ethical discussion in an undergraduate science course. *Biochemistry and Molecular Biology Education, 35*(5), 332–336.

Society for Ethics Across the Curriculum. (2000). Society for ethics across the curriculum. http://www.rit.edu/~w-ethics/seac/.

Steneck, N. H. (1999). Designing teaching and assessment tools for an integrated engineering ethics curriculum. In *Frontiers in Education Conference*, 1999. FIE '99. 29th Annual (Vol. 2, pp. 12D6/11–12D6/17). San Juan, Puerto Rico: Stripes Publishing L.L.C. https://doi.org/10.1109/FIE.1999.841685.

Steneck, N. H., & Bulger, R. E. (2007). The history, purpose, and future of instruction in the responsible conduct of research. *Academic Medicine, 82*(9), 829–834.

Tate, P. D., Denecke, D. D., Council of Graduate Schools. (2006) *Graduate education for the responsible conduct of research*. Washington, DC: Council of Graduate Schools.

Weil, V. (2003). Ethics across the curriculum: Preparing engineering and science faculty to introduce ethics into their teaching. In *Emerging technologies and ethical issues in engineering* (pp. 117–124). Washington, DC: National Academies Press.
van der Burght, S., & Kalichman, M. (2017). Train the trainers: RCR courses and workshops. 5th World Conference on Research Integrity. Amsterdam, 28 May.
Zaikowski, L., & Garrett, J. (2004). A three-tiered approach to enhance undergraduate education in bioethics. *BioScience, 54*(10), 942–949.

The Impact of Ethics Across the Curriculum at Union College, 2006–2017

Robert Baker

To every thing there is a season, and a time to every purpose under the heaven.
A time to plant, and a time to reap; A time to keep, and a time to cast away.

Ecclesiastes 3.1-2.

Abstract Union College's Everyday Ethics Across the Curriculum project was initiated in 2006 and immediately became a vital presence on campus, funding over one hundred ethics course segments or stand-alone ethics courses in departments other than Philosophy, over one hundred lunchtime workshops, symposia and other events, and a program of outside speakers running the gamut from white-collar felons to eminent historians of the Holocaust. Among those featured were Kwame Anthony Appiah (*Cosmopolitanism: Ethics in a World of Strangers* and *Honor Code*), Stephanie Bird (editor of the journal, *Science and Engineering Ethics*), Bernard Gert (*Morality* and *Common Morality*), and Eyal Press (*Beautiful Souls*). The initiative played a pivotal role the development of the college's honor code and also hosted four national conferences. It caught the attention of the *Chronicle for Higher Education*, (Union College Website: https://www.union.edu/news/stories/2011/07/the-challenge-of-putting-a-grade-on-ethical-learning.php.) received shout-outs from public intellectuals like Martha Nussbaum, and was the focus of an issue of the journal, *Teaching Ethics* (*Teaching Ethics*, Vol. 9, No. 2, Spring 2009.). In this account the initiative's founding director reflects on factors that contributed to its flourishing, its lasting accomplishments, and his decision to terminate it.

Keywords Donor fatigue · Economics education · Eliphalet nott
Ethics across the curriculum · Ethics education · Honor code · Philosophy
Union college

R. Baker (✉)
Union College, New York, USA
e-mail: bakerr@union.edu

E. E. Englehardt and M. S. Pritchard (eds.), *Ethics Across the Curriculum—Pedagogical Perspectives*, https://doi.org/10.1007/978-3-319-78939-2_22

Origins of Union College's Ethics Initiatives

Upset and frustrated over the accounting scandals that led to the fall of ENRON and several other companies, in 2003 Union College alumnus Michael Rapaport (Class of 1959) approached Professor Harold Fried of Union's Economics Department with a question, "Why didn't a bell go off in the heads of these very clever people to warn them that they were crossing a line, that ethical issues were involved?" Was there some way, Rapaport wondered, to insure that Union College alumni did not turn into morally-obtuse felons like Harvard MBA, Jeffrey Skilling, who "cooked the books" at ENRON? To transform Rapaport's question into an actionable idea, Fried designed a program to entice and educate Economics faculty into introducing ethics into economics courses. This may not seem like a major innovation but at that time no Union College economics course included discussions of ethics in its syllabus. In fact, a review of the college's 2003 Catalogue indicated that, except for a few Philosophy courses, not a single course description mentioned "ethics" or "morality."

The college's founders would have been shocked. Founded in 1795 as America's first non-denominational college, Union, like the eight other colleges founded in colonial America, required all students to take a capstone course emphasizing moral philosophy and paid close attention to students' moral development. In 1804 it even recruited as its fourth president one of America's most famous antebellum moralists, Eliphalet Nott (1773–1866), whose influential sermon "On the Death of Hamilton" led to the outlawing of dueling in the Northeastern United States.[1] By the twentieth century, however, the college had become less focused on students' moral development, terminating its student honor court and honor code in 1925 and rejecting attempts to reinstate an honor code in 1965, 1981 and 1985. As late as the 2004–2005 academic year the college's mission statement made no mention of ethics or of students' moral development, and these words to not appear in any course description in its catalogue—aside from the few in Philosophy. Moreover, when the Philosophy Department proposed an ethics-across-the-curriculum initiative to the college's Academic Affairs Committee, it was rejected on the grounds that "ethics had no place in art, math, science or social science courses."

Swimming against this tide, from 2003 to 2005 Professor Fried's ethics initiative engaged nine of the eleven members of the Economics Department.[2] They had been enticed by course development funding for introducing ethics segments into their courses, and engaged by on-going workshops on moral philosophy led by Professors Baker and Mathias of Union's Philosophy Department, and by Aine Donovan, Director of the Institute for the Study of Applied and Professional Ethics at Dartmouth College. Buoyed by the success of the Economics initiative, Fried and Rapaport approached Baker with the idea of initiating a campus-wide ethics-across-the curriculum initiative.

---

[1]Eliphalet Nott, 1804, "On the Death of Hamilton" http://www.bartleby.com/268/8/24.html accessed August 16, 2017.

[2]Fried (2009).

The Organizational of Union's Ethics Across the Curriculum Initiative

Baker had directed several innovative educational initiatives at Union including one establishing an on-line bioethics masters program[3] and an NIH-funded initiative to use on-line e-education to train research ethicists in Central and Eastern Europe.[4] Drawing on this experience and on elements of Fried's successful Economics project, Baker and Rapaport agreed on a three-year grant whose purpose was to

- **Demonstrate** that "colleges can stimulate students understanding and appreciation of everyday ethics as it affects people in their jobs, professions and daily activities so that if they encounter ethical issues they will recognize them 'as if a little bell rang inside their head to warn them.'"
- **Administrating the Initiative would be an**: Executive committee, a one-half-time director (Baker), and an assistant director.
- **Space and Services** would be provided by Union College and the
- **Faculty would receive** monetary compensation and public commendation for course/segment development (as in Fried's initiative);
- **A Grants Program** would fund development of course segments, speakers, conference and workshop attendance (as in Fried's initiative);
- **Measurable objectives would be**:
  - **Ethics courses or segments** in 50 non-philosophy departments;
  - **Numbers of regular faculty workshops and symposia held**;
  - **Campus-wide extracurricular programs** for student- and faculty-initiated speakers and events;
  - **Dissemination of information** about the initiative nationwide;
  - **Documentation of Ethics Across the Curriculum activities;**
  - **Assessment** of progress regularly.

Implementing the Rapaport Ethics Across the Curriculum Initiative

After an initial six-month set-up period the initiative was launched in the 2006–2007 academic year. During that year the ethics initiative held ten on-campus workshops (about twenty faculty attending per session) on subjects ranging from pedagogical demonstrations, to reports from participating faculty, to sessions on

---

[3]Now the Bioethics Program of Clarkson-University and the Icahn School of Medicine at Mount Sinai http://bioethics.uniongraduatecollege.edu.

[4]Now part of the Clarkson-Mt. Sinai bioethics program http://bioethics.uniongraduatecollege.edu/nih-grants/europe.

such topics as engineering ethics and "Teaching About Race and other 4-letter topics."[5] The Initiative received 24 proposals for ethics segments in courses (from over 10% of Union's 200 faculty members), 14 of which were taught in 2006–2007;[6] it sponsored seven campus-wide speakers/events and it supported faculty

---

[5]Rapaport Ethics Across the Curriculum Faculty Development Workshops

1. *Ethics Workshop*, Aine Donovan, Dartmouth College, April 27, 2006
2. *Ethics Workshop*, Robert Baker and Michael Mathias, May 16, 2006
3. *Ethics faculty meeting*, October 4, 2006
4. *Ethics Workshop*, Robert Baker, October 18, 2006
5. *Scientific Integrity Workshop*, John Kaplan, Professor, Albany Medical College, November 1, 2006
6. *Teaching Race and other 4-letter topics*, Robert Baker, Michael Mathias, January 10, 2007
7. *Ethics Workshop*, Robert Baker, January 24, 2007
8. *Ethics Lunch with Grant Recipients*, Robert Baker, February 7, 2007
9. *Ethics Workshop*, Robert Baker, April 25, 2007
10. *Engineering Ethics Workshop*: "Exploring New Curricula in Engineering and Liberal Education" Deans Cherrice Traver and Douglass Klein, June 15, 2007.

[6]**Course Ethics Segments Supported by the Initiative (listed by trimester)**

Ethics Segments Implemented, Fall '06:
Prof. David Baum (English) FYP "Freedom on Trial"
Prof. Kelly Black (Mathematics) "Introduction to Statistics"
Prof. George Gmelch (Anthropology) "Sports, Society, and Culture"
Prof. Gregory Reid (Chemistry) "Organic Chemistry I"
Prof. Channette Romero (English) "American-Indian Women Writers"
Ethics Segments Implemented, Winter '07:
Prof. Jeffrey Corbin (Biology) "Introduction to Environmental Studies"
Prof. Ashraf Ghaly (Engineering) "GIS for Humanity"
Prof. Peter Heinegg (English) FYP "Secular Humanism"
Prof. Gregory Reid (Chemistry) "Organic Chemistry II"
Prof. Jordan Smith (English) "Introduction to the Study of Literature: Poetry"
Ethics Segments Implemented, Spring '07:
Prof. Christopher Pizzino (English) "Science Fiction"
Prof. Kenji Tierney (Anthropology) "Food and the Self"
Prof. Mark Walker (History) "Science, Medicine, and Technology in Culture"
Prof. Tom Werner (Chemistry) "Chemistry and Athletic Performance"
Ethics Segments/Events *Planned*, Fall '07
Prof. Martin Benjamin (Visual Arts) "Photographing Another Culture: Vietnam"
Prof. Brendan Burns (Computer Science) "Computer Games"
Prof. Katherine Lynes (English) FYP "Literature and the Environment"
Profs. Seyfollah Maleki (Physics) & Walker (History) "Physics and Politics"
Prof. Victoria Martinez (Modern Languages)
"Human & Civil Rights: Literature & Film of the Mexican-AmericanBorder"
Ethics Segments *Planned*, Winter '08
Prof. Martin Benjamin (Visual Arts) "Photography I" and "Photography III"
Prof. Anupama Jain (English) "Utopian Philosophies and Fictions"
Prof. Sharon Gmelch (Anthropology) "The Museum: Theory and Practice"
Prof. Bonney MacDonald (English)
"Literature, Culture, & Ethics in the Rangeland West" (miniterm)
Prof. Joyce Madancy (History) "Opium: East and West," Sophomore Seminar.

and students attendance at off-campus conferences or workshops.[7] Among the events hosted for'06-'07 were lectures on "Business Ethics from the Perspective of A Convicted White-Collar Criminal," and a full day of seminars on "Engineering Ethics" led by Dr. Stephanie Bird, editor of the journal, *Science and Engineering Ethics*.[8] Union's Ethics Across the Curriculum was also preparing to host the American Society for Bioethics and Humanities' National Undergraduate Bioethics Conference (NUBC) in 2008.

Reinstituting Union's Honor Code and Student Honor Council

In 2005–2006 several engineering students who, in their view, had been improperly convicted of illicit collaboration by an all-faculty panel, were given a penalty assignment of researching academic integrity. In their report they recommended

---

[7]**Conference Attendance Supported by the Initiative 2006–07**

1. Academic Integrity, University of Colorado, Boulder, CO, October 19–21, 2006 attended byAli Gardezi and Vishal Patel, student members of the AAC Subcouncil on Academic Integrity.
2. Ethics across the Curriculum, Dartmouth College, Hanover, NH, Nov. 17–18, 2006 attended by Dr. Anastasia Pease.
3. "Reading Human Rights" Workshop, the Institute for Writing and Thinking, Bard College, Annadale-on-Hudson, NY, December 8–10, 2006, attended by Mary Mar, Director, Writing Center.
4. American Society for Engineering Education, New England Section Conference, Rhode Island University, Kingston, RI, April 20–21, 2007, attended by Prof. Ashraf Ghaly, Engineering.
5. Biennial Conference for the Study of Literature and the Environment, Wofford College, Spartanburg, SC, June 11–16, 2007, attended by Prof. Katherine Lynes, English Department.
6. American Literature Association Conference, Boston, MA, May 25–26, "Ethics and Native American Literature," presentation by Prof. Chanette Romero, English Department.

[8]**Ethics Events Organized and/or Supported 2007**:

1. *Ethics of Public Spaces Lecture "A Century of Theme Park Utopias"* (public lecture), Professor Michael Pinsky, University of South Florida, May 10, 2007.
2. *Ethics and the Face of Difference* (public lecture), Professor Bonnie TuSmith, Northeastern University, May 15, 2007.
3. *Business Ethics from the Perspectives of A Convicted White-Collar Criminal* (public lecture), Walt Pavlo, MBA former MCI business executive, May 22, 2007.
4. *Sports Ethics*: *What Everyone Should Know about the Anabolic Steroids Abuse Crisis* (public lecture) Professor Bruce Svare, University at Albany, May 24, 2007.
5. *Engineering Ethics*: *Dr. Stephanie Bird, Editor of the journal Science and Engineering Ethics*, Prof. Stephanie Bird, MIT, June 14, 2007.
6. *Engineering Ethics*: *Engineering Principles for a Living Planet*, Professor William Vitek, Clarkson University, June 14,'07.
7. *A Brief Introduction to the History of Modern Research Ethics,* Presented to Union Summer Research Scholars, Robert Baker, July 19, 2007.

that the college join the Center for Academic Integrity[9] and consider implementing an honor code with a student majority honor court. They gave a copy of the report to Professor Baker who made the initiation of an honor code a major objective of the newly created ethics initiative. The college administration soon appointed Baker to chair a committee to assess whether to reinstitute Union's honor code. After two years of study, which involved a survey campus opinion, the committee recommended the restoration of an academic honor code and honor court. Union's faculty, students and trustees approved their proposal.

During the same period Union underwent a strategic planning process. Acknowledging the idea behind the ethics initiative, the opening lines of the planning document offered the following guidance to the planners.

> The strategies we will pursue all have the purpose of moving toward fulfillment of our vision for the College: **Union College will be a leader in educating students to be engaged, innovative, and *ethical* contributors to an increasingly diverse, global, and technologically complex society.**[10]

The bolded words were later integrated into Union's 2010 mission statement and for the first time in at least a half-a-century, perhaps for the first time since 1925, the college officially regarded ethics education as integral to its mission. In 2012 Union reinstated its student honor council and honor code.

External Review 2009–2010

Michael Rapaport continued to fund the project in 2009–2010 academic year during which Union's administration supported an external review of the initiative by three leading ethics educators: Daniel E. Wueste, President of the Society for Ethics Across the Curriculum, Aine Donovan and Stephanie J. Bird. The review committee summed up their analysis as follows

> Since its inception in 2006, Union's Ethics Across the Curriculum has provided more than 40 lunch workshops guided by external speakers bringing their expertise in the fields of ethics or other related fields to help faculty and staff learn about incorporating ethics teaching and concepts into the classroom. Ethics Across the Curriculum has also funded more than 70 different proposals by faculty members in most every department on campus to incorporate ethics segments into their own courses, bring in speakers, buy supplementary materials or attend conferences for further ethics-related training. Ethics Across the Curriculum has funded proposals in the following departments between 2006 and 2009: Anthropology (3), Arts (5), Biology (3), Chemistry (3), Classics (5), Computer Science (2), Economics (9), Engineering (3), English (19), History (4), Physics (1), Math (1), Political Science (1), Psychology (5), Sociology (2), Spanish (2), Writing Center (3)....

---

[9]Center for Academic Integrity (established 1992) http://www.academicintegrity.org/icai/home.php accessed August 17, 2017.

[10]Union College Strategic Plan—bolding in original proposal, italics added. Current version of mission statement https://www.union.edu/about/mission/ accessed August 17, 2017.

We were very impressed by Union's Ethics Across the Curriculum Initiative. We were especially impressed with the high level of interest and enthusiasm exhibited by the faculty, administrators, students, and staff with whom we met. The self-report and other materials we received document the many accomplishments of the Initiative. Notable among the materials we received is an issue of *Teaching Ethics*, (Volume 9, Number 2-Spring, 2009), some 82 pages of which are about or illustrative of the ethics across the curriculum work that has been going on at Union College. An article in this issue by Anastasia Pease and Robert Baker, "Union College's Rapaport Everyday Ethics Across the Curriculum Initiative," summarizes the Initiative's goals and measureable objectives (pp. 6–7) and reports the accomplishments of the Initiative over a three-year period, March 2006–March 2009 (pp. 14–15). It is clear that in the terms set out at the beginning of the Initiative, the program we were asked to evaluate is a genuine success; it is something of which Union College and all those who have been active in it should be very proud.

We believe that the program is well positioned to move to the next level, though as the reader will be aware, we see some challenges ahead.

> …. The program faces some challenges, of course. For example, the fact that the program has been formed by and is heavily dependent on the vision of its founding director led us to wonder about a plan of succession in leadership… at present, there is no such plan [and] … we were surprised to find that the new chair of the philosophy department is not quite certain about the details of the program or what sort of connection it should have with philosophy. Moreover, the program did not figure in his recruitment.

The External Review Committee also noted that although the administration was supportive, it had not expressly recognized the tie between the ethics initiative and Union's educational mission.

The Apogee and Perigee of Union's Ethics Initiative: 2010–2016

From 2010 to 2013 the initiative continued with a mix of support from the original donor and from the administration, enabling it to support two national conferences in 2010–2011, (an externally funded national conference on Ethics and Disability, and the national conference of the Society for Ethics Across the Curriculum). With over twenty-four course segments being created in each year this became the high point, the project's apogee. After that point, and especially after the passage of Union's honor code, space and other resources available to project from the college were thinned. In 2012–2013 the dedicated space was removed and the assistant director was replaced by shared use of administrative services. After September 2014, the director ceased to receive release time for administering the initiative.

During the same period donor fatigue set in and the amount of external funding diminished substantially. Nonetheless major speakers like the moral philosophers,

Bernard Gert (2011) and Kwami Anthony Appiah (2012), the crusading medical reformer, Stephen Miles MD, and two Holocaust related events drew substantial audiences: one was offered jointly by a physician from the American Medical Association and a staff member of the National Holocaust Museum (2014); the other, by the journalist Eyal Press author of *Beautiful Souls*, was a documentary account of people who aided Jews and other victims during the Holocaust (2015). Creation of new course segments, however, fell to fewer that five per year, and many were from previous grantees.

Reflections on the Initiative and Its Impact

The ethics Initiative would probably have limped on had not a personal event intervened. In the fall of 2015 the director's wife was diagnosed with an aggressive form of uterine cancer and the director informed Union's administration that he could not fulfill any extracurricular obligations, including directing the ethics initiative. This history and assessment is being written in the last days of the summer of 2017, as the director's wife, who is now a cancer survivor, emerged from treatment and its aftereffects (the so-called "chemo-fog" of fatigue and clouded memory that lingers long after chemo and radiation therapy cease). As the initiative's director I decided to take the opportunity offered by this essay to reflect on the initiative and its impact.

As I stared at the blank computer screen thinking about how to open and organize this essay, words from *Ecclesiastes* 3.1 came to mind, "To every thing there is a season." The ethics initiative was conceived as a response to the erosion of personal morality and professional ethics in the American business community that underlay and culminated in the housing bust of 2006 and the great recession of 2007–2009. Appropriately to that season, the initiative started in the Economics Department and then went college wide. As a response to failures of personal morality, from its early embrace of the idea of an honor code to first-year talks by convicted white collar felons, to Eyal Press's final presentation which, to quote the subtitle of *Beautiful Souls,* was about "the courage of ordinary people in extraordinary times," the initiative's focus was on individual moral responsibility. Yet in the run-up to the election of 2016 the focus of America's national debate was on our collective responsibility for such things as climate change, income disparities, mass incarceration and whether Black lives matter, terrorism, immigration, and the exodus of White working-class jobs. The season had changed, and the initiative had fallen out of step with the dominant narrative of the day. Not surprisingly, therefore, faculty lost interest.

Closer to my academic home, as the external reviewers feared the newly hired chair of the Philosophy Department turned out to be benignly indifferent to the ethics initiative: willing to support specific endeavors (e.g., funding Kwame Anthony Appiah's speaker's fee) but unwilling to encourage junior faculty to become involved in the initiative. Consequently, the project has no successor. The administration's interest also seemed to wane after the reintroduction of the honor code—although at the point that I terminated the project I still had a 5-figure budget and shared administrative assistance. Perhaps appropriately, the initiative's last financial transaction involved funding related to a survey to assess the efficacy of Union's honor code.

In the end, the donor and I "burned out." Nonetheless our initiative left an enduring impact on Union College, stirring it to include a commitment to educating ethical alumni in its mission statement and reinstituting a student majority honor council and an honor code. Less concrete but no less important was our ability to engage more than half of Union's faculty in an ethics initiative, so that, for the better part of a decade, most Union students were immersed in discussions of ethics not only in humanities courses but also in engineering, science and social science courses. We will never learn whether, as Michael Rapaport hoped, this will affect their future conduct; but it should better prepare alumni to recognize and understand moral challenges as they confront them.

On a personal note: stirring a campus to engage in ethical reflection was a rare pleasure for someone like me, who has spent a lifetime reflecting on moral philosophy. Had I known at the outset what the initiative's life and half-life would turn out to be, I would happily do it over again. Yet, to return to *Ecclesiastes*, the time to keep the initiative has passed, now is the time to cast it away.

Lessons Learned

What might others learn from my experience? Perhaps most importantly that the initiative succeeded in fulfilling the original aims of Michael Rapaport's demonstration project: that with an modest allocation of money, some course releases and administrative support it is possible to create courses and course segments that engage students in discussions of ethics across a college's curriculum, even in a college like Union that initially confined ethics to its Philosophy Department, with some spill over into other humanities courses. The project succeeded beyond anyone's expectations, leaving a permanent imprint on the college's professed mission and reinstituting a student-led honor code/honor council. It was so successful that it morphed into an effort to become a permanent program. However that proved a step too far since neither the donor nor the college was willing to commit resources on a permanent basis. Instead they deployed resources elsewhere so that the initiative dwindled to one-person's extracurricular activity supported by college funds and administrative assistance.

In the end, for both personal and professional reasons, I too decided to deploy my time and energy elsewhere. Had I kept the initiative going I would have rekindled it with events more in tune with a national conversation about a divided country, climate change, anti-science/anti-elitism and similar topics to pique faculty interest, find alternative streams of funding, and tap new sources institutional and external support. I believe that anyone trained in moral philosophy or moral theology willing to commit time and energy to initiating an ethics across the curriculum initiative can be successful: should such an opportunity present itself their motto should be *carpe diem*: seize the moment.[11]

Reference

Fried, H. (2009). The Michael S. Rapaport initiative to introduce ethics into the economics curriculum at Union College. *Teaching Ethics, 9*(2), 25–50.

[11]From Horace's *Odes* (1.11): the Latin reads "carpe diem, quam minimum credula postero," in John Conington's 1882 translation, "Seize the present; trust tomorrow e'en as little as you may," or as I put it, "seize the moment."

The Ethics Across Campus Program at the Colorado School of Mines

Sandy Woodson and Qin Zhu

Abstract By taking the Ethics Across Campus Program (EAC) at the Colorado School of Mines (CSM) as an example, this paper discusses how institutional, social, and policy contexts provide opportunities—and challenges—for ethics centers and programs to create effective moral learning experiences for students. This paper begins by depicting the historical background and institutional context within which this EAC program was founded and has evolved. It then introduces the major campus programs and curricular initiatives (e.g., the Daniels Fund Faculty Fellows Program) that aim to enhance students' learning opportunities. Finally, this paper reflects on some major challenges faced by the program, including how to bridge physical and intellectual distances between faculty, and how to best cultivate an ethical climate for STEM education and research.

Keywords Ethics education · Professional development · Ethics across the curriculum · Science and engineering ethic

"[The e]ngineering curriculum addresses ethics in the context of safety and the code of ethics because these are perceived as quantifiable ways of analyzing complex decisions....This leads to a diluted approach to ethics in engineering education.... The cyclical nature of engineering education results in professors who are unable or unwilling to have these multi-dimensional discussions with their students and to assess their ability to analyze ethical dilemmas. Therefore, to change the way that engineers approach ethics, a change must be made in the culture that surrounds ethics and other so-called 'soft' topics, as perceived by STEM majors." Mechanical Engineering Student Paper, Fall 2015.

S. Woodson (✉) · Q. Zhu
Colorado School of Mines, Golden, USA
e-mail: swoodson@mines.edu

Q. Zhu
e-mail: qzhu@mines.edu

E. E. Englehardt and M. S. Pritchard (eds.), *Ethics Across the Curriculum—Pedagogical Perspectives*, https://doi.org/10.1007/978-3-319-78939-2_23

Background

The Colorado School of Mines is a public university of approximately 5700 students that focuses on engineering and applied science. Mines opened its doors in 1879, focusing on gold and silver mining and evaluation. Over time its curriculum expanded to include a wide spectrum of engineering, applied science and mathematics. Currently, the top three undergraduate majors are Mechanical Engineering, Petroleum Engineering, and Chemical and Biochemical Engineering (Colorado School of Mines 2015). Of the 5700 students matriculating at Mines, approximately 1200 are graduate students. Mines offers 30 different master's programs, and 24 PhD programs. In May 2017, Mines conferred 741 bachelor's degrees, 163 master's degrees and 39 doctoral degrees (Colorado School of Mines 2017).

Mines has distinguished itself by developing a curriculum and research program geared towards responsible stewardship of the earth and its resources. In addition to strong education and research programs in traditional fields of science and engineering, Mines is one of a very few institutions in the world having broad expertise in resource exploration, extraction, production and utilization. Mines is emerging as one of the nation's university leaders in the critical areas of water and the environment, unconventional and renewable energy, and strategic and critical minerals. A Mines education is dually-focused, emphasizing both practical problem-solving skills and creative critical thinking abilities.

Mines enjoys high rankings from a variety of sources, including the following: #1 from USA Today's "Top Ten Engineering Schools in the US" (Stockwell 2016), #1 Public University in Colorado and #53 in the country by The *Business Journals* (The Business Journals 2015), and #33 in US News and World Reports "Top Public Schools" in 2017's Best Colleges (U.S. News & World Report 2017). CSM has the highest admission standards of any college in Colorado and a much higher than average freshman-to-sophomore retention rate of 94%. CSM has exceedingly strong ties to industry (of the $63 million in research funding in 2015, almost half came from non-governmental sources). In fact, companies approach CSM faculty for help with issues surrounding Corporate Social Responsibility, Social License, and Professional Ethics. These ties allow for an alternative entry point for ethical considerations, possibilities extending beyond more traditional business ethics, bioethics, etc.

The focus on science and engineering ethics education at Mines reflects professional and national priorities: (1) the Accreditation Board for Engineering and Technology (ABET) requires students who graduate from accredited programs to achieve eleven learning outcomes, of which three are directly relevant to ethics (ABET 2016). (2) The National Academy of Engineering (NAE) has a center devoted to ethics education—the Center for Engineering Ethics and Society, and it also manages the internationally known Online Ethics Center. Mines faculty have been involved in the Center in a variety of ways, including by participating in workshops and consulting on research projects (3) Additionally, the National Science Foundation (NSF) funds a National Center for Professional and Research Ethics, as well as sponsors a major program, "Cultivating Cultures for

Ethical STEM." In fact, faculty at Mines have received two separate CCE-STEM grants. (4) Engineering societies are also raising awareness about ethics; for example, the Institute for Electrical and Electronics Engineers (IEEE) Standards Association introduced the Global Initiative for Ethical Considerations in the Design of Autonomous Systems (IEEE Standards Association 2016).

The national dialogue around engineering education has been led by NAE, and it is deeply invested in engineering ethics. NAE produced *The Engineer of 2020*, which has been a guiding document for all engineering programs in the US since its publication in 2004. The text states that upholding "high ethical standards" and developing "a strong sense of professionalism" are vital attributes for the engineer of 2020 (National Academy of Engineering 2004). Mines has been recognized by NAE Center for Engineering Ethics: a 2016 NAE Report named 25 "Exemplary Engineering Ethics Programs," honoring three "exemplars" from CSM (National Academy of Engineering 2016). Only one other institution had more than one program honored—MIT. The National Academy continues in its efforts to provide incentives and guidance for engineering pedagogy.

The NSF provides funding for programs to implement the pedagogical goals articulated by the NAE, as well as innovations that researchers develop on their own. The NSF stipulates that all institutions that receive funding must verify that undergraduate students, graduate students, and postdoctoral researchers participate in RCR (responsible conduct of research) training. Furthermore, NSF "anticipates that institutions will develop their RCR training programs in a manner that helps prepare the next generation of researchers, including the consideration of risks or other factors associated with student and postdoctoral researcher participation in research" (National Science Foundation 2009) Each institution has the responsibility to "determine both the content and the delivery method for the training that will meet the institution's specific needs for RCR training in all areas that institution for which NSF provides support" (National Science Foundation 2009) As an applied science and engineering institution, Mines has the obligation to design ethics education learning outcomes that best serve its own institutional missions and campus culture, and ensure that students and postdocs funded by NSF meet these ethics education learning outcomes.

Organizational Context

The original impetus for building an EAC program at Mines emerged in 1997 (Mitcham and Englehardt 2016). Over the last 20 years, commitment from the institution's administration has waxed and waned, which has inevitably affected attempts to establish a robust EAC presence at Mines. As it was originally envisioned, the EAC program continues to be a campus-wide effort, with the director reporting directly to the provost. The EAC program as it now stands is the product of many years of effort and administrative shifts, and is poised to make new inroads on the Mines campus.

One major milestone in the Mines EAC program's evolution was the establishment of the Boettcher Foundation Endowment for Distinctive Educational Programming Fund. Since 1952 the Boettcher Foundation has provided merit-based scholarships to Colorado residents. The Mines Programming Fund came online in 2010, and focuses on supporting "innovative ways to enhance the educational experience for exceptional undergraduate students at Colorado School of Mines." To that end, monies were directed to support EAC initiatives, including faculty support to develop a graduate-level Research Ethics course, support for faculty workshops, as well as other activities. For example, from 2013–14 EAC provided funds to develop and moderate an on online ethics discussion platform, "Collaborative Online Discussion on Ethics" (CODE), which was designed as an outlet for graduate students. CODE was envisioned as a way to "facilitate confidential discussions of ethics in relation to students' academic and career fields together with easy access to relevant resources via and anonymous, moderated online exchange (Mitcham and Englehardt 2016, p. 18). In the third semester of the pilot program, the moderators invited students from the undergraduate ethics class to participate, an activity that highlighted some of the particular strengths and weaknesses of Mines students.

As indicated above, Mines awards degrees only in areas of engineering and applied science. According to the most recent statistics (2016–17), Mines students are academically very strong, with new freshmen having average ACT scores of 31, SAT scores of 1344, and high school GPA's of 3.9. Our undergraduate population is skewed heavily toward men, with a 72% male-identifying population. Ethnic minorities make up 17% of the student body (Colorado School of Mines 2016a, b). Thus, the undergraduate student population is markedly homogenous: white, male, and high-achieving. During the semester that undergraduate students participated in the CODE program, for example, several female students described their trepidation about working on-site in remote locations, with no other women around. Several other female students described their similar fears. The male students who responded to the post were all remarkably hostile, either telling the women to get different jobs, or to stop blaming men for their problems. The anonymity, perhaps, seemed to be used to different effect, depending on gender. This example, which included a tiny fraction of the student body, nevertheless raises important issues for EAC's consideration.

In the first author Woodson's experience in the classroom, Mines undergraduates tend to be exceeding literal, linear thinkers, so in a fundamental way they are well-suited to philosophical argumentation couched in terms of premises and conclusions. On the other hand, that linearity lends itself to "flowchart" thinking, where processes are broken into discrete steps that ignore the big picture. As one HASS professor describes it, engineers and applied scientists are by nature problem-solvers. We humanists problematize, which is antithetical to them. Indeed, much of our work consists in helping these particular students see the big picture, to see the world through the eyes of another and recognize the intrinsic and instrumental value in doing so.

In 2015, Dr. Paul Johnson was named the 17th President of the Colorado School of Mines. A chemical engineer by training, he came to Mines from Arizona State University where he was the Dean of the Ira A. Fulton Schools of Engineering, and professor in the School of Sustainable Engineering and the Built Environment (Ira A. Fulton Schools of Engineering 2015). As the dean leading ASU's engineering college, which is one of the most innovative engineering schools in the country, Johnson focused on improving student outcomes and promoting innovative programming, "[arguing] that students are to be engineers from day one" (Ira A. Fulton Schools of Engineering 2015). During his time at ASU, majors in engineering jumped from 6000 to 19,000 (Colorado School of Mines 2016a, b), a fact that surely didn't escape the notice of the Mines Board of Trustees. At Mines, President Johnson has been explicitly committed to creating diverse and professional learning experience for students, including building a program on entrepreneurship and mentoring that emphasizes learning outside traditional class work. Early in his tenure at Mines, Johnson appointed a new Provost—Dr. Tom Boyd—who is deeply committed to promoting EAC, and Johnson himself has demonstrated his commitment to EAC programs by attending numerous lectures by visiting ethics education scholars, and inviting the Mines students who competed in Intercollegiate Ethics Bowl to his office for pizza. Virtually all of the upper administration–the President, the Provost, Deans and Department Chairs—have regularly attended EAC lectures and workshops, and the Provost committed institutional funds and other resources to support a (sadly unsuccessful) bid to bring the Association for Practical and Professional Ethics headquarters to Mines. Clearly, we have entered a new era of EAC support.

Also in 2015, the EAC program saw other major organizational shifts. In Spring 2015, Carl Mitcham (co-director since the inception of EAC) began transitional retirement. Roel Schneider (the other co-director) stepped back from EAC to act as Interim Department Chair of Geology, and starting in fall 2017, he will lead a center devoted to Professional Development Education. Sandy Woodson was named director for EAC in January 2016, with her particular strength in undergraduate ethics education. The composition of the EAC committee membership changed as well: Another committee member began transitional retirement, so two members needed to be replaced, all the while keeping in mind the necessity for broad representation from across campus. (Mitcham and Englehardt 2016, p. 24) In our case, "broad representation" has been interpreted to include faculty—both tenure-line and teaching–, disciplinary variety, administration, and students. From the pre-2016 committee, a teaching professor in Physics and the Dean of Students remain. We added the Director of the Title IX Program, a tenured professor in petroleum engineering, and two undergraduate students. Each person at the table now brings unique perspectives that will inform our work moving forward: administrators who are privy to important information about the campus climate, students who are a part of that climate, and faculty who have the power to affect change within the curriculum.

Given input from each of our members, this iteration of the EAC committee has identified the following focus areas:

- Improving the campus culture for women
- Assessing and promoting academic honesty
- Evaluating ethics instruction in STEM courses
- Establishing a lecture/film series
- Providing a central clearinghouse for ethics initiatives on campus, a resource for all faculty on campus

These focus areas may well change, and in fact we anticipate that they will. These focus areas, however, have been chosen as avenues to improve the entire campus, both inside and outside the classroom. Ultimately, this version of EAC is an attempt to help and coordinate all the efforts at improving the campus culture, as envisioned by students, faculty and administrators, or as stated on the Mines EAC website:

The Ethics Across Campus (EAC) program at Colorado School of Mines serves as an umbrella for multiple ethics-related teaching, research, and outreach activities. As such, it seeks:

- to promote, extend, and deepen the understanding of ethical issues in relation to applied science and engineering education and research;
- to coordinate ethics teaching, learning, and practice;
- to serve as a consultative body and resource for any group or organization whose policies and/or procedures affect the ethical aspects of life at CSM.

Finally, in 2016 two other significant events occurred: (1) EAC received a $65,000 grant from the Daniels Fund to promote ethics education on campus, and (2) a postdoctoral position in engineering education was established. A postdoc was hired to help administer the Fund and to teach the graduate Research Ethics course. In 2017, the postdoc position was renewed, with the expectation that research and funding proposals continue apace. These resources—both financial and intellectual—are being employed in a variety of initiatives, which are detailed below.

Campus Programs

Broadly speaking, there are two classes of initiatives currently undertaken by EAC: Campus Programs, which seek to engage students and faculty across campus and in a variety of venues, and Curricular Programs, which as the name implies, focus on the 4-year curriculum as well as specific courses within the curriculum. Here we describe the programming developed for the broader campus.

Campus programs seek to engage and develop synergies for student and faculty engagement with ethics and ethical issues. For example, the Intercollegiate Ethics Bowl—developed and sponsored by the Association for Practical and Applied Ethics (APPE)—is an activity available to all undergraduate students. The

Intercollegiate Ethics Bowl program was created 20 years ago, and is sponsored and organized by APPE: Teams of students analyze and respond to ethical dilemma scenarios, many of which are actual cases culled from ongoing controversies. At the Ethics Bowl, teams face each other and must answer a question connected to the case. The teams do not know the question or which case that will be used in any given round. A panel of judges, selected from faculty, local business leaders, graduate students and alumni, assesses the quality of each team's analysis on the bases of how well team members organize and present their approach, attend to and analyze the morally relevant features of the case, and anticipate and preemptively respond to commentary and questions. The Ethics Bowl is designed to develop and showcase moral reasoning and civil discourse.

The EAC Program sent Woodson to a June 2014 workshop for the Intercollegiate Ethics Bowl program, where she began the process of bringing Mines into the competition. She worked with another faculty member (Toni Lefton, Humanities, Arts & Social Sciences) and the McBride Honors Program in Public Affairs to field the first Mines Ethics Bowl team in the Rocky Mountain Regional competition during the fall semester 2014. EAC sent Woodson back to the 2015 Ethics Bowl Workshop to help prepare Mines to host the Rocky Mountain Regional competition in November 2015. The McBride Program; the Division Director for Humanities, Arts and Social Sciences; the Dean of the College of Earth Resources, Science and Engineering; and EAC have joined together to provide the funding to send the Mines Ethics Bowl teams to the national competition.

Winners of the Regional Ethics Bowl Competitions qualify to compete at the annual National Ethics Bowl Tournament, held at the annual APPE conference. (For more information, see http://appe.indiana.edu/ethics-bowl/ethics-bowl.) In the second year of participation in the Ethics Bowl program, the Mines team from the honors program won the Rocky Mountain Regional, securing a bid to the National Ethics Bowl competition that took place in Reston, VA. Mines fielded a second team in 2015, coached by Rachel Osgood and Cortney Holles. The second team placed in the bottom half of the Regional competition, but EAC and Mines paid for members of that team to attend the APPE conference. Each of the Mines students attended various sessions of the conference and either participated in or observed the National Ethics Bowl. In our first trip to the National Ethics Bowl, the CSM competitors represented Mines with distinction, demonstrating the characteristics Ethics Bowl is designed to foster: moral reasoning and civil discourse.

For many years the Rocky Mountain Regional Ethics Bowl was hosted by the University of Colorado-Boulder, but Mines hosted the Regional Bowl in 2015 and 2016. Institutions that have competed in regional competition include the following: University of Colorado-Denver, University of Colorado-Boulder, University of Nebraska-Lincoln, University of Winnipeg, Arizona State University, Air Force Academy, University of Northern Colorado, Colorado State University, and Mines. Hosting the Bowl entails recruiting competition judges, and that process provides excellent opportunities for community engagement with campus and students, opportunities that will be discussed in somewhat more detail below.

Beginning with the first Ethics Bowl team in 2014, EAC has actively pursued a collaboration with the McBride Honors Program in Public Affairs. The Honors Program at Mines was established in 1978, and since that time it has attracted gifted students and faculty. As stated on the McBride website, "Advocating a broad-ranging education for high achieving Mines students, President Guy T. McBride was an unseen hand behind the creation of the program in 1978. 'He recognized,' as Dr. Dendy Sloan observed, 'that leading a good life required connecting with a broader context than technical engineering and science could provide. It meant addressing larger human problems, ones that were not answered in the back of the book, such as the questions encountered in classic literature and the liberal arts'" (McBride Honors Program n.d.). Given the values implicit in the McBride Program, there are many synergies that have emerged and will continue to emerge: with McBride, EAC co-sponsored a public lecture by Jonathan Haidt, a social psychologist and Professor of Ethical Leadership at New York University's Stern School of Business. Two of the Daniels Fund Faculty Fellows are incorporating ethics content into the McBride curriculum, which be discussed in additional detail, below.

Similarly, EAC has begun to collaborate with the Hennebach Program in the Humanities. The Hennebach Program in the Humanities was built on the Hennebach Visiting Professorship endowment that was established in 1991. Since 1995, the Hennebach Program has supported a series of visiting scholars to help make the humanities an essential component of a Mines education. Scholars can visit for a semester, a week, or a class, but each visitor is chosen for their capacity to bring important opportunities for Mines students to encounter exceptional practitioners in their fields. For example, the Hennebach Program sponsored a visit by Mines alumnus George Saunders, the author of *Lincoln in the Bardo*, the winner of a MacArthur Fellowship and many other awards. Saunders gave a public reading and visited several classes. The spring semester 2017, Hennebach sponsored an Artist-In-Residence, and recently, the Hennebach Program has established a 2-year position in Environmental Humanities. EAC provided some funds to the George Saunders visit, and there are plans to work together on the film series, as well as to develop programming for Black History Month, Women's History Month, and Earth Day.

Initiatives are also underway to collaborate with institutional programs that revolve around governance. For example, the first "Mines Leadership Institute" is underway, and the Associate Vice President of Human Resources contacted EAC to provide support for their efforts to include ethics in their program. In another example, the Title IX Coordinator sits on the EAC committee, and regularly updates us about issues on campus, e.g., data from student surveys on sexual harassment/violence. The members of the committee then brainstorm about how to address these concerns via faculty, educational programming, etc. This data—combined with ongoing institutional efforts to increase the number of female students at Mines—led to the EAC committee's prioritization of improving the campus culture for women. We anticipate significant overlaps with the Society for Women Engineers as well as the Minority Engineering Program.

An explicitly "ethical" faculty/staff program was held in fall 2015. EAC sponsored an activity organized by the Association for Practical and Professional Ethics (APPE), where faculty and administrators participated in an "Ethics Roundtable." At that meeting faculty from across campus gathered to discussed the following:

- What ethical issues are either implicitly or explicitly at play in developing and submitting funding proposals?
- What ethical issues do you see emerging in engineering and science?
- What sorts of resources would help you include ethics and ethical considerations in your classes?
- What are ethical issues you face as a professor?

There were 22 attendees, including the Provost, and the discussion was wide-ranging. Some of the observations included concerns about universities losing their status as unbiased research institutions, and having to "sell their souls" to industry to keep their doors open, aka "Frackademia." The culture of cheating on the Mines campus was discussed, as well as the idea that faculty and students see ethics as a lecture and not a practice. For this reason technical faculty avoid talking about values, and one commented, "How do you act as an agent of change and survive?" Clearly, EAC has a multitude of concerns and issues to consider.

Some of the campus culture issues have been addressed head-on by student groups, and EAC will support them whenever possible and appropriate. For example, "Equality Through Awareness" (ETA) is a student run organization that facilitates discussions related to social justice and diversity in engineering. Other students are involved in oSTEM, the LGBT student organization on campus, and in spring 2016 the Title IX Office sponsored a film and panel discussion for Women's History Month. EAC is committed to supporting these groups/events, advertising their meetings and providing resources (financial and intellectual) for their projects.

One way to add to the intellectual support on campus is to build academic resources. To that end, we have been working with the library to extend and diversify ethics education opportunities on campus. For instance, in collaborating with library, we have created a book display featuring specific applied ethics issues every one or two months. We have also developed two online electronic libraries for ethics education. One is for the general audience across the campus and the other one is for the Daniels Fund Faculty Fellows Program. The first library site is mainly a portal that categorizes all resources on ethics education held by the library into different themes such as geoengineering and environmental ethics, engineering ethics, computer and cyber security ethics, biomedical ethics, etc. Faculty and students are thus able to efficiently identify resources useful for their courses and make good use of these resources. The second online library is devoted to the Daniels Fund Faculty Fellows Program. In addition to the ethics education resources included in the first library, the second one includes useful tools and methods for designing effective curriculum plans, assessment strategies, and pedagogical activities in ethics education formal and informal settings.

Finally, EAC has participated in more traditional activities, *viz*, bringing guest speakers to campus. Given the campus culture, we focus our efforts on attracting lecturers who address contested applied ethical topics, e.g., geoengineering or nanotechnology. Most recently, we invited Dr. Jack Stilgoe, Senior Lecturer in the Department of Science and Technology Studies, University College London. He gave a lecture titled "Machine Learning, Social Learning and Self-Driving Cars." Mines students and faculty are not particularly interested in philosophical jargon, so we work to engage these constituencies with ethical issues more than the lexicon of moral theory. These lectures have successfully engaged our engineering faculty in joining the conversation about ethics of technological design. Our long term goal is to create a lecture series focusing on ethical and political issues of emerging technologies. We are hoping to further invite our engineering faculty to expand their research agenda by incorporating broader ethical and political considerations into their proposal-writing and everyday decision-making.

Curricular Initiatives

For more than 20 years, HASS has housed the two humanities Core Courses required of every student matriculating at Mines: **Nature and Human Values** (NHV) and **Human Systems** (Systems). NHV is a 4-credit hour hybrid course of freshman composition plus environmental/professional ethics; in essence, students encounter, discuss and write about ethics. In 2016 NHV was recognized as an "Exemplary Education or Program Activity" by the National Academies of Science, Engineering and Medicine. As stated in the NAE report, "Nature and Human Values gives students ethical preparation for their engineering practice by highlighting ways that new technologies and engineering feats are changing people, society, and culture; exploring the evolving definitions of nature and the environment and how they impact human interactions and occupations; and emphasizing the obligation to forge ethical solutions to debates that acknowledge the values of all stakeholders. The class stresses written and oral communication as a crucial component of professional and civic dialogue, and encourages critical reading, thinking, and conversation about engineers' specific ethical obligations as professionals and their broader moral, social, and environmental responsibilities as world citizens" (National Academy of Engineering 2016).

The McBride Honors Program received $2.5 million in 2016 in order to develop an alternative to NHV, a course specifically designed for prospective Honors students. This process resulted in a signature year-long honors class titled "Innovation and Discovery in Engineering, Arts and Sciences (IDEAS) that "uses real-world problems to prepare students to be successful scientists, engineers, and citizens. It integrates open-ended design problems and discussion of ethical dilemmas in science and technology with core skills needed for success: communication, teamwork, problem-solving, and critical thinking" (McBride Honors Program n.d.) AY 2016–17 saw the launch of the course, enrolling 90 students.

Human Systems is a 3-credit hour course that examines the political, economic, social and cultural systems on a global scale during the modern era. It introduces global topics such as the development patterns in key regions of the world, the causes and outcomes of globalization, and the influence of energy, technology, and resources on the development. The specific contents of each section are often designed according to the expertise of each instructor. Nevertheless, global ethics and development issues (e.g., immigration, global justice, international development, corruption, world religions and conflict) can be found across most sections of the course. For instance, the section taught by the second author of this chapter spends two thirds of the course teaching global ethics and justice theories (e.g., human capabilities, lifeboat ethics, and the tragedy of the commons, etc.). These theories help students better understand the ethical and political issues arising from the design and implementation of engineering projects in the developing world.

Beyond the Core requirements, we offer four higher level ethics courses: (1) LAIS 198 "Ethics Bowl," (2) LAIS 321 "Ethics," (3) LAIS 425 Environmental Philosophy and Public Policy, and (4) SYGN 502 "Introduction to Research Ethics." There are other courses that do or soon will incorporate ethics into their curricula, e.g., Corporate Social Responsibility, classes offered in the Humanitarian Engineering program, as well as courses from Daniels Faculty Fellows. The Daniels Program will be described in some detail, below, but here we will examine the four stand-alone ethics courses offered at Mines.

LAIS198 "Ethics Bowl" was formed in response to student interest in Ethics Bowl. This 1-credit hour course was piloted in fall 2016, and will continue to be offered every fall semester. As the title indicates, the class is designed to prepare students to compete in the Regional Ethics Bowl competition held every November. Not every student enrolled in the course will be "at the table," i.e., face off against the opposition in the competition, but each student is required to do research about cases and/or ethical implications engendered by them. The course is a seminar, driven by discussion and wrestling with complex moral questions. In this course the cases come first, and then we learn about moral theories that may apply.

A different, 3-credit hour course in the McBride Program was offered in fall 2015. That class was structured as a standard honors ethics course, using primary texts almost exclusively, where the ethics bowl cases were added in. This approach to preparing for Ethics Bowl was not particularly effective—we were occasionally working at cross-purposes: working with primary texts from the moral theory cannon, e.g., *Groundwork of the Metaphysics of Morals*, was a wonderful foundation for competing in Ethics Bowl, but we didn't have time in class to both "go deep" into the philosophy and research and prepare the cases. Hence, the 1-credit hour, free elective course, described above. That said, the final project for the honors course was excellent, and was the reason we added students to the EAC committee. Each student was required to write a research paper that evaluated how to improve ethics education at Mines, and a quote from one of those papers opens this chapter.

LAIS320 "Ethics" was proposed in 2005. This course was piloted as a senior-level seminar. (Before 2005, no stand-alone ethics courses were offered at

Mines.) Given that Mines students have relatively little background in the humanities—no less in academic philosophy—the decision was made to use moral theory as the organizing principle for a course introducing a history of (western) philosophy. For example, students are required to read a chapter on Subjectivism in James Rachels' *Elements of Moral Philosophy*, an excerpt from David Hume's *Treatise of Human Nature*, and Karl Popper's "Science and Falsification" from *Conjectures and Refutations.* From these texts students encounter not only moral theory, but epistemology and the philosophy of science. This approach diverges rather dramatically from a more traditional undergraduate course, which typically focuses on the "Big Three": Aristotle, Kant and Mill. The Mines Ethics students don't encounter the "Big Three" until half-way through the course: by that time they have studied Moral Relativism, Subjectivism, and Ethical Egoism. Later in the course we examine Feminist Ethics and Existentialism.

LAIS320 is further organized around a bookended assignment: an "Ethics Autobiography." The assignment has two sections: (1) A "diary" of their actions, and (2) a list and analysis of their personal moral code/principles. At the very beginning of the semester, we ask students to keep a diary of an actual day, including every, single action they take. As stated on the assignment: "I'm not so interested in detail about bodily functions, but do include information, e.g., about whether your breakfast was food you 'borrowed' from your roommate, etc." Students are then asked to evaluate each entry for ethical implications, e.g., if they snoozed their alarm 25 times and they have a roommate whose sleep was interrupted by this. In the second section, students are asked to list their most important ethical principles, rank those principles, and the source(s) of those principles.

The second iteration of the "Ethics Autobiography," is due at the end of the semester, and is in lieu of a final exam. For this assignment students are required to revisit the first autobiography and reevaluate the ethics present in the diary of their day, connecting up to three entries to a theory learned in class. Likewise, students are asked to reevaluate their own ethical principles and relate their principles to ethical theories studied in the course. The final section of the essay requires students to choose the "best" theory, explain it, and defend it against at least one objection. Students are strongly encouraged to assign "best" to the ideals presented in the theory, not the one that most closely resembles what they already hold (although that's possible as well). In short, we ask student to analyze how they behave, what they believe, and who they aspire to be. This process is clearly an exercise in self-reflection, and can be seen as an exercise in developing moral sensitivity.

Most engineering educators recognize the necessity—and challenges—of teaching students moral sensitivity. As recently pointed out by some scholars, along with moral sensitivity, promoting "self-knowledge" is significantly lacking in engineering curricula (Mitcham 2014; Walling 2015). We suggest that a version of the "Ethics Autobiography" employed in some health and psychological science programs (Bashe et al. 2007; Hardwig 1997) can serve as a useful tool for teaching engineering students moral sensitivity and self-reflective competencies. To that end, we have begun to assess the Ethics Autobiography as a pedagogical tool.

LAIS 425 "Environmental Philosophy and Public Policy" has not been offered in several years; it is set to be revived in 2018. The course's objectives are to understand and analyze competing claims that underlie environmental issues, as well as examine the complexity of the issues themselves. Students work with fundamental concepts that drive environmental arguments, and apply ethical and political theory to concrete cases. Students are expected to use the full repertoire of their educational experience—technical and non-technical knowledge, research experience and intellectual curiosity—to evaluate their own and others' arguments. A new programmatic emphasis on "environment" within the HASS division is likely to make this a regularly offered course going forward.

SYGN 502 "Introduction to Research Ethics" is a one-credit hour research ethics course SYGN 502 was designed to fulfill the ethics education requirement for graduate students on campus who receive funding from the National Science Foundation. The course starts by introducing three major ethical theories: consequentialism, deontology, and virtue ethics. It then helps students identify and respond to some major issues in research ethics, including norms of science, mentoring and advising, the treatment of data, human subjects, authorship, research misconduct, and science and society. It is our hope that building a foundation with the three ethical theories can help students make better sense of the ethical significance of these issues. During the course, each student is asked to select a most recent case or story relevant to research ethics from *Science* magazine, share with the rest of class, and facilitate class discussion on the case or story. The final assignment for each student to create a personal ethics statement that he/she is committed to live by.

Departments on campus are allowed to develop their own educational programs that can be used to substitute for SYGN 502. Early on the EAC program was tasked with the responsibility to review these alternatives to SYGN 502 and make sure they meet the learning outcomes Mines students and postdoctoral researchers are supposed to achieve. The course requirements for SYGN 502 are

- compare, contrast, and evaluate at least three basic ethical theories;
- address a range of ethical issues they may confront in their professional lives; and
- articulate their ethical ideals and commitments to science, society and the environment.

Departments that propose to develop substitutes for SYGN502 are required to submit the following information to EAC:

- mode(s) of delivery
- materials used
- description how the three basic outcomes are to be met
- methods of assessment of student learning outcomes
- name of the responsible faculty member and commitment or background of that person to ethics relate activities (e.g., EAC events, McBride, ethics activities of professional societies, etc.)

One of our future goals is to compare the effectiveness of three major modes of delivering research ethics on the Mines campus: stand-alone "face-to-face" course such as SYGN 502, online learning, and ethics component or module in a technical course. This research may include some of the projects developed in the Daniels Fund Faculty Fellows Program.

The Daniels Fund provides support for a variety of nonprofit entities located in Colorado, Utah, New Mexico and Wyoming. The Fund focuses on several areas, including aging, addiction, early childhood education, and others (Daniels Fund n. d.). The collegiate program of the Daniels Fund Ethics Initiative strives to "deliver principle-based ethics education and reinforce the value of ethical conduct" at higher learning institutions. We received $65,000 grant (2016–2017) from the Daniels Fund and created the **Daniels Fund Faculty Fellows Program** on campus, which is dedicated to providing support for faculty to integrate principle-based ethics into their curricula. Mines is the first STEM-focused university to receive a Daniels Fund grants for ethics, and we have discovered that many faculty are interested in this opportunity. On August 5, 2016, seven Mines faculty were awarded the inaugural Daniels Fund Faculty Fellows award for their proposals on how to incorporate ethics into their courses. The Faculty Fellows are expected to develop lessons, modules or projects that incorporate ethical considerations as a central focus. Upon completing their courses, the Faculty Fellows assess the activity/ies and develop strategies for adaptation, the goal being to create strategies that can be applied to a variety of courses. The faculty are required to share their results with the campus community.

In order to support faculty in their efforts to incorporate ethics into their course curricula, Daniels Fund monies have been used to purchase materials (textbooks) and bring in leading scholars in engineering ethics and engineering ethics pedagogy. Each of the following guests gave a public luncheon lecture, where all faculty and staff were invited, and then the following day there was a 2-hour workshop. While these workshops were designed for Daniels Fund Faculty Fellows, they were open to any faculty interested in learning specific pedagogical and assessment strategies for incorporating ethics into their classes, and how to assess their success at the same. We were fortunate to have the following visitors.

- **Elaine Englehardt** (Distinguished Professor of Ethics, Utah Valley University) and **Michael Pritchard** (Willard A. Brown Professor of Philosophy and Co-Director of the Center for the Study of Ethics in Society, Western Michigan University). Their public lecture drew insights from moral psychology and business to discussed obstacles to effective ethical decision-making in the organizational context. Their faculty workshop focused on how to teach moral decision-making procedures to engineering students.
- **Michael Davis** (Senior Fellow at the Center for the Study of Ethics in the Professions and Professor of Philosophy, Illinois Institute of Technology). Davis' public lecture discussed the scandal at Volkswagen and how their corporate code of ethics partly led to the scandal. Davis' faculty workshop

introduced a variety of tools for teaching (e.g., test questions for moral decision-making; micro insertion) and presented strategies for how to turn "common" classroom assessment efforts (e.g., quizzes, exams) into publishable research.

- **Michael Loui** (Dale and Suzi Gallagher Professor of Engineering Education, Purdue University). Loui's public lecture introduced the historical development of professional engineering ethics in the United States and surveyed the most important topics and prevalent pedagogies in the field. Drawing on theories in learning sciences, Loui's faculty workshop introduced the so-called "backward design" approach to teaching engineering ethics, which aligns curriculum with pedagogy and assessment.

Again, the Daniels Faculty Fellows were strongly encouraged to attend these lectures/workshops, but each event was open to any faculty on campus. The Daniels Fund Fellows did attend, but faculty from across campus also joined both the luncheon and workshop programs, with an average of 35 participants for lunch, and 25 for the workshops. The vast majority of attendees were from STEM (vs. humanities), and either the Provost, President and/or Deans attended each luncheon lecture.

To date, the Daniels Fund Faculty Fellows have introduced new ethics content into the following: CSCI 475/598: Information Security and Privacy, CSCI 261: Programming Concepts, CSCI370: Computer Science Field Session, CEBN310: Introduction to Biomedical Engineering, GEGN 469: Geology Design, GEGN 673: Geological Engineering Design, and the new 1st Year McBride Honors course, "Integration and Discovery in Engineering, Arts & Sciences." Faculty have incorporated ethics content in a variety of ways, from including discussions of the ethical implications of the science, e.g., head transplants in the Biomedical Engineering course, to requiring additional readings, to "micro-insertions" (Davis 2006) of ethics into cybersecurity engineering design and computer programming problems. Faculty Fellows in computer engineering integrated Daniels Fund ethics principles and professional codes of ethics into field sessions and used pre- and post-surveys to assess improvement in moral knowledge among students. Each of these projects is ongoing, and we will share results with the wider campus via the library website, mentioned above.

Perhaps the most ambitious project emerged from the 1st Year McBride course. According to Daniels Faculty Fellow Toni Lefton, "Our…activities sought to overcome two specific challenges related to infusing ethics into the first-year engineering curriculum: (1) the tendency of ethics instruction to separate the personal (values and moral philosophy) from the professional (policies and codes of conduct); and (2) the tendency of case studies/assignments/activities to focus on either the 'engineering' aspects of the case or the 'moral' aspects of the case, but not both simultaneously." Professors Lefton and Hitt will continue to modify this course, and in fact hope to expand their work to "Ethics Across the Honors Curriculum."

Community Engagement

Every spring semester—for the last decade—students in LAIS320 have chosen the winner for the Golden Rotary Club's Ethics in Business Award. The Rotary Club solicits nominations for exemplary businesses in the Golden area, both nonprofit and for-profit. Each nominee is asked to provide information on a questionnaire about their business practices, and then the students compare the nominees on their success in community-building, environmental ethics, and treatment of their customers, clients and employees. Students take the process remarkably seriously, arguing vociferously for "their" nominee. In addition to the opportunity do engage in a very different class activity, students learn about their community, and what it actually means to conduct business ethically. This takes ethics out of the realm of theory and into practice, and it is an excellent opportunity for students to see that operating ethically can also be profitable.

Similarly, EAC has started working with the Mines Career Center: student behavior with prospective employers and during internships has real implications for the institution. If, for example, a student reneges on a job offer, the company may refuse to recruit on campus again. In these kinds of cases, again, ethics moves from the abstract into the "real world," with real consequences. In some ways, working with the Career Center is "low hanging fruit," given the seriousness of the consequences. However, we shouldn't dismiss "easy" ways to raise moral sensitivity, and as stated above, Mines students are exquisitely attuned to practical matters.

Although the Ethics in Business Award is clearly an activity designed to connect students to the wider community, and students interacting with recruiters and companies have direct impacts beyond campus, hosting the Rocky Mountain Regional Ethics Bowl also provided a different kind of community engagement: The Bowl requires people to judge the contest, and a variety of individuals were recruited for that purpose. For example, members of the Golden Rotary Club agreed to send judges, including the President, Vice President, and Vocational Services Chairman. Lockheed Martin sent two Ethics Officers (students didn't know such a job title existed!), an alumna from Ball Aerospace, the President of Red Rocks Community College, faculty from Mines, and other Colorado universities, a colonel from the Marines, a clinical ethicist, and several attorneys. This competition allows STEM faculty and the wider community to see the seriousness with which we take ethics and ethical issues, and may disabuse some people of the common attitude that young people don't care about ethics, or that Mines students are only good at math and science.

EAC also has supported activities explicitly aimed at the wider, non-student, community, for example "citizen & science" events: (1) WorldWide Views on Global Warming (2009), and (2) WorldWide Views on Biodiversity (2012). WorldWide Views is a program developed by the Danish Board of Technology, and it seeks to educate and poll "average" citizens about their attitudes concerning big issues. In each of the two meetings hosted by Mines, 50–100 citizens from the greater Denver area spent the day together, learning about climate change or biodiversity, and developing recommendations for policy, moving forward. All the results were compiled and distributed to negotiators at the United Nations, and the results were generated from sites across the globe. This program connects Mines to people and institutions all over the world, and the next WorldWide Views program will address "Oceans and Seas" hosted by the Permanent Mission of Germany to the United Nations.

Finally, individual Mines (ethics) students have engaged and impacted the wider community. For example, a chemical and biochemical engineering major received the city of Golden's first-ever GoldMine Award, part of the Mayor's Awards for Excellence and established this year to honor a project that highlights the close ties between the city and Colorado School of Mines. This student "matched" Mines students with volunteer opportunities in the city of Golden, and she was a member of the Ethics Bowl team. An engineering physics major was interviewed by Colorado Public Radio about her experiences competing in the Ethics Bowl. These two students exemplify some of the possibilities we will be exploring in order to become more fully integrated into the Golden community.

Institutional Challenges and Negotiating the Future

While EAC is certainly on an upward trajectory, all is not perfectly rosy. As an institution, there exists deep uncertainty about our financial future: given that STEM schools pull a significant chunk of their funding from the federal government, whether it be DOD, DOE, NSF, etc., big changes in federal funding can have a disproportionately serious impact on an institution like Mines; a dramatic reduction in federal funding would certainly affect the entire institution, and EAC with it. The current national, political climate highlights an ongoing issue: support for EAC at Mines has not been consistent, and we must prepare for an eventual, less-favorable administration.

Beyond the broad political climate, there remain other issues, and many of them were identified in the Ethics Roundtable: (1) disciplinary silos, aka, how to bridge physical and intellectual distances between faculty; (2) STEM faculty fears about discussion of ethical issues; (3) how to improve a culture, whether in terms of cheating or sexual harassment; (4) how to weave ethics throughout the curriculum; (5) how to keep our momentum going, knowing there are limits on resources and time.

Here is where we learn a something from the linear, literal engineers and scientists: large-scale projects must be broken down into their constituent parts, and not everyone will do every thing. We will need to make choices: as enticing and exciting as all of our opportunities are, current resources (staff and income) do not allow for a dramatic expansion of EAC programs. Hence, we've identified priorities, e.g., the campus culture for women, and we are partnering with the Title IX office to address an issue that has vexed this campus for years. We will submit grant proposals, including to NSF, to secure funding for both programming and hires, through which we will (hopefully) be less dependent on revolving administrations. We will support faculty through the Daniels Fund Faculty Fellows program and provide a clearinghouse for sharing ideas, and conduct research on the efficacy of these innovations. Finally, EAC will address the graduate student experience via Roel Snieder: he has been tasked with developing the Center for Professional Enhancement and Innovation, and he is slated to lead that Center once it is established. As a full professor in a STEM field, Snieder's support for EAC has been—and will continue to be—an enormous asset for ethics initiatives at Mines. Our ongoing task will be to discover and promote the innovations rising across the Mines campus, and provide leadership on the issues that EAC was created to address: to promote, extend, and deepen the understanding of ethical issues in relation to applied science and engineering education and research; to coordinate ethics teaching, learning, and practice; and to serve as a consultative body and resource for any group or organization whose policies and/or procedures affect the ethical aspects of life at Mines.

References

ABET. (2016). *Criteria for accrediting engineering programs*. http://www.abet.org/wp-content/uploads/2016/12/E001-17-18-EAC-Criteria-10-29-16-1.pdf. Accessed June 10, 2017.

Bashe, A., Anderson, S. K., Handelsman, M. M., & Klevansky, R. (2007). An acculturation model for ethics training: The ethics autobiography and beyond. *Professional Psychology Research and Practice, 38*, 60–67.

Colorado School of Mines. (2015). Colorado School of Mines 2015–2016 mini fact book. http://www.minesnewsroom.com/sites/default/files/wysiwyg-editor/docs/2015-16-Mini-Fact-Book.pdf. Accessed June 10, 2017.

Colorado School of Mines. (2016). *President Paul C. Johnson*. http://www.minesnewsroom.com/president. Accessed June 10, 2017.

Colorado School of Mines. (2016). *Students—2016/2017*. https://inside.mines.edu/Accountability-Students. Accessed June 10, 2017.

Colorado School of Mines. (2017). *Mines celebrates its new graduates*. Mines newsroom. http://www.minesnewsroom.com/news/mines-celebrates-its-newest-graduates. Accessed June 10, 2017.

Daniels Fund. (n.d.). *Daniels Fund grants program: An overview*. http://www.danielsfund.org/Grants/index.asp. Accessed June 10, 2017.

Davis, M. (2006). Integrating ethics into technical courses: Micro-insertion. *Science and Engineering Ethics, 12*, 717–730.

Hardwig, J. (1997). Autobiography, biography, and narrative ethics. In H. L. Nelson (Ed.), *Stories and their limits: Narrative approaches to bioethics* (pp. 50–64). New York, NY: Routledge.

IEEE Standards Association. (2016). *IEEE Standards Association introduces global initiative for ethical considerations in the design of autonomous systems*. http://standards.ieee.org/news/2016/ieee_autonomous_systems.html. Accessed June 10, 2017.

Ira A. Fulton Schools of Engineering. (2015). Paul C. Johnson, Ph.D. http://faculty.engineering.asu.edu/pauljohnson/. Accessed June 10, 2017.

McBride Honors Program in Public Affairs. History of the McBride program. http://inside.mines.edu/MCB-history. Accessed June 10, 2017.

McBride Honors Program in Public Affairs. Why first year honors? http://inside.mines.edu/MCB-whyfirstyear. Accessed June 10, 2017.
Mitcham, C. (2014). The true grand challenge for engineering: Self-knowledge. *Issues in Science & Technology, 31,* 19–22.
Mitcham, C., & Englehardt, E. E. (2016). Ethics across the curriculum: Prospects for broader (and deeper) teaching and learning in research and engineering ethics. *Science and Engineering Ethics*. https://link.springer.com/article/10.1007%2Fs11948-016-9797-7.
National Academy of Engineering. (2004). *The engineer of 2020: Visions of engineering in the new century*. Washington, D.C.: The National Academies Press.
National Academy of Engineering. (2016). *Infusing ethics into the development of engineers: Exemplary education activities and programs*. Washington, D.C.: The National Academies Press.
National Science Foundation. (2009). *RCR frequently asked questions*. https://www.nsf.gov/bfa/dias/policy/rcr/rcrfaqs.jsp. Accessed June 10, 2017.
Stockwell, C. (2016). *The top 10 schools in the U.S. for an engineering degree*. USA today. http://college.usatoday.com/2016/11/18/top-engineering-colleges/. Accessed June 10, 2017.
The Business Journals. (2015). *Exclusive: Here are the best public colleges in America*. http://www.bizjournals.com/bizjournals/news/2015/02/12/michigan-earns-top-honors-in-rankings-of-public.html. Accessed June 10, 2017.
U.S. News & World Report. (2017). *Top public schools: National universities*. https://www.usnews.com/best-colleges/rankings/national-universities/top-public. Accessed June 10, 2017.
Walling, O. (2015). Beyond ethical frameworks: Using moral experimentation in the engineering ethics classroom. *Science and Engineering Ethics, 21,* 1637–1656.

Ethics Across the Curriculum at Dartmouth College

Aine Donovan

Abstract This essay presents an overview of one of the earliest ethics across the curriculum programs, designed specifically to enhance liberal arts teaching and research. The Dartmouth academic community has long had an emphasis on moral meaning-making, but the threads of that interest were varied and not connected to the mission of the college. The EATC program sought to integrate the mission and the educational objectives into the framework of ethics education as a necessary component of a robust liberal arts education. EATC at Dartmouth, now in its 15th year, has become a much sought after opportunity for faculty and staff.

Keywords Ethics · Experiential learning · Faculty development

Introduction

Dartmouth College is a small liberal arts school with a rich heritage of ethics education. Established in 1769 by Rev. Eleazar Wheelock, a Congregational minister, who sought to educate and convert Native Americans. The college took root with a mission that had explicit ethical norms for the student body, and those goals tied into a social vision of the college graduate as a socially responsible agent. Dartmouth's mission was, in the beginning, similar to the other institutions of higher education in the Northeastern United States: shaping young men into Christian adults who would foster social ideals of hard work, responsibility and prudence. The methods for achieving that goal was a weaving of ethical intention into every aspect of the curriculum—from language classes to the dining hall. And, importantly, students accepted the mission and enthusiastically entered into the process of adult formation.

By the 19th century the Dartmouth College curriculum had shifted to include explicit courses on character formation, as was the case with many liberal arts

A. Donovan (✉)
Dartmouth Ethics Institute, Hanover, USA
e-mail: Aine.Donovan@Dartmouth.EDU

E. E. Englehardt and M. S. Pritchard (eds.), *Ethics Across the Curriculum—Pedagogical Perspectives*, https://doi.org/10.1007/978-3-319-78939-2_24

colleges at the time. Indeed, as Douglas Sloan points out, "the most important course in the college curriculum was moral philosophy, usually taught by the college president and required of all seniors."[1] This change toward a more explicit form of character education reflected the rapid changes that were taking place in American society, from westward expansion to the industrial revolution. The purpose of moral education was to provide an anchor of stability in a changing world.[2] Additionally, moral philosophy was the unifying thread throughout the college curriculum. Yet by the end of the 19th century another educational shift had occurred with the rise of the modern university and a greater emphasis on the training for the newly forming industrial-managerial society. Educational leaders such as Daniel Coit Gilman, President of Johns Hopkins, took the lead in developing academic departments and specialized fields of study that would usher in an entirely new type of educational experience.

By the 20th century the highly specialized university model of education had swept the United States. Dartmouth College, on the other hand, had resisted the urge to specialize and continued to educate young men (and, by the 1970s, women) in a manner that reflected the holistic approach of its founder. Gone, of course, were the required chapel attendance and the ethics courses, but they were replaced with a robust program of social engagement that held service as the ultimate goal of every Dartmouth graduate. Much of that emphasis came from Dartmouth president John Sloan Dickey, serving from 1945 until 1970. Dickey's commitment to the liberal arts took shape in the introduction of a "great issues" course for all undergraduate Dartmouth students, with a much repeated goal for the students he served—that 'the world's problems are your problems'. Ahead of his peers, Dickey piloted a program that granted academic credit for participation in social activism for undergraduate students. This insight of blending experiential learning to academic course work would stand alone as a unique educational approach until the 1980s when many colleges and universities initiated similar programs.

Dartmouth's conscious decision to remain a premiere liberal arts *college*, as opposed to a research university, has made the continued efforts to integrate ethics into the curriculum a somewhat simpler process than might be found at a larger university. In 1982 three Dartmouth thought leaders, Dean of the Tuck School of Business John Hennessey, Philosophy Department Chair Bernard Gert, and Dartmouth Medical School Professor Charles Culver collaborated to establish a center for the study of applied and professional ethics. The purpose of the center was, according to Dean Hennessey, to foster interdisciplinary dialogue and research that would enhance understanding and provide moral guidance for the rapidly changing needs of society.

In 2001 the Provost of Dartmouth College, the Faculty Director of the Ethics Institute (Ronald M. Green) and the Ethics Institute Faculty Advisory Board invited me to join the institute as the Director. One of the first initiatives that I developed as

---

[1]Sloan (1980).

[2]ibid.

Director was an ethics across the curriculum program. In consultation with Philosophy Professor Bernard Gert I created a year-long program for faculty that is now in its 13th year. The program was funded by a Dartmouth graduate, A. George Battle who, like John Sloan Dickey, believed that ethics education was critical to the adult formation of young men and women. The Battle fund provides the foundation for a program that has been exceptionally well received.

A pilot program for ethics across the curriculum (EATC) was launched in the fall of 2002 with six faculty members from a variety of disciplines. The program has expanded over the thirteen years of its existence, but the decision was made early on to keep the number of participants to 15, or a maximum of 20. The small group dynamic has proven critical to the success of the program and the opportunity to provide individualized instruction to faculty on their projects.

The foundation of the EATC program at Dartmouth lies in what Dennis Thompson (former Safra Center Ethics Director, Harvard University) called a "linking discipline"—it bridges theory and practice. Theory, in itself, often conflicts with other seemingly suitable theory; therefore the theory needs to be understood through the lense of the conflict itself. This, in essence, is the field of practical ethics, and it draws on a variety of disciplines to inform the moral decision-maker about choices and consequences. But whichever theory is invoked, the most important element of a successful ethics-across-the-curriculum program is identifying the reason for the program. Is it intended to inspire citizenship? Bolster the honor code? Counter the upswing in undergraduate cheating? Or merely to present ethical options in a world of cafeteria choice lifestyle? This fundamental aspect of EATC is often missing in secular liberal arts institutions, and it surely was at Dartmouth in the early years of our program. The advisory board for the ethics institute spent two years at monthly meetings wrestling with the question of purpose, and the common good. Though a deeper understanding of the complexities of the task of educating college students was attained, we did little to settle the question of *why* we were involved in the project of ethics education. Not so when I taught at a Jesuit University, or at the United States Naval Academy; both educational institutions promoted their mission as tethered to their moral foundations. Therefore we were faced with a dilemma: teach ethics as a normative discipline, with assessable goals in student behavior and belief; or teach ethics as a descriptive exercise in cultural relativism. The program has vacillated between the two orientations, with the latter being reported as more "comfortable" for faculty who are attempting to introduce ethical concepts into their specific discipline. This is an issue that warrants far greater exploration than this space permits, but it is safe to say that faculty are increasingly voicing concern about promoting any one value over another. In a 2006 study conducted by the Higher Education Research Institute, 62 percent of the college juniors surveyed reported never having experienced a professor who encouraged discussion about religion, spirituality or the meaning of life.[3] The Dartmouth "ethics across the curriculum" program was

[3]Rainey (2006).

established to spark precisely these types of discussions, within a framework that respects diverse opinions, reasoned arguments and intellectual curiosity.

This hesitation in serving as moral guide and mentor flies in the face of accepted psychological theory that shows the college years to be especially rich for moral development. Erik Erikson's theory of stage development presents college educators with ample evidence that the moral disruptions of the college years offer a pivotal opportunity for mature ethical growth. The two ways of perceiving moral/emotional growth in college students tends to be broken down into two camps: developmental and/or environmental.[4] The *developmental* focus relies on explorations of phases of individual growth and concentrates on outcomes, or the nature of student change. On this view 'identity formation' is often cited as the ultimate goal of EATC. The *environmental* model identifies variables that are presumed to exert an influence on one or more aspects of student change, focusing often on institutional characteristics, programs and services. Yet these two perspectives are merely faces of a complex theory of moral education, they are not oppositional.

The Dartmouth Program

The Dartmouth EATC program is a year-long program for tenure-track, or tenured faculty. This qualification is, of course, ripe for questioning. But the Dartmouth administration determined that resources of this sort, and stipends, should be restricted to faculty that they deemed long term and invested in Dartmouth's future. As a note, adjunct faculty, of course, are often longer term than many tenure track faculty, and often have greater interest in the student community; but this is a separate issue to be decided on an individual basis.

The program is a year-long seminar, with advertising beginning in the preceding spring term through every academic department. The call for participation offers seminars in moral psychology, ethical theory, and case teaching. Additionally, the faculty participants are required to complete a "project" that, upon completion, will garner a stipend to their research accounts. The call for participation is sent again in the early fall term when, in addition, a "kick-off" event is scheduled. The rationale for a kick-off event is simple—to inspire interest in the subject of ethics in college education while a large group gathers for a lively presentation and social event. These events are designed with a marketer's eye toward gathering a clientele: timely, provocative, fun, thought provoking. An example of a particularly popular event was a faculty workshop on "The Ethical Code in the Sopranos". The panel consisted of two of the Soprano writers and two academics who had written on the ethics of popular television programming. The afternoon event (3–5:30) was open to all faculty and addressed the questions of whether Tony Soprano was a moral exemplar for his particular community, and if so what that meant for the larger

[4]Pascarella and Terenzini (1991).

society. The afternoon was a lively discussion and exploration of the intersection of popular culture and moral theory. It concluded with a reception and cocktail hour of socializing for attendees. This element of socializing should not be underestimated—it is critical for the fostering of community in a profession that is increasingly isolated and disconnected. Aside from the obvious enjoyment of a lively intellectual discussion, it also provides an opportunity for advertising the year-long seminar on ethics that are showcased in the kick-off event.

The applications for the seminar are simple, but detailed enough to demonstrate commitment on the part of applicants. The group is selected so that they represent a cross section of academic disciplines; this approach is, of course, a choice—equally compelling is the argument for providing discipline specific exclusivity. Twice in the history of the program I have done this—once with medical school faculty because their concerns are so vastly different from arts and sciences faculty, and once with business school faculty because the physical location of the schools made it difficult to combine with arts and sciences. Both approaches have advantages, but given a choice the interdisciplinary approach seems to offer greater depth of discussion and richer understanding of the complexities of moral decision-making.

One textbook is used for the seminar, James Rachel's Moral Theory. The advantage to this slim volume is the general overview of complex matters for non-philosophers, and the short cases studies included in the book. Originally, when the seminar was first taught, I co-taught with colleague and noted philosopher, Bernard Gert and we opted to use his groundbreaking book, Moral Rules. But while the book is a significant contribution to the field of ethics, it does not easily lend itself to an adult learner who has limited time to understand complex issues and apply them to their own narrow area of expertise. Therefore, we abbreviated much of Gert's theory into the introductory session as a launch to understanding the varieties of ethical theory. The occasional seminar participant will utilize the resources of the Ethics Institute library and extend their knowledge of some aspect covered in the session, but most of the faculty are content to receive the general overview from the Rachel's text as a base line. They are given a four page annotated bibliography that is updated very new academic year as well.

The curriculum for the seminar consists of the following seven, three-hour sessions. All sessions are held in the evening, with dinner provided and a short (half hour) social/wine time factored in for community building.

1. *Session One*—Introduction; the question of what is meant by the term "ethics" and how the discipline is viewed either normatively or descriptively and what that distinction means for higher education. An emphasis is placed on the Greeks and the distinction between the Aristotelian and Platonic idea of the good life. The much abbreviated history of important philosophical contributions to the fields of ethics is explored, including Eastern philosophers. The run up to the contemporary thought is, of course, difficult to condense into a short session, but the post-Enlightenment is highlighted and serves as a foundation to a discussion about the emerging role of relativism and its influence on a normative approach to ethics education. An exercise that I developed for

participants to match up colleges with different mission statements has been instrumental in generating discussion about the role of purpose in higher education and, further, how purpose and meaning are (or are not) intertwined.

2. *Session Two*—moral psychology is the emphasis on the second session, with additional readings from Piaget, Kohlberg, Erikson, Gilligan and Haidt. What do students need, developmentally, is the underlying question for this session.
3. *Session Three*—Ethical theory/Deontology is the focus of this session, with a particular emphasis on Kant.
4. *Session Four*—Ethical theory/Utilitarianism is the focus of this session with a particular emphasis on Bentham and Mill, as well as Peter Singer.
5. *Session Five*—Pedagogical approaches to ethics education; the method for teaching ethics matters greatly, and this session focuses on the case study method as a preferred method for ethics instruction. Additional readings from the Harvard Business school case study expert, C. Roland Christensen, serve as a foundation to the discussion. The seminar is a more applied session, with participants required to take an active part in the demonstrations. A case that I co-wrote on the General Patton "slapping incident" is often used as an example of a longer, multi-part, case. It demonstrates the occasional need for background information (in this case, world war two) and a very real dilemma for Supreme Allied Commander, General Dwight D. Eisenhower. The case lends itself to role play, a difficult challenge for many students who fear speaking in public, because the roles are clearly defined and the choices not readily evident. The session is a demonstrated pedagogical exercise, where I "freeze frame" the discussion at several points to ask the seminar participants to notice what has just happened, or to highlight a position that exemplifies one of the theories, or to analyze the framework for a position that has been articulated. A ten-minute video clip of the Patton film is a rich addition to the discussion.

The role of film as a case study is included in this session, and how clips of pivotal decision-forcing moments can bring an ethics discussion to life for students in a way that text sometimes cannot.

6. *Session Six*—Assessing student learning. The participants spend time reading through student essays (names omitted) and evaluating the strength of their moral arguments. The assigned reading is my five page "primer" on how to grade philosophical position papers and avoid the trap of biased interpretation. This session also includes a self-assessment tool for evaluating our biases in teaching.
7. *Session Seven*—Conclusion and project sharing. Each of the participants are required to complete a project that will enhance their teaching. The possibilities include, but are not limited to, case studies, new course syllabus, a book, or video game. The faculty who have participated in the Dartmouth "EATC" program have produced a wide range of 'deliverables' in a variety of fields. Several projects have resulted in published journal articles, and there are currently three new courses being taught at Dartmouth that were generated from the seminar project.

Upon completion of the project, and sharing with fellow participants, the seminar participants receive a stipend and are eligible for a small grant for conference presentation.

The Dartmouth program for integrating ethics into the college curriculum has expanded over the years to include a separate program for graduate students. And that program is now being replicated at the Dartmouth hospital for all research faculty. The need continues to evolve and expand, especially as politics and policy collide with the mission to educate the whole person: mind, body and spirit. The divisive state of affairs in higher education make our task the highest priority in liberal education.

References

Pascarella, E. & Terenzini P. (1991). How college affects students. Jossey Bass.

Rainey, A. (2006, March 10) Professors favor building moral values in students. In *The Chronicle of higher Education*, p. A12.

Sloan, D. (1980). The teaching of ethics in the american undergraduate curriculum 1876–1976. In D. Callahan & S. Bok (Eds.), *Ethics teaching in higher education* (pp. 1–57). New York: Plenum Press.

Ethics Across the Curriculum at UPRM: A Roadmap for STEM Integration

William J. Frey and José A. Cruz-Cruz

Abstract This chapter presents over three decades of EAC (ethics across the curriculum) experience at UPRM, the second largest campus of the University of Puerto Rico. An introductory section outlines our evolving concept of EAC. The second part outlines four initiatives, three of which were funded by the National Science Foundation: interdisciplinary faculty workshops in EAC, GERESE (Graduate Experience in Research Ethics in Science and Engineering), and the EAC Toolkit. The final section covers strategies for sharing EAC best practices and methods for refashioning materials from faculty workshops into EAC modules and activities. In short, this paper hopes to provide readers from similarly situated programs with a roadmap for developing a successful EAC program.

Keywords Ethics · Ethics across the curriculum · EAC · Interdisciplinary Business · Engineering

Introduction

Ethics across the curriculum (EAC) has been sustainably practiced at UPRM (University of Puerto Rico at Mayagüez) for over thirty years now. This paper will use the UPRM experience to construct a roadmap to EAC for those similarly situated. First, it will engage in what Gilbert Ryle terms, "conceptual cartography" by outlining important characteristics of EAC. Second, it will add stations to this roadmap by describing four UPRM EAC initiatives: (1) interdisciplinary workshops and retreats, (2) using EAC to integrate research ethics into UPRM's Science, Technology, Engineering, and Math (STEM) graduate curriculum, (3) an EAC Toolkit consisting of a repository best practices in EAC, and (4) a Statement of

W. J. Frey (✉) · J. A. Cruz-Cruz
University of Puerto Rico, Mayagüez, PR, USA
e-mail: freyuprm@gmail.com

J. A. Cruz-Cruz
e-mail: j.cruz@upr.edu

E. E. Englehardt and M. S. Pritchard (eds.), *Ethics Across the Curriculum—Pedagogical Perspectives*, https://doi.org/10.1007/978-3-319-78939-2_25

Values drafted by UPRM's College of Business Administration stakeholders. Finally, the roadmap will outline our journey from a fragmented, disengaged culture toward an increasingly integrated and engaged culture. Because this agenda is ambitious, the reader will be directed "off map" to resources published over the years that disseminate initiatives and practices in EAC carried out at UPRM.

The EAC (Ethics Across the Curriculum) Concept at UPRM

Nearly 25 years ago, the authors characterized EAC as interdisciplinary, holistic, and directed toward "ethical empowerment" (Cruz and Frey 2003). This core still holds but the EAC concept has also been expanded and enriched by many initiatives between then and now. Here is a summary.

1. *EAC is interdisciplinary.* To begin, EAC requires synthesizing good pedagogy and content from both ethicists and STEM specialists; technical knowledge is necessary to formulate ethical challenges while ethical frameworks, principles, and concepts help STEM students take a broader, more critical perspective on their learning and research. One of us has been working for nearly fifteen years with STEM graduate students to help them identify ethical issues embedded in their research (Frey and Rivera-Vélez 2014). This has ranged from promoting responsible research practices to uncovering ethical issues that attend the application of their research. Uncovering (or discovering) ethical issues embedded in theoretical research requires moral imagination, and EAC creates opportunities to teach and practice this skill (Frey 2015a).
2. *EAC at UPRM relies on a "co-integration" strategy* based on (1) integrating STEM expertise into ethics courses and modules and (2) integrating ethical analysis and methodology into STEM teaching and practice. Just as Kant said that concepts without percepts are empty and percepts without concepts are blind, ethics courses without STEM content become abstract and lack relevance while STEM teaching and practice without ethical analysis and moral imagination becomes myopically technical and disengaged.

Faculty development workshops carried out at UPRM over the years have been designed to promote co-integration: STEM teachers and practitioners have taken the lead in identifying ethical issues and challenges that arise in day-to-day practice; they have also taken the lead in modeling the EAC activities they have integrated into their own classes. Ethicists have shared with their STEM colleagues their ethics teaching practices. This co-integration strategy allows ethics and STEM faculty members to collaborate as equals. The outcome is a jointly designed program that both can fully embrace.

3. *EAC is holistic and provides a largely untapped resource for integrating the university curriculum and making it more comprehensive.* At UPRM, EAC has integrated (1) *freestanding courses* such as business and engineering ethics, (2) *micro-insertions* of ethics into the mainstream STEM curriculum, (3) *special activities* like the Ethics Bowl which is deployed at UPRM as the capstone activity in introduction to ethics classes, (4) *faculty development workshops* that facilitate compliance with ABET ethics standards, and (5) *EAC resource repositories* (cases, explanatory materials, exercises, assessment instruments) that together form a "toolkit" designed to promote the development and sharing of best practices.

EAC can coordinate and integrate different components of the STEM curriculum by emphasizing (through workshops and other activities) the following:

- **Recognize**. EAC is built on a concerted effort to recognize and document ongoing ethics integration activities carried on by both ethics and STEM faculty. For example, some STEM faculty already integrate ethics into their teaching and practice; they understand that practical and professional ethics is inseparable from technical, STEM studies and practice. They also understand that ethics integration is possible because one can teach STEM and ethics *at the same time.* Ethical issues in the STEM curriculum can be brought to the surface by reframing the technical, professional, or occupational. A professor in business at UPRM uses a local bakery as a case study to teach entrepreneurship.[1] This bakery loses money because it has to throw away the pastries it has failed to sell during the day. This opens the way towards a "pivot" to the ethical. Can it sell these pastries the next day as day-old pastries? Should it reduce the price since these are less fresh? Should it use aromatic chemicals (which may be carcinogenic) to create the (subliminal) impression of just-baked freshness? These questions make students aware of moral issues present in day-to-day professional and occupational activities. EAC should deliberately and systematically recognize such "pivots to the ethical" and bring them to the foreground of curricular development.
- **Disseminate and Model**. STEM faculty repeatedly raise two objections to doing EAC: no space and no expertise. Disseminating recognized ethics integration activities carried out by other STEM faculty cuts off both of these objections at their source. Modeling demonstrates that EAC is possible by showing that it is already taking place. STEM professors who give demonstrations at EAC workshops and retreats show concretely how to make time to cover ethics in the classroom. These demonstrations also show that ethics integration exercises can be effectively carried out by those who are not ethics "experts." Examples from past EAC demonstrations include redesigning textbook exercises as ethical integration activities, using cartoons (such as Calvin and Hobbes) to stimulate reflection on classroom ethical issues, discussion exercises tied to current events (e.g., newspaper articles), and

[1] http://www.uprm.edu/etica/PDF/Web/FreshlyBakedGoods.pdf.

"Cine Ética" activities that use movies to dramatize ethical scenarios. Having STEM faculty model their EAC modules demonstrates that EAC is possible because actual; STEM professors become convinced that they can do ethics in their classes when they see how their colleagues do this.
- **Coordinate**. Coordination begins when one shares ethics integration activities with colleagues. A textbook used by an Information Systems professor has an activity where students use Kant's categorical imperative to analyze a case involving the use of Microsoft Excel to produce a deceptive graph, one that makes a negligibly successful marketing campaign appear highly successful (Kroenke 2015). The information systems professor consults with a colleague in business administration on how to present the categorical imperative more effectively. Together they develop and teach an activity that improves upon the textbook's explanation (Cruz and Frey 2015). After this, the colleague adds a short presentation on the categorical imperative in his classes because his students often take his colleague's Information Systems classes. Sharing an ethics integration exercise refines these activities and positively influences other activities. When colleagues begin to coordinate their teaching, these collective efforts help solidify an emerging community in EAC teaching.

EAC at UPRM continues to be holistic and integrative. This holistic approach helps STEM professors integrate ethics without forcing them to redesign their courses, crowd out essential technical content, or become ethics experts.

4. *EAC proceeds by creating "ethical empowerment."* Cruz and Frey (2003) propose five skills of ethical empowerment and argue that these can help recognize ongoing ethics integration activities and guide the development of new EAC models. Here these skills are recapitulated and then reformulated to show how they have been developed through two ongoing ethics initiatives at UPRM, *GREAT IDEA* and *CRWS*.[2]

    - The skill of *ethical awareness* consists of the "ability to perceive ethical issues in complex, concrete situations" (Cruz and Frey 2003). A "pre-test" activity promotes this skill; students are presented with short scenarios and are asked whether these raise ethical issues, are controversial, and represent common occurrences. Recently in the CRWS initiative (Cultivating Responsible Well-being in STEM) this skill has been applied to community development initiatives. Reformulated in the CRWS proposal, it is the "ability to pick out social, ethical, and global relevance in technical artifacts and socio-technical systems."

[2]The ethics empowerment skills have been used in both GREAT IDEA and CRWS. Christopher Papodoupolus was the PI of *Graduate Research and Education for Appropriate Technology: Inspiring Direct Engagement and Agency* (GREAT IDEA, NSF/EESE 1033028) and Marcel Castro-Sitiriche is the PI of *Cultivating Responsible Wellbeing in STEM: Social Engagement through Personal Ethics* (CRWS, NSF 1449489). Quotes of the reformulated empowerment skills come from the CRWS proposal.

- *Ethical evaluation* consists of the "ability to assess a product or process in terms of different ethical approaches or tests" (Cruz and Frey 2003). This skill can be developed through the well-known Gray Matters activity that forms a part of the ethics training programs of many multinational corporations (Whitbeck 2011). Students are provided with a scenario that ends abruptly at a point of decision; they bring the story to a close by choosing from among different "possible endings," i.e., possible solutions to the problem raised in the scenario. Reformulated in the CRWS proposal, ethical evaluation consists of the skill of "evaluating and ranking different appropriate technology solutions to problems of community development."
- The skill of *prevention* consists of an "ability to anticipate ethical/social problems and design counter-measures" (Cruz and Frey 2003). An example of this skill and an exercise to develop it would be the Social Impact Statement activity developed by Chuck Huff at St. Olaf University (Huff 2001). This activity encourages students to use socio-technical system analysis to identify ethical problems embedded in socio-technical systems and practice solving these problems upstream in the design process before they manifest themselves in the form of full-blown ethical dilemmas. In the grant projects, GREAT IDEA and CRWS, participants have pursued the skill of prevention by deploying frameworks taken from Appropriate Technology and the Capabilities Approach (Papadopoulos et al. 2014). They use these approaches to envision integrating a design into a given socio-technical system, anticipating possible value conflicts, and building changes into the design to mitigate or eliminate these problems.
- The skill of *ethical integration* consists of the "ability to integrate ethical considerations into an activity [such as a product, service, design prototype, or technology]" (Cruz and Frey 2003). This skill anticipates different design approaches such as value sensitive design, participatory design, and capability sensitive design (Oosterlaken 2015). These empower students to integrate different moral values into design and problem-solving. Reformulated in CRWS, this skill was described as the "ability to treat social, ethical, and global values as ends in the designing activity such that one is able to generate designs that translate or realize these values." This skill overlaps with "techno-socio sensitivity" as characterized by Harris: "critical awareness of the way technology affects society and the way social forces in turn affect the evolution of technology (Harris 2008). As students learn that technical designs can alter the socio-technical systems into which they are enacted, they learn to develop designs that avoid destructive impacts. As they learn that socio-technical systems effect the successful operation of an artifact, they tailor technical designs to specific socio-technical systems.
- Finally *value realization*, consists of the "ability to recognize and exploit opportunities for realizing moral value." This ethical empowerment skill targets the exemplary. First enacted in community development projects, this skill is promoted when STEM students learn to deploy their technical skills

and knowledge in the service of the values such as social responsibility and social justice. Value realization also provides an excellent vehicle for learning about moral exemplars. At UPRM, EAC faculty designed ethics integration modules where students study moral exemplars in business and engineering. These studies focus on the supererogatory and supplement "Big News/Bad News" cases in Puerto Rico (Pritchard 1998).[3]

The skills of ethical empowerment have anchored many successful EAC initiatives at UPRM. They provide assessment heuristics that recognize and highlight ongoing ethics integration activities in the STEM curriculum. They have helped guide the design of faculty development workshops and concentrate the efforts of STEM professors.

5. *EAC at UPRM has taken the form of "15/85": 15% of STEM faculty empowered in ethics can reach 100% of their students and bring about significant improvements in 85%.* 15/85 focuses on recruiting a core of committed STEM faculty to integrate ethics modules into mainstream STEM classes. To support these "ethics champions," it creates a repository of resources including case studies, tested classroom activities, and self-explanatory lesson plans, all embedded in a straightforward assessment process. Here is 15/85 broken down into its parts.

- Faculty development workshops recruit and empower STEM teachers interested in EAC who are willing to champion this effort in their discipline. EAC prospers when there is a highly motivated and committed core of faculty; it falters when faculty feel that they have been drafted into the movement. EAC works best when voluntary.
- The efforts of this core group (call them EAC "veterans" or "champions") are magnified and strengthened by providing them offline and online support; *offline*, faculty development workshops build skills, identify issues, and create enthusiasm; *online*, a toolkit populated with resources (cases, exercises, pedagogical strategies, assessment instruments) advances the infusion of ethics content and activities throughout the STEM curriculum. The right kind of support magnifies the efforts of this committed core of STEM professors.
- Key points can be identified in the STEM curriculum where the insertion of ethics modules would have substantial impact. For example, EAC veterans could teach ethics modules in portal or gateway courses through which all or most STEM students must pass; placing modules in these required courses ensures that STEM students will be exposed to ethics presented in the context of everyday STEM teaching and practice.
- An assessment model could be developed (an informal one making use, perhaps, of the five empowerment skills) to document the impact of these ethics insertions. This assessment approach should be modest; it could focus on ethical awareness, ethical evaluation, practicing decision-making

[3]See also http://legacy.cnx.org/content/m60084/1.2/.

(in pre-test and Gray Matters formats), and carrying out socio-technical analyses (This list is by no means exhaustive.).

- The target at UPRM has been set at 15–85: 15% of the STEM faculty trained in ethics instruction and strategically placed at gateway points in the STEM curriculum could reach 100% of the students with ethics integration exercises and have a positive impact on 85% of these students. There is nothing sacred or magical about the number 15%, but it does represent a reasonable estimate of the minimum faculty commitment needed to achieve positive outcomes. For example, in a recent survey taken by the Center for Ethics in the Professions at the Illinois Institute of Technology, it appears that 15% of that university's STEM faculty has been able to anchor a sustainable EAC program (Davis et al. 2016). (Especially when one recognizes the excellent leadership exercised by the members of that university's Center for the Study of Ethics in the Professions.) Fifteen percent, thus, represents a plausible, testable hypothesis.[4]

Ethics Across the Curriculum Grants and Projects

Over the years, EAC researchers at UPRM have carried out a series of projects (or initiatives) designed to enact different components of the EAC strategy described above. This section will outline four: (1) interdisciplinary retreats with UPRM faculty members from the ethics and STEM areas; (2) GERESE, an initiative designed to tackle research ethics using an EAC strategy; (3) the EAC Toolkit, an online repository of EAC resources and materials for generating, sharing, and disseminating EAC best practices; and (4) the Statement of Values process carried out in UPRM's College of Business Administration culminating in the drafting and implementation of a Statement of Values by this college's stakeholders. While the previous section outlined key constituents of EAC, this section will show efforts at UPRM to embody these concepts in an ongoing EAC program.

1. The "*Interdisciplinary Research and Training Program in Ethics for Business, Science, and Engineering in the Puerto Rican Context*" (NSF SBR 9810253) sponsored four faculty retreats and four dissemination workshops between 1996 and 2000. The main activity was a five-day retreat held in Guanica, Puerto Rico in 1998 with approximately 25 participating faculty members from UPRM and other island universities. Highlighting this retreat were (1) issue identification activities (based on presentations from representatives of the surrounding Puerto Rican business community), (2) workshops on ethical theory, and (3) case writing activities carried out in small, interdisciplinary teams.

[4]Davis et al. (2016): "On the most conservative reading, the survey tells us that at least 15.5% of IIT faculty are involved in some way in teaching ethics at the undergraduate level (70 of 452) and at least 13.7% at the graduate level (62 of 452)".

This retreat generated many useful cases and exercises.[5] Along with follow-up retreats held from 1999–2000, it provided an opportunity to design and refine a retreat methodology. But its focal achievement was the emergence of a community at UPRM united around the goal of integrating ethics across the Business, Science, and Engineering curriculum. Cruz and Frey (2003) summarize the emergence of this community, its case writing activities, the emerging concept of EAC, and assessment instruments and activities.

2. *Graduate Experience in Research Ethics for Science and Engineering* (GERESE—NSF SES 0629377) provided a multi-dimensional EAC approach to research ethics. From 2006–2009, GERESE developed and assessed one- and three-hour standalone courses designed to cover the nine ORI issues in research ethics (Steneck 2004). These courses were team-taught by an ethicist and a STEM specialist along with graduate assistants from the STEM areas. Both versions were assessed to see if research ethics could be taught better in one- or three-hour formats.

A series of four research ethics workshops was also designed and tested. The series took a year to complete, and the grant tested two cycles. Each workshop targeted a specific ethical empowerment skill. Here is a short description:

- The GAW (*Graduate Awareness Workshop*) targeted ethical awareness by organizing themes in research ethics along a "double axiological axis" (Ferrer-Negrón 2007). An axis of truth proscribed practices that blocked the pursuit of truth and prescribed those that empowered the realization of epistemic research values such as honesty, accuracy, objectivity, and efficiency (Steneck 2004). An axis of social responsibility collected and organized practices that promoted or prevented the pursuit of social responsibility. For example, this axis covered the responsible treatment of humans and animals as research subjects (Frey and Rivera-Vélez 2014). Several cases were outlined that fell along this double axiological axis including the Tuskeegee case.
- The MDW (*Moral Deliberation Workshop*) targeted ethical evaluation by first outlining the four principles from bioethics (autonomy, justice, beneficence, and non-maleficence). These principles were then integrated into a decision-making procedure termed "moral deliberation."[6] Workshop leaders presented cases in research ethics to help students apply the principles of bioethics and to practice the seven step decision procedure of Moral Deliberation. This workshop was led by bioethicist, Jorge Ferrer-Negrón.

[5]See http://uprm.edu/etica.

[6]The steps to this process consist of the following: (1) determining the facts, (2) identifying the morally problematic situation, (3) identifying possible courses of action, (4) identifying the disagreements and moral problems in each one of the possible courses of action, (5) values in play and the hierarchy of duties, (6) deliberating on the consequences, and (7) justification of the decision taken (Ferrer-Negrón 2007).

- The CAW (*Case Analysis Workshop*) targeted ethical integration. It began with a review of the materials covered in the previous two workshops then continued by discussing more interactive cases including a role-play where, instead of working with a dominant narrative, students took up different participatory perspectives and presented the case through several participant narratives. The goal was to integrate the conflicting narrative perspectives and practice-decision making using the tools presented in the Moral Deliberation Workshop (Valdes et al. 2009). This workshop also provided students with the opportunity to practice moral deliberation with other ethics cases (Rios-Velazquezs et al. 2013).
- The capstone workshop, the *Research Ethics Banquet*, consisted of student groups presenting posters which depicted ethics cases in their area of research; they examined their cases using the four principles of bioethics and the method of moral deliberation. To prepare for the banquet, students constructed their posters under the supervision of faculty mentors. This workshop concluded with an awards ceremony where each participant who completed the workshop series received a certificate in research ethics.

GERESE also addressed other components of EAC including faculty development. This was built around (1) an issues identification workshop where faculty and graduate participants discussed short scenarios in research ethics and then set about identifying and prioritizing research ethics issues, (2) a workshop that disseminated GERESE initiatives and discussed the implementation of research ethics across the STEM graduate curriculum, and (3) a two-day retreat held in Western Puerto Rico where STEM faculty from universities across the island participated in module and case development activities.[7] Recently (2015), a workshop was held that integrated the four workshops developed through GERESE into a one-day activity. Participants found this workshop experience valuable, and there are plans to make it a regular, required activity for STEM graduate students.

3. The EAC (ethics across the curriculum) Toolkit[8] was designed to facilitate the sharing of EAC resources and cases through the creation of a "content commons" (Cruz et al. 2010). It was also proposed to provide continuity between different EAC events and activities; instead of jump-starting EAC with high profile workshops and retreats, the Toolkit was designed to provide continual support to the key activities carried out in workshops: issue identification, conceptual analysis, recognizing and coordinating EAC best practices, and solidifying an EAC community. This project required identifying existing online platforms that would support following three activities:

   (1) *Browsing*. The first stage was to provide a "commons" or repository of EAC materials and resources which individuals could explore and from which

[7]www.cnx.org/content/m32949/latest/.

[8]"Collaborative Development of Ethics Across the Curriculum Resources and Sharing of Best Practices" NSF SES 0551779.

they could extract useful materials. Connexions®, a content sharing platform developed and maintained by Rice University, supported this activity.[9] A "content commons" housed cases, classroom exercises, and lesson plans and made them accessible to everyone with access to the Internet. To facilitate browsing, Connexions® provided a module taxonomy (built out of keywords), search features (title, author, and topic searches), and what can be termed as "grouping" capabilities that drew individual modules together into larger functional groups called courses or collections. Connexions® modules also supported links to other websites allowing individual modules to function as nodes in networks of interrelated EAC resources and materials.

(2) *Commenting*. One of the challenges facing an online content repository is to bring about continuous content improvement. The EAC Toolkit singled out commenting as a route to this goal because here the experience of the EAC community could be brought to bear on content evaluation and improvement. Commenting is enacted online through such features as rankings and commentaries. Connexions® supports these activities but the most successful online platform found was the Knowledge Hub (later Temoa) housed in MTI (Monterrey Technical Institute).[10] Temoa rates content for the Open Education Resources Forum and has provided ratings and commentaries on 41 EAC Toolkit modules and cases. For example, in one module commentary, a high school teacher rated her experience using a toolkit module in her *English as a Second Language* class. She used the module to give her students an opportunity to practice discussion in English but also noted that the activity touched upon ethical awareness. Interestingly enough, this reverses the activity of recognition described above. Now an ethics module has been recognized (and redescribed) as a language-building activity. Documenting this expanded use in Temoa provided the module authors with valuable feedback. It has also validated the potentially valuable role commenting can play in the development of an ethics toolkit.

(3) *Authoring*. The biggest challenge to the Toolkit was to generate a forum in which potential authors could meet to collaborate and to develop new content in EAC. To function as a commons, the EAC Toolkit would have to generate new as well as continually refine existing modules. Connexions® provides several features to support individual and group authoring: (1) a private forum where authors can meet to generate content; (2) authoring roles (authors, maintainers, copyright holders, editors, and translators) embedded in a template for new modules and derived copies; (3) a Creative Commons Attribution License, a broad copyright protection that promotes sharing but requires attributing the contribution of each author. An important feature (related to the Attribution License) is the "derived-copy

[9]Connexions® has evolved over time and is now known as OpenStax CNX (https://cnx.org/).

[10]http://www.temoa.info/search/apachesolr_search/Frey%20ethics.

feature": an author can check out modules from the Connexions® content commons then add or mix new content to contextualize the module. The derived-copy template nicely divides attribution by recognizing the layer contributed by the original author and acknowledging the new content added by the collaborating, new author.

The EAC Toolkit project was funded by the NSF from 2006 until 2010 and has produced lasting results including 96 modules, 16 collections, and an EAC Toolkit Lens (Collaborative Development of Ethics Across the Curriculum Resources and Sharing of Best Practices) that has drawn together 7 collections and 33 modules. The EAC Toolkit module concept has been used in another grant carried out at UPRM, CIVIS,[11] which has incorporated modules from the EAC Toolkit but has also developed General Studies modules (O'Neill-Carrillo et al. 2009). CIVIS has its own collection in the Connexions® Commons called the "CIVIS Project—UPRM", and this includes modules grouped under the following headings: Professional Ethics, Information Literacy, Sustainability, Writing in the Disciplines, Social Impact/Global Issues, World Cultures, and Financial Literacy—some 25 modules altogether.

Recently STEM graduate students at UPRM have contributed to the EAC "content commons." *Bio-sand Filters in Haiti*, *Tropical Bamboo for Construction*, *The River Plata Bridge*, *Cube Satellite*, and *TMDLs in Water Resources in Puerto Rico* are cases developed by graduate students to identify and discuss the ethical issues embedded in their research. Publishing these cases in the EAC Toolkit makes it possible for undergraduate students in business administration to access, study, and learn from their discussions; in this way, the EAC Toolkit helps develop mentoring relations between graduate and undergraduate STEM students at UPRM. The EAC Toolkit has proven to be remarkably fertile in its ability to generate learning opportunities based on EAC.

4. In 2006, the College of Business Administration at UPRM approved a "Statement of Values" that was drafted by the college's stakeholder groups: civil service workers, administration, students, and teachers (Frey and Cruz-Cruz 2013). The Statement of Values was originally developed in compliance with accreditation standards of the AACSB. However, it also stands as an EAC best practice. It has served as a reminder of the aspirations of the business administration community, helped to generate classroom activities that integrate ethics into the business administration curriculum, and provided a focal point for stakeholder community building by stimulating a dialogue on ethics.

The Statement of Values process followed value sensitive design methodology (Flanagan et al. 2008; Oosterlaken 2015). Values were **discovered** during workshops in which stakeholders crafted rules of conduct, identified the values embedded in these rules, and abstracted from these rules a master list of five values: *justice*, *responsibility*, *respect*, *trust*, and *integrity*. The second process, value

[11]Center for Resources in General Education: A New Perspective in General Education at UPRM, U.S. Department of Education Award, No. P031S080124.

translation, took place when a committee of stakeholders rewrote the value profiles provided by an outside ethics consultant; they took the ethicist's theory-dense value characterizations and rewrote them in language more friendly to stakeholders. The process of writing and rewriting value profiles is itself a valuable ethics integration exercise; writing about values empowers those seeking to understand and deploy them in practical and professional contexts.

The third stage of value sensitive design, **verification**, consists of determining if the targeted values have been realized. Taking the lead from Johnson & Johnson, business administration decided to verify its Statement of Values by enacting a series of challenges. The first consisted of translating the Statement of Values into Spanish. Students in business ethics classes, working in small groups of five or six, prepared several draft translations. Then, in a *translating workshop*, stakeholders integrated the student translations into one master translation. Translating the Statement of Values raised crucial ethical and pedagogical issues. For example, participants debated the differences between "confianza" and "trust" as well as "equidad" and "fairness". A second series of challenges concentrated on the content of the Statement of Values. Participants questioned whether the value profiles were clear, whether the values list was comprehensive, and whether the profiled values had been properly ranked. (Stakeholders had previously carried out a prioritization exercise in which the values were ranked in the following order from most to least important: justice, responsibility, respect, trust, and integrity.) This challenging process has become a regular ethics integration module for two required courses: The Environments of the Organization (ADMI 4016) and Introduction to Business, Management, and Ethics (ADMI 3009).

Codes of ethics traditionally target member conduct. But the Statement of Values has picked up on other functions that contribute to community development: education, stimulating dialogue, articulating aspirations, and solidifying identity. It has promoted intra-community discussion, provided an ongoing ethics integration exercise, and has guided business administration through its day-to-day ethical challenges.

An EAC Roadmap to the Future

This section aims to complete the EAC roadmap by providing information on how to get from **here** to **there**. To start, what is meant, in the context of this map, by "here" and "there"?

"**Here**" denotes, first of all, a STEM curriculum that is fragmented and deals with ethics—if at all—outside the mainstream STEM curriculum. Ethics at this point is an extracurricular activity. Second, the STEM faculty is ethically disengaged; they are reluctant to discuss practical and professional ethical issues in their areas of expertise because of reasons cited above, namely, the crowded curriculum and the perceived need to delegate ethics to the "experts." Third, this stage draws on materials and resources that are derived from external sources and generally focused on abstract issues and activities. In a recent call for proposals, the NSF cited studies that claimed this fragmented approach actually caused STEM students to become disengaged, not only from ethical issues, but also from social and global concerns (Cech 2014).

"There" denotes a curricular approach where ethics is integrated with STEM curricular content, so that students come to see ethics as a necessary, integral part of day-to-day activities. Furthermore, participants in EAC represent many disciplines and, through EAC faculty development retreats and workshops, have become ethically empowered. This community now sees ethics, not as the province of experts, but as an interdisciplinary adventure where individuals from different disciplines teach one another how to approach embedded ethical challenges. Finally, "there" denotes an approach to ethics integration built upon an EAC Content Commons, a repository of materials developed by those who will use them and refined to be responsive to socio-technical surroundings.

EAC can take us from here to there by promoting interdisciplinary teaching and research, co-integration, a holistic approach to curricular development, a bottom-up strategy for designing and implementing ethics integration activities, and an emphasis on ethical empowerment in teaching and research. In what follows, a path to an integrated STEM curriculum will be outlined that emphasizes faculty empowerment, the successful "hand-off" of EAC best practices between members of the EAC community, and building a repository of EAC best practices to support the EAC community.

Faculty Development A fundamental component of this roadmap consists of the faculty development workshops and retreats that target the constituents of EAC outlined in Section "The EAC (Ethics Across the Curriculum) Concept at UPRM". Davis (1999) provides a detailed description of these workshops and retreats. Because we have followed this model in Puerto Rico, we will not repeat it here. However, those interested in looking at sample workshop and retreat agendas used at UPRM can find examples in the module, *Faculty Retreat in Research Ethics—Modules and Issues*, published in Connexions.[12]

Train-the-Trainer At UPRM, Train-the-Trainer has been adopted by the College of Engineering to introduce ethics into the freshman Graphics Design course, a course that all first year students are required to take. An engineering ethics module was developed by members of the engineering faculty who then trained their colleagues to teach this module. TTT_1 takes an integration exercise through three stages to bring about a module "hand-off" where the module's author passes off that module to colleagues who then teach it in their classes.

- First, the module's author visits a colleague's class and teaches the module. After this "guest lecture" they discuss the activity, the pedagogical strategies it

[12]See Connexions® https://cnx.org/contents/Im9U_550@4/Faculty-Retreat-in-Research-Et or https://legacy.cnx.org/content/m32949/1.4/?legacy=true.

employs, the background knowledge required and different pedagogical situations that may arise. The colleague learns by observing the module author's teaching.

- Second, the colleague adopting the module teaches it to another section while the module's author observes. After class, the module's author offers friendly, non-confrontational advice on how to teach the module better. Both further discuss issues pertinent to the teaching of the module with special emphasis on how one goes about adapting the module to one's own teaching style.
- In the final stage, the module adopter "goes solo." This three-stage process renders a STEM teacher capable of integrating an ethics module into his or her class. Having all professors who teach a given course teach the same ethics integration exercise in the same way supports accreditation and assessment requirements. It provides solid, uniform data on how effectively a program has integrated ethics into its curriculum. But it may also overly constrain those teachers who already have their own ethics integration exercise or find it awkward to modify their teaching style toward teaching a module developed by others.

TTT_1 thus raises two problems. First, some professors had already developed their own ethics integration exercises for their sections of the course and are, understandably, reluctant to set their module aside and adopt one developed by colleagues. Second, even though they had gone through the three steps of TTT_1, some still find it difficult to teach the module solo in their classes. This has led some to drop the ethics module and substitute other activities more directly related to the course's technical content. While TTT_1 has mitigated some standard objections to EAC, it has not eliminated all; citing the crowded curriculum and lack of expertise, many STEM professors still requested that the module be taught by its authors as guest lecturers.

To get around these and other problems, we have developed a different approach to Train-the-Trainers called TTT_2. The difference is that TTT_2 has incorporated two lessons from EAC: (1) EAC works best if it deploys a co-integration strategy (where both ethicists and STEM teachers contribute to a module's content) and (2) those who already have EAC modules in their classes are due recognition and credit for these modules and not be required to set aside their activity in order to adopt one authored by someone else. Furthermore, TTT_2 allows those in position to adopt an ethics module to choose how to build the module and how—and whether—to integrate it into the classroom.

The new ethics integration activity centers on a case entitled "Mega Weapons" where a pacifist engineer, who is unemployed, has to deliberate on whether to apply for a job with a weapons manufacturer. A business administration graduate student produced a short, animated video to present the case; students could watch this animation online before class or during the actual class. Then an ethics professor was video-taped leading a class discussion on the case followed by a ten minute lecture introducing three ethics tests (reversibility, harm, publicity).

The objective in TTT_2 is to give the STEM professor building blocks for constructing an ethics activity around the instructor's preferences. In TTT_2 the STEM professor can choose from the following options:

- He or she can make the entire class virtual by having the students go to the LMS (Learning Management System), view the video, watch the taped discussion and ethics presentation, then write an essay discussing this experience. This accommodates professors who, for various reasons, choose not to do EAC during regular class time.
- The STEM professor can take over the entire class by showing the video, having her students discuss the problems it raises, present a problem-solving framework (using ethics codes, applying ethical theory, or deploying ethics tests), and conclude with an evaluative, reflective activity, all carried out in the classroom. Here the ethically empowered STEM professor becomes an ethics mentor to his or her students.
- STEM professors can also take a hybrid approach by having, say, an in-class discussion of the animation and then assign viewing the taped presentation online after the classroom activity. In this way, the STEM professor partners with the ethics professor to provide a balanced, in-depth activity which integrates the expertise of both.
- If a STEM professor already has an ethics integration exercise in use, he or she can continue with their own activity and have the students do the new module online and outside of class. It might even be possible to integrate the two activities. The point here is to give the professor maximum freedom in deciding how to integrate this content into his or her course.

Sharing EAC modules (module hand-offs) can be difficult and, from time to time, may go sideways because they go against the interdisciplinary, co-integration, and community dimensions of EAC. On the other hand, TTT_2 preserves these characteristics of EAC by allowing both ethicists and STEM teachers to collaborate in bringing about a successful ethics-integration.

Building a repository of EAC resources Here is where the roadmap metaphor is most pertinent. Building a repository of ethical issues, cases responsive to these issues, and classroom exercises that integrate these issues and cases into classroom activities is the best and most direct path to EAC. This part of the roadmap will focus on developing issue lists and case taxonomies and having ethicists and STEM professors collaborate to write ethics scenarios and cases.[13]

- At UPRM, STEM faculty write cases in small groups in faculty development workshops. These workshops provide an excellent forum in which cases can be crafted and refined. Usually there is only limited time and participants are reluctant to continue after the workshop. But within the time frame of a day-long

[13]Davis (1999) again provides excellent suggestions on how to write cases, ethics integration exercises, and how to rewrite textbook exercises to include an ethics component.

workshop, there is time to outline scenarios that pose ethical issues. For example, during a workshop, a participant wrote a short scenario where the EPA shut down a Puerto Rican aquaculture business for violating EPA regulations. This case was short and posed from the evaluator's perspective. But a group of students in an engineering ethics class took this scenario and rewrote it from a participatory standpoint. In the new version, the EPA cited the aquaculture facility and gave it six months to reach compliance with regulations. The students role-played as consultants hired to help the aquaculture facility comply with EPA regulations. Cases provide narratives that can be approached from different standpoints (evaluative and participant) which allow for targeting a whole range of pedagogical and ethical objectives.

- The scenarios developed by faculty groups in workshops can be correlated with the ethics issues lists. This provides yet another basis from which a scenario can be rewritten and refined. Different issues lists developed by STEM faculty have helped EAC stakeholders to modify cases to fit the Puerto Rican context. Bringing an issues list to a case also helps pose useful questions: Does the case address ethical issues highlighted and prioritized by the EAC community? Can a given case or scenario be "tweaked" to address one or more of these vital issues? Finally, scenarios developed by faculty in a workshop can be assessed in terms of whether they address key issues; those covered in the scenario are hits while those not covered are misses. Issues lists, thus, provide specifications for modifying existing scenarios or devising new ones to address uncovered issues.
- Scenarios can also be expanded changing them from thin to thick cases. For example, complicating circumstances can be layered onto a core scenario to address the Hastings Center objective touching on moral ambiguity and non-agreement (Rios-Velazquez et al. 2013). Or cases can be expanded through further technical research. An engineer passes by a laminating press room and sees that the press operator is covered with a white powder and is not using protective gear. Does this pose a problem? What should the engineer do? This case could be expanded by doing research on laminating presses, identifying their chemical and waste by-products, and the long term effects of exposure on health. At UPRM, a group of graduate students adopted this scenario and expanded it into a thick, technically complicated case that they shared with others.
- Scenarios can be integrated into different activity platforms to produce a variety of ethics integration exercises. Huff and Frey (2005) provide a list of things to do with cases that includes (1) structured and unstructured discussion, (2) role-plays, (3) taking the evaluator's point of view, (4) role-playing from a participant's perspective, and (5) writing out dialogues to rehearse how a course of action might play itself out (Dewey 1922). Each of these serves as an activity platform into which a scenario or case can be integrated to create an EAC module. For example, a scenario developed by STEM faculty could be deployed in the Pre-Test platform; here the scenario would serve as the occasion for an unstructured discussion asking whether it raises an issue that is ethical, common, or controversial. One could add to this an introduction to ethics tests

(reversibility, harm, publicity) to provide more structure to Pre-Test discussions. Or the same scenario could be inserted into a Gray Matters format to practice decision-making. Students could choose from among "alternative endings" to bring the narrative to a successful conclusion. Or they could use ethical considerations to guide them in designing their own endings. The scenarios could provide the occasion for practicing moral imagination by identifying participatory perspectives and having students develop role-plays from within these perspectives. In historical cases with "bad endings," students could construct "What if" dramas where they imagine participants taking different decisions at key points (Frey 2015b). They could then dramatize these decisions and see if they would lead to different outcomes. This, by no means exhaustive, list makes a central point: bringing cases to activity platforms can generate any number of ethics integration activities which can be fine-tuned to bring about specific pedagogical objectives.
- Students at UPRM—especially graduate students—have played a key role in expanding the EAC content commons. They have turned their own research into ethics case studies by (1) zooming in on their research projects and plans, (2) zooming out to the socio-technical context of research or application, and (3) discussing the ethical issues that arise when research projects are projected into specific socio-technical systems (Frey and Rivera-Vélez 2014). These cases have been assigned to undergraduates who make posters presentation. All of this facilitates the sharing of EAC practice and research.

An EAC community using interdisciplinary collaboration and co-integration can write its own ethics integration materials. These can be designed to be responsive to community concerns and help students and faculty develop skills of ethical empowerment. This bottom-up, participatory approach to EAC advocated throughout this essay makes it likely that these materials will hit their mark and provide powerful ethics integration exercises.

Conclusion

This chapter has approached ethics across the curriculum through the authors' experience at UPRM from 1990 to 2017. It has concentrated, first, on conceptual cartography; EAC is interdisciplinary, deploys co-integration between STEM and ethics faculty, takes a holistic approach to curricular development, proceeds by means of workshops and retreats that ethically empower participating faculty members, and deploys the strategy outlined above as 15/85. Several activities carried out at UPRM have given body to this EAC concept: a series of interdisciplinary workshops and retreats in EAC, workshops targeting ethical empowerment skills for integrating research ethics into the STEM graduate curriculum, an EAC Toolkit which has brought online the cases, lesson plans, and resources in EAC generated at UPRM, and a Statement of Values process in which the College of Business Administration at UPRM created a value statement that provided form and content to their emerging ethical community. This roadmap has been set forth to show the pathway from a fragmented, multi-disciplinary curriculum to an integrated, interdisciplinary curricular approach, one that reengages faculty and students around ethical best practices in professional and occupational studies.

References

Cech, E. (2014). Culture of disengagement in engineering education? *Science, Technology and Human Values, 39*(1), 42–72.

Cruz, J., & Frey, W. (2001). Ethics and empowerment: An ethics module for introduction to computers. In *Proceedings of the 2001 American Society for Engineering Education Annual Conference & Exposition*, Copyright © 2001, American Society for Engineering Education. Session 1520.

Cruz, J., & Frey, W. (2003). An effective strategy for integrating ethics across the curriculum in engineering: An ABET 2000 challenge. *Science and Engineering Ethics, 9*(4), 543–568.

Cruz, J., & Frey, W. (2015). Dual Approaches: Integrating ethics into the information systems curriculum. In *Proceedings: ACBSP 2015.* https://www.slideshare.net/ACBSPAccreditation/cruz-frey-acbsp2015v2.

Cruz, J., Aurbelo, A., & Frey, W. (2010). Doing ethics across the curriculum: The EAC toolkit. *Teaching Ethics, 10*(2), 47–69.

Davis, M. (1999). *Ethics and the university* (pp. 111–174). London, UK: Routledge.

Davis, M., Hildt, E., & Lass, K. (2016). Twenty-five years of Ethics Across the Curriculum: An assessment. *Teaching Ethics, 16*(1), 55–74.

Dewey, J. (1922/1988). Human nature and conduct. Carbondale, IL: Southern Illinios University Press, p. 216.

Ferrer-Negrón, J. (2007). *Deber Y Deliberación: Una Invitación a la Bioética* (pp. 83–110). Mayagüez, PR: Centro de Publicaciones Académicas, Facultad de Artes y Ciencias, UPR-RUM.

Flanagan, M., Howe, D., & Nissenbaum, H. (2008). Embodying values in technology: Theory and practice. In J. van den Hoven & J. Weckert (Eds.), *Information technology and moral philosophy* (pp. 322–353). Cambridge: Cambridge University Press.

Frey, W. J., & Cruz-Cruz, J. A. (2013). Value integration: From educational computer games to academic communities. *IEEE Technology and Society Magazine, 32*(1), 31–35.

Frey, W. & Rivera-Vélez, C. (2014). Bringing responsible research into engineering ethics: Responsible research in appropriate technology. In *2014 IEEE International Symposium on Ethics in Science, Technology and Engineering.*

Frey, W. (2015a). Training engineers in moral imagination for global contexts. In C. Murphy, P. Gardoni, H. Bashir, C. E. Harris, & E. Masad (Eds.), *Engineering ethics for a global world* (pp. 229–247). Dordrecht: Springer.

Frey, W. (2015b). Teaching responsibility: Pedagogical strategies for eliciting a sense of moral responsibility. *Teaching Ethics, 15*(2), 217–336.

Harris, C. (2008). The good engineer: Giving virtue its due in engineering ethics. *Science and Engineering Ethics, 14*(2), 153–164.

Huff, C. (2001). Introduction to social impact analysis. ComputingCases.org. www.computingcases.org.

Huff, C., & Frey, W. (2005). Moral pedagogy and practical ethics. *Science and Engineering Ethics, 11*(3), 389–408.

Kroenke, D. M. (2015). *MIS Essentials* (4th ed., pp. 16–17). Boston: Pearson.

O'Neill-Carrillo, E., Collins, D. L., López-Garriga, J., Macchiavelli, R., & Cruz, J. A. (2009). Center for resources in general education (CIVIS): Towards student success in general and STEM education. In *Proceedings of the 39th ASEE/IEEE Frontiers in Education Conference.* San Antonio, TX, October 2009, M3F1–M3F6.

Oosterlaken, I. (2015). *Technology and human development* (pp. 46–77). London, UK: Routledge.

Papadopoulos, C., et al. (2014). Sponsoring research in appropriate technology. In *121st ASEE Annual Conference and Exposition*, p. 10625.

Pritchard, M. (1998). Professional responsibility: Focusing on the exemplary. *Science and Engineering Ethics, 4*(2), 215–234.

Rios-Velazquez, C., Frey, W., Jarramilli-Giraldo, E., & Echeverry-Solarte, M. (2013). Accepting the likelihood of ambiguity and disagreement on moral matters: Transitioning into the "Gray World". *Teaching Ethics, 13*(2), 55–72.

Steneck, N. H. (2004). ORI: Introduction to the responsible conduct of research. University of Michigan Library.

Valdes, D., Jaramillo Giraldo, E., Ferrer, J., & Frey, W. (2009, June). Case analysis: A tool for teaching research ethics in science and engineering for graduate students. Paper presented at 2009 Annual Conference & Exposition, Austin, Texas. https://peer.asee.org/5729.

Whitbeck, C. (2011). Ethics in engineering practice and research, (2nd ed., pp. 254–258) Cambridge, UK: Cambridge University Press.

Correction to: Ethics Across the Curriculum—Pedagogical Perspectives

Correction to:
Chapters 1 and 2 in: E. E. Englehardt and M. S. Pritchard (eds.), Ethics Across the Curriculum—Pedagogical Perspectives, https://doi.org/10.1007/978-3-319-78939-2

In the original version of the book, the following belated corrections have been incorporated in chapters 1 and 2.

The author's name stated in chapter 1 has been corrected from "Karlana June" to "Karlynd June".

The co-author name and her email address in chapter 2 was corrected from "Karlana June, kmorgan@usf.edu" to "Karlynd June, karlyndjune14@gmail.com".

The correction chapters have been updated with the changes.

The updated version of these chapter can be found at
https://doi.org/10.1007/978-3-319-78939-2_1
https://doi.org/10.1007/978-3-319-78939-2_2

E. E. Englehardt and M. S. Pritchard (eds.), *Ethics Across the Curriculum—Pedagogical Perspectives*, https://doi.org/10.1007/978-3-319-78939-2_26

MIX
Papier aus verantwortungsvollen Quellen
Paper from responsible sources
FSC® C105338

If you have any concerns about our products,
you can contact us on
ProductSafety@springernature.com

In case Publisher is established outside the EU,
the EU authorized representative is:
Springer Nature Customer Service Center GmbH
Europaplatz 3, 69115 Heidelberg, Germany

Printed by Libri Plureos GmbH
in Hamburg, Germany